TRANSLATIONS SERIES IN MATHEMATICS AND ENGINEERING

TRANSLATIONS SERIES IN MATHEMATICS AND ENGINEERING

M.I. Yadrenko
Spectral Theory of Random Fields

1983, viii + 259 pp.
ISBN 0-911575-00-6 Optimization Software, Inc.
ISBN 0-387-90823-4 Springer-Verlag New York Berlin Heidelberg Tokyo
ISBN 3-540-90823-4 Springer-Verlag Berlin Heidelberg New York Tokyo

G.I. Marchuk
Mathematical Models In Immunology

1983, xxv + 353 pp.
ISBN 0-911575-01-4 Optimization Software, Inc.
ISBN 0-387-90901-X Springer-Verlag New York Berlin Heidelberg Tokyo
ISBN 3-540-90901-X Springer-Verlag Berlin Heidelberg New York Tokyo

A.A. Borovkov, Ed.
Advances In Probability Theory: Limit Theorems and Related Problems

1984, xiv + 378 pp.
ISBN 0-911575-03-0 Optimization Software, Inc.
ISBN 0-387-90945-1 Springer-Verlag New York Berlin Heidelberg Tokyo
ISBN 3-540-90945-1 Springer-Verlag Berlin Heidelberg New York Tokyo

V.A. Dubovitskij
The Ulam Problem of Optimal Motion of Line Segments

1985, xiv + 114 pp.
ISBN 0-911575-04-9 Optimization Software, Inc.
ISBN 0-387-90946-X Springer-Verlag New York Berlin Heidelberg Tokyo
ISBN 3-540-90946-X Springer-Verlag Berlin Heidelberg New York Tokyo

N.V Krylov, R.S. Liptser, and A.A. Novikov, Eds.
Statistics and Control of Stochastic Processes

1985, xiv + 507 pp.
ISBN 0-911575-18-9 Optimization Software, Inc.
ISBN 0-387-96101-1 Springer-Verlag New York Berlin Heidelberg Tokyo
ISBN 3-540-96101-1 Springer-Verlag Berlin Heidelberg New York Tokyo

Yu. G. Evtushenko
Numerical Optimization Techniques

1985, xiv + 561 pp.
ISBN 0-911575-07-3 Optimization Software, Inc.
ISBN 0-387-90949-4 Springer-Verlag New York Berlin Heidelberg Tokyo
ISBN 3-540-90949-4 Springer-Verlag Berlin Heidelberg New York Tokyo

Continued on page 559

Yurij G. Evtushenko

NUMERICAL OPTIMIZATION TECHNIQUES

Optimization Software, Inc.
Publications Division, New York

Author
Yu.G. Evtushenko
Computing Center
The USSR Academy
of Sciences
Ulitsa Vavilova 40
Moscow B-333
USSR

Series Editor
A.V. Balakrishnan
School of Engineering
University of California
Los Angeles, CA 90024
USA

Translation Editor
Prof. Dr. J. Stoer
Institut für Angewandte Mathematik
und Statistik der Universität Würzburg
Am Hubland
D-8700 Würzburg
The Federal Republic of Germany

Library of Congress Cataloging in Publication Data

Evtushenko, IU.G. (IUrii Gavrilovich)
Numerical optimization techniques.

(Translations series in mathematics and engineering)
Translation of: *Metody resheniia ekstremal'nykh zadach i ikh primenenie v sistemakh optimizatsii.*
Bibliography: p.
Includes index.
1. Mathematical optimization. 2. Maxima and minima.
I. Title. II. Series.
QA402.5.E9713 1985 519 85-7230
ISBN 0-911575-07-3

Worldwide Distribution Rights by Springer-Verlag New York Inc.,
175 Fifth Avenue, New York, New York 10010, USA and
Springer-Verlag Berlin Heidelberg New York Tokyo,
Heidelberg Platz 3, Berlin-Wilmersdorf-33, The Federal Republic
of Germany.

ISBN 0-911575-07-3 Optimization Software, Inc.
ISBN 0-387-90949-4 Springer-Verlag New York Berlin Heidelberg Tokyo
ISBN 3-540-90949-4 Springer-Verlag Berlin Heidelberg New York Tokyo

ABOUT THE AUTHOR

Yurij Gavrilovich Evtushenko is Deputy Director of the Moscow Computing Center of the USSR Academy of Sciences and Head of the Laboratory of System Optimization. He is one of the leading Soviet specialists in Numerical Methods of Optimization as well as Optimization Software. He received his D.Sci. degree from the Moscow Physico-Technical Institute in 1980.

CONTENTS

FOREWORD

The book of Professor Evtushenko describes both the theoretical foundations and the range of applications of many important methods for solving nonlinear programs. Particularly emphasized is their use for the solution of optimal control problems for ordinary differential equations. These methods were instrumented in a library of programs for an interactive system (DISO) at the Computing Center of the USSR Academy of Sciences, which can be used to solve a given complicated problem by a combination of appropriate methods in the interactive mode. Many examples show the strong as well the weak points of particular methods and illustrate the advantages gained by their combination. In fact, it is the central aim of the author to point out the necessity of using many techniques interactively, in order to solve more difficult problems.

A noteworthy feature of the book for the Western reader is the frequently unorthodox analysis of many known methods in the great tradition of Russian mathematics.

J. Stoer

PREFACE

Optimization methods are finding ever broader application in science and engineering. Design engineers, automation and control systems specialists, physicists processing experimental data, economists, as well as operations research specialists are beginning to employ them routinely in their work. The applications have in turn furthered vigorous development of computational techniques and engendered new directions of research. Practical implementation of many numerical methods of high computational complexity is now possible with the availability of high-speed large-memory digital computers. Indeed, experience has shown that the most efficient way of solving optimization problems is by interactive man-machine mode, allowing the use of a variety of optimization techniques for any given problem.

This book deals with computational techniques (including interactive man-machine methods) for solving nonlinear programming problems as well as constrained optimal control problems.

The book has seven chapters. Chapter 1 reviews relevant parts of convex analysis and derives necessary and sufficient conditions for optimality in nonlinear methods for solving systems of nonlinear equations, including minimax solutions; these methods of independent interest are used later on. Chapters 3, 4 and 5 treat numerical methods for solving nonlinear programming problems. Chapter 3 considers various modifications of the penalty-function method. Chapter 4 deals with methods using several modifications of the Lagrangian. Relaxation methods are described in Chapter 5. Chapter 6 presents numerical methods for solving optimal control problems with state constraints (including nondifferentiable functionals), drawing on nonlinear programming--in particular, modified Lagrangians, constrained gradients, gradient projections, Newton-Raphson, among others.

Chapter 7 is new--added in the English edition. It contains new results on global numerical methods for a variety of problems (e.g., global minimization of multivariable functions, nonlinear programming, multicriteria optimization, and solution of nonlinear equations) based on the method of nonuniform covering. While available in the Soviet literature, they are almost unknown in the West. Various versions of the nonuniform-covering method can also be used for parallel computers. Appendices I and II provide a review of relevant results from Analysis, Linear Algebra, and Point-to-Set Mapping Theory, used in the book. The bibliography makes no claim of completeness (the number of extant references exceeds well over a thousand) and lists only those papers and monographs used directly in writing the book. Special attention has been given to those methods which had been tested extensively in recent years at the USSR Academy of Sciences Computing Center, and improved upon.

The author expresses his deep gratitude to N.N. Moiseev for his encouragement in writing this book and his continued interest. Extensive access to the Western literature on this subject was made possible through the courtesy of Professor O. Hellman and the Turku University Library. The author is grateful to his co-workers at the Computing Center, O.P. Burdakov, A.I. Golikov, N.I. Zhadan, and V.A. Purtov for reading the manuscript and making many useful comments.

O.P. Burdakov assisted in writing Section 2.6, and Section 6.7 was written in collaboration with N.I. Grachev.

Notation

Theorems, Lemmas and Definitions are labelled by a triple number: the first digit stands for the chapter number, the second digit for the section number, and the third digit is the ordinal number of the theorem, lemma or definition in the section. The formulas are labelled by two digits: the first digit stands for the section number, the second digit is the number of the formula in the section. If reference is made to a formula in another chapter, one more digit is added to indicate the number of the chapter.

$x \in X$: x is an element of the set X;

$x \notin X$: x is not an element of the set X;

$X \subset Y$: the set X is a subset of Y, $X = Y$ is not excluded;

$X = Y$: the sets X and Y coincide;

$X \cup Y$: the union of the sets X and Y;

$X \cap Y$: the intersection of the sets X and Y;

$X \setminus Y$: the difference between the sets X and Y, i.e., the set of all $x \in X$ such that $x \notin Y$;

$X \times Y$: the Cartesian product of the sets X and Y, i.e., the set of pairs (x,y) where $x \in X$, $y \in Y$;

$\operatorname{int} X$: the set of interior points of the set X (see Definition 1.1.2);

$\bar{X}$: the closure of the set X (see Definition (1.1.3);

(2) NOTATION

$X = \emptyset$: the set X is empty;

$\{x:T\}$: the set of all elements x satisfying condition T;

$\lambda \in (a,b)$: λ satisfies the condition $a < \lambda < b$;

$\lambda \in [a,b]$: λ satisfies the condition $a \leq \lambda \leq b$;

$f: X \to Y$: a one-to-one mapping of X onto Y;

$W: X \to 2^Y$: a multivalued mapping of X onto Y (see Appendix III);

$\forall x$: reads "for all x";

$\exists x$: reads "there is an x...";

dis (x,y): the distance between the two points x and y;

$\text{dis}(x,X) = \inf_{y \in X} \text{dis}(x,y)$: the distance between the point x and the set X;

$G(q)$: the open neighborhood of the point q;

$G(X)$: the open neighborhood of the set X;

$G_\varepsilon(X)$: the ε-neighborhood of the set X; $G_\varepsilon(X) = \{x: \text{dis}(x,X) < \varepsilon\}$;

R^n: the real linear (normed) n-dimensional space;

E^n: the Euclidean n-dimensional space;

E^n_+: the nonnegative orthant of E^n, i.e., the set of all vectors of E^n all the coordinates of which are nonnegative;

$x \in E^n$: the vector x is an element of the space E^n

x^i: the ith coordinate of the vector x; in some places, $x^{(i)}$ is written for greater clarity;

(3) NOTATION

e^i: the ith axis, column vector, the ith coordinate of which is unity, the remaining, zeros;

p^T: a transposed vector;

B^T: a transposed matrix;

$\|x\|$: the norm of the vector x, in most cases in the text, the Euclidean norm is meant (for more detail, see Appendix II);

$\|A\|$: the norm of the matrix A, adapted to the norm of the vectors;

$|A|$: the determinant of the matrix A;

$p \geq 0$: all coordinates of the vector p are nonnegative;

$B > 0$: the symmetric matrix B of order n is positive definite, i.e., for any $x \in E^n$ and such that $\|x\| \neq 0$, $x^TBx > 0$;

$B \geq 0$: the symmetric matrix B of the order n is positive semi-definite, i.e., for any $x \in E^n$, $x^TBx \geq 0$;

$D(z)$: a diagonal matrix whose i^{th} diagonal element is the ith coordinate of the vector z, the dimension of the matrix D is determined by the dimension of the vector z;

I_s: the identity matrix of the order s;

I_+: the set of nonnegative real numbers;

O_{nm}: the zero matrix $n \times m$ (in many places in the text, where this does not cause any misunderstanding, the subscripts nm are omitted);

$(a,b) \neq 0$: at least one of the vectors $a \in R^c$ or $b \in R^m$ is nonzero;

$z = [a,b]$: shorthand notation for $z = [a^T, b^T]^T$, $a \in R^c$, $b \in R^m$,

$\langle a,b \rangle$: the scalar product of the vectors a and b;

$h_+(x)$, $h_-(x)$: the vector functions whose i^{th} components are defined by the formulas:

$$h_+^i(x) = \max\,[0, h^i(x)] \quad ,$$

$$h_-^i(x) = \min\,[0, h^i(x)] \quad ,$$

if $h, v \in E^c$, then

$$(h+v)_+^2 = \sum_{i=1}^{c} [\max\,[0,\ h^i + v^i]]^2 \ ;$$

$f_x(x)$: the n-dimensional column vector the ith component of which is $\frac{\partial f(x)}{\partial x^i}$;

$f_{xx}(x)$: the square matrix of the order n whose $(i,j)^{th}$ element is $\frac{\partial f^2(x)}{\partial x^i \partial x^j}$;

$g_x(x)$: the rectangular matrix whose $(i,j)^{th}$ element is $\frac{\partial g^j(x)}{\partial x^i}$;

$S'(q)$: a derivative of the function S in the scalar argument q;

$\mathrm{Re}\, z$: the real part of the complex number z;

$\mathrm{Im}\, z$: the imaginary part of the complex number z;

$\bar{z}$: the conjugate of z;

$|z|$: the modulus of z;

$\overline{\lim\limits_{k\to\infty}}\, x_k$: the upper limit of the sequence $\{x_k\}$;

$\underline{\lim\limits_{k\to\infty}}\, x_k$: the lower limit of the sequence $\{x_k\}$;

$\dot{x}$: the differentiation with respect to the independent variable t;

$\partial f(x)$: the set of subgradients of the function f at the point x (see Definitions 1.2.1 and 1.2.2):

$Q_i(x)$: estimates of the rate of convergence (the definition of them is given in Section 2.3);

$\sigma(x)$: the set of active bounds of the inequality type at the point x (see the formula (1.7.3));

$L(x,u,v)$: the Lagrange function in the nonlinear programming problem (1.6.1) (see the formula (1.6.4));

$\operatorname{Arg}\min\limits_{x\in X} f(x)$: the set of all those points $x \in X$ at which the minimum of the function f is attained on X;

sup: supremum;

inf: infimum;

///: the proof is completed.

Chapter 1

AN INTRODUCTION TO OPTIMIZATION THEORY

In this chapter we present definitions and basic theoretical results to be used in developing and justifying numerical methods for solving extremal problems. We also give necessary and sufficient conditions for the extremum in various optimization problems. The material of this chapter is essential for understanding the subsequent chapters.

1. CONVEX SETS AND CONVEX FUNCTIONS

1. BASIC DEFINITIONS

Let $x, y \in E^n$. By the closed line segment joining the points x and y we mean the set of all points representable as $\lambda x + (1-\lambda)y$, where $0 \leq \lambda \leq 1$.

DEFINITION 1.1.1. The set $X \in E^n$ is convex if the closed line segment joining every two points of X belong to X.

DEFINITION 1.1.2. The point $x \in X$ is an interior point of the set X if for any $y \in E^n$ we can find $\bar{\lambda} > 0$ such that $x + \lambda y \in X$ for all $0 \leq \lambda \leq \bar{\lambda}$. We denote the set of all interior points of X by $\text{int}\, X$.

The convexity of the set X implies the convexity of $\text{int}\, X$.

We say that a set is open if each of its points is an interior point. The set X is said to be closed in E^n if the complement of X in E^n, that is the set $E^n \setminus X$, is open.

DEFINITION 1.1.3. We say that the point x is a limiting point of the set X if there exists a sequence of points $x_k \in X$, converging to x. The aggregate of all limiting points of the set X is said to be its closure and is denoted by $\bar{X}$.

DEFINITION 1.1.4. A set X is said to be compact if any sequence of its points contains a subsequence converging to some point of X.

In the space E^n the term "compact set" is synonymous with "bounded closed set."

DEFINITION 1.1.5. A function $f: X \to R \cup \{\pm \infty\}$ defined on a set $X \subset E^n$ is called convex on X if for any two points $x, y \in X$ and $\lambda \in E^1$ such that $0 \leq \lambda \leq 1$ and $\lambda x + (1-\lambda)y \in X$, the condition

$$f(\lambda x + (1-\lambda)y) \leq \lambda f(x) + (1-\lambda)f(y) \tag{1.1}$$

is satisfied.

If we require in addition that for $\lambda \neq 0$, $\lambda \neq 1$ and $x \neq y$ the sign $\leq$ be replaced by $<$ in (1.1), the function $f(x)$ is said to be strictly convex on X.

If instead of $\leq$ we take $\geq$ in (1.1), the function $f(x)$ is said to be concave on X. If the function $f(x)$ is convex, the function $-f(x)$ is concave. Thus all the properties of convex functions are easily applicable, when appropriately modified, to concave functions.

A function $f(x)$ defined on the convex set X is convex if

for any $x, y \in X$ and $\lambda \in E^1$ such that $0 \leq \lambda \leq 1$, (1.1) is satisfied.

If $f(x)$ is a convex function of x on the entire space E^n, we say simply that the function $f(x)$ is convex. The Euclidean norm of the vector $||x|| = \sqrt{\langle x,x\rangle}$ is a simple example of such a convex function on E^n. Indeed, using the triangle inequality

$$||\lambda x + (1-\lambda)y|| \leq ||\lambda x|| + ||(1-\lambda)y|| \quad ,$$

the property of the norm $||\lambda x|| = |\lambda|||x||$, $||(1-\lambda)y|| = |1-\lambda|||y||$, we obtain for $0 \leq \lambda \leq 1$

$$||\lambda x + (1-\lambda)y|| \leq \lambda||x|| + (1-\lambda)||y|| \quad ,$$

implying in turn the convexity of the function $||x||$.

An example of an open convex set is the set of points interior to the n-dimensional sphere of radius ε centered at the point $\bar{x}$:

$$G_\varepsilon(\bar{x}) = \{x \in E^n : ||x-\bar{x}|| < \varepsilon\} \quad .$$

If $x, y \in G_\varepsilon(\bar{x})$, we obtain by the triangle inequality:

$$\begin{aligned} \|\lambda x+(1-\lambda)y-\bar{x}\| &= \|\lambda(x-\bar{x})+(1-\lambda)(y-\bar{x})\| \leqslant \\ &\leqslant \lambda\|x-\bar{x}\|+(1-\lambda)\|y-\bar{x}\| < \lambda\varepsilon+(1-\lambda)\varepsilon=\varepsilon, \end{aligned}$$

implying in turn the convexity of the set $G_\varepsilon(\bar{x})$.

In the sequel we shall call the set $G_\varepsilon(\bar{x})$ the ε-neighborhood of the point $\bar{x}$.

DEFINITION 1.1.6. The function $f(x)$ defined on E^n is said to be infinitely large, if for any positive M there exists $R(M)$ such that for any x satisfying $||x|| > R$ the inequality

$f(x) > M$ holds.

We denote by dis (x,X) the distance between the point x and the set X:

$$\text{dis}\,(x,X) = \inf_{p\in X} \|x-p\| .$$

If the set X is convex and closed, in the last equality the minimum is attained at a unique point $p(x) \in X$, which is called the projection of the point x onto the set X and is found from the condition

$$p(x) = \operatorname{Arg}\min_{p\in X} \|x-p\| . \tag{1.2}$$

We can show that dis (x,X) is a convex function of x. Let x_1 and x_2 be arbitrary points of E^n and $0 \leq \lambda \leq 1$. Then for any $z \in X$ we have the inequality

$$\text{dis}\,(\lambda x_1 + (1-\lambda)x_2,\ X) \leq \|\lambda x_1 + (1-\lambda)x_2 - z\| . \tag{1.3}$$

Using the definition (1.2), we write $p_1 = p(x_1)$, $p_2 = p(x_2)$, $p_1, p_2 \in X$. Due to the convexity of the set X, all the points in the line segment joining p_1 and p_2 belong to X; hence, putting in (1.3) $z = \lambda p_1 + (1-\lambda)p_2$, we obtain

$$\begin{aligned}\text{dis}\,(\lambda x_1 + (1-\lambda)\,x_2,\ X) &\leqslant \\ \leqslant \|\lambda\,(x_1-p_1) + (1-\lambda)\,(x_2-p_2)\| &\leqslant \\ \leqslant \lambda\,\text{dis}\,(x_1,\ X) + (1-\lambda)\,&\text{dis}\,(x_2,\ X)\end{aligned}$$

implying in turn the convexity of the function dis (x,X) on E^n. It is also obvious that for a compact set X this function is infinitely large.

<u>DEFINITION 1.1.7.</u> By the epigraph (hypograph) of the convex function $f(x)$, epi f, we mean the set of points of R^{n+1} on the

graph and above the graph f:

$$\text{epi } f = \{x \in E^n,\ \mu \in E^1 : f(x) \leq \mu\} .$$

DEFINITION 1.1.8. By the effective domain of the convex function f, dom f, we mean the set of points of E^n, in which f attains finite values or the value $-\infty$:

$$\text{dom } f = \{x \in E^n : f(x) < +\infty\} .$$

The set dom f is the projection of epi f onto E^n, i.e., dom f = $\{x \in E^n : \exists \mu \in E^1$ such that $(x, \mu) \in \text{epi } f\}$.

DEFINITION 1.1.9. The convex function f is called proper if the set dom f is not empty and $f(x) > -\infty$ $\forall x \in \text{dom } f$.

In other words, a proper convex function does not attain the value $-\infty$ and is not identical to $+\infty$. In the sequel, we shall only consider proper convex functions, without specifying this property each time.

A convex function is continuous at all interior points of dom f.

DEFINITION 1.1.10. The nonempty set $K \subset E^n$ is said to be a convex cone, if the following conditions are satisfied:

- 1. $x + y \in K$ for any $x, y \in K$;
- 2. $\lambda x \in K$ for any $x \in K$ and each $\lambda \geq 0$.

These conditions are equivalent to the requirement that $\alpha x + \beta y \in K$ for any $x, y \in K$ and each $\alpha \geq 0$, $\beta \geq 0$.

The nonnegative orthant E^n_+ is a convex cone. If A is a $m \times n$ matrix and $x \in E^n$, or $x \in E^n_+$, the set of all solutions of the system $Ax \leq 0$ is a convex (polyhedral) cone.

By a hyperplane in E^n we mean the set

$$\Gamma = \{x \in E^n : \langle c, x \rangle = \alpha\} ,$$

where $c \in E^n$, $\|c\| \neq 0$, α being real. This set is convex and always nonempty. If $x_* \in \Gamma$, the hyperplane Γ can be represented as

$$\Gamma = \{x \in E^n : \langle c, x - x_* \rangle = 0\} ,$$

i.e., Γ consists of only those points x for which the vector $x - x_*$ is orthogonal to the vector c. The vector c is said to be the vector "normal" to the hyperplane Γ.

We say that the hyperplane Γ with normal vector c separates two nonempty sets X and Y, if there exists some γ such that for any $x \in X$, $y \in Y$ the inequalities

$$\langle c, x \rangle \leq \gamma \leq \langle c, y \rangle$$

are satisfied.

DEFINITION 1.1.11. The hyperplane Γ is said to be a "support" to the set X if $\langle c, x \rangle \geq \alpha$ for all $x \in X$ and $\langle c, y \rangle = \alpha$ for some point $y \in \bar{X}$. If $y \in X$, the vector c is said to be the support vector of the set X at the point y, c being, at the same time, the vector normal to the hyperplane which is a support to X and passes through the point y.

THEOREM 1.1.1 (on Separability). Let X be a nonempty, convex set in E^n, not containing the origin. Then there exists a hyperplane separating the set X from the origin.

The proof of this theorem can be found in many works on Convex Analysis (see, for instance, Nikaido [1], Mangasarian [1], Rockafellar [1].

2. PROPERTIES OF CONVEX FUNCTIONS WITHOUT DIFFERENTIABILITY

THEOREM 1.1.2. Let the strictly convex function $f(x)$ attain the minimum value on E^n at some point x_*. Then $f(x)$ is an infinitely large function.

Proof. Let S be the surface of the sphere of unit radius centered at the point x_*, and let x_1 be an arbitrary point outside S. We draw the straight line between x_* and x_1, and denote by $\tilde{x}$ the intersection point of this line with S. We then write

$$\tilde{x} = \lambda x_* + (1-\lambda)x_1 , \qquad 0 < \lambda < 1 .$$

We define the value of the coefficient λ, depending on the choice of the point x_1, by the condition $\|\tilde{x} - x_*\| = 1$, and obtain

$$0 < 1 - \lambda(x_1) = \frac{1}{\|x_1 - x_*\|} < 1.$$

The convexity condition of $f(x)$ in x implies that $f(\tilde{x}) \leq \lambda f(x_*) + (1-\lambda)f(x_1)$. Next we find

$$0 < \frac{1}{1-\lambda(x_1)}[f(\tilde{x}) - f(x_*)] \leqslant f(x_1) - f(x_*).$$

If $\|x_1\| \to \infty$ then $1 - \lambda(x_1) \to 0$; from these inequalities we conclude that $f(x_1) \to \infty$, i.e., $f(x)$ is an infinitely large function. ///

THEOREM 1.1.3. A real-valued function f defined on the convex set X, is convex on X iff its epigraph is a convex set in E^{n+1}.

Proof. Let the set epi f be convex, $x, y \in X$. Then

$$[x, f(x)] \in \text{epi } f, \quad [y, f(y)] \in \text{epi } f .$$

The convexity condition for the epigraph implies that

$$[\lambda x + (1-\lambda)y, \lambda f(x) + (1-\lambda)f(y)] \in \text{epi } f$$

for any $0 \le \lambda \le 1$.

By the definition of the epigraph this implies in turn that

$$f(\lambda x + (1-\lambda)y) \le \lambda f(x) + (1-\lambda)f(y) ,$$

i.e., f is convex on X.

To prove necessity, let f be convex on X; $[x,\mu] \in \text{epi } f$, $[y,\nu] \in \text{epi } f$. From the convexity property of f we have

$$f(\lambda x + (1-\lambda)y) \le \lambda f(x) + (1-\lambda)f(y) \le \lambda\mu + (1-\lambda)\nu$$

for any $0 \le \lambda \le 1$. Therefore,

$$[\lambda x + (1-\lambda)y, \lambda\mu + (1-\lambda)\nu] \in \text{epi } f$$

for any $0 \le \lambda \le 1$, i.e., the set epi f is convex in E^{n+1}. ///

THEOREM 1.1.4. The function $f(x)$ is convex on E^n iff for any $x, y \in E^n$ the function

$$\psi(\lambda) = f(\lambda x + (1-\lambda)y)$$

is convex for any $0 \le \lambda \le 1$.

Proof. Let f be convex, and let x, y be arbitrary points of E^n. We show that the epigraph of the function $\psi(\lambda)$ defined by

$$\text{epi } \psi = \{\lambda \in E^1, \alpha \in E^1 : 0 \le \lambda \le 1, \psi(\lambda) \le \alpha\}$$

is a convex set. Let $[\lambda_1,\alpha_1] \in \text{epi } \psi$, $[\lambda_2,\alpha_2] \in \text{epi } \psi$,

$$z_1 = \lambda_1 x + (1-\lambda_1)y \quad , \qquad z_2 = \lambda_2 x + (1-\lambda_2)y \quad .$$

We then have

$$f(z_1) = f(\lambda_1 x + (1-\lambda_1)y) = \psi(\lambda_1) \leqslant \alpha_1,$$
$$f(z_2) = \psi(\lambda_2) \leqslant \alpha_2.$$

Therefore, $[z_1, \alpha_1]$, $[z_2, \alpha_2] \in \text{epi } f$. The set epi f is convex; hence for any $0 \leq \nu \leq 1$ we have

$$[\nu z_1 + (1-\nu)z_2, \ \nu\alpha_1 + (1-\nu)\alpha_2] \in \text{epi } f \quad ,$$

yielding

$$f(\nu z_1 + (1-\nu)z_2) \leq \nu\alpha_1 + (1-\nu)\alpha_2 \quad .$$

By the definition of the function ψ we obtain

$$f(\nu z_1 + (1-\nu)z_2) = \psi(\nu\lambda_1 + (1-\nu)\lambda_2) \leq \nu\alpha_1 + (1-\nu)\alpha_2 \quad .$$

Hence the epigraph of the function ψ is a convex set and, therefore, the function ψ is convex. The converse is proved similarly. ///

<u>THEOREM 1.1.5</u>. If the convex function f attains a finite value at a point $a \in E^n$, the set X defined by

$$X = \{x : f(x) \leq f(a), \ a \in E^n\}$$

is convex and not empty.

<u>Proof</u>. The set X is not empty, since it a priori contains the point a. For any points $x, y \in E^n$ and $0 \leq \lambda \leq 1$ (1.1) is satisfied. We assume in addition that $x, y \in X$. Then

$$f(\lambda x + (1-\lambda)y) \leq \lambda f(x) + (1-\lambda)f(y) \leq f(a) \quad .$$

By the arbitrariness of λ of the interval $[0,1]$ we conclude that the closed line segment joining two points x and y in X belongs to X. Hence the set X is convex. ///

<u>THEOREM 1.1.6</u>. Let $f_i(x)$ be a family of convex functions $i \in [1:m]$. Then for any $\alpha_i \geq 0$, $i \in [1:m]$, the following functions also are convex:

$$\varphi_1(x) = \sum_{i=1}^{m} \alpha_i f_i(x), \quad \varphi_2(x) = \max_{i \in [1:m]} f_i(x),$$
$$\varphi_3(x) = \max_{i \in [1:m]} [0, f_i(x)].$$

<u>Proof</u>. For any $x, y \in E^n$, $0 \leq \lambda \leq 1$ we have the inequalities

$$\varphi_1(\lambda x + (1-\lambda) y) = \sum_{i=1}^{m} \alpha_i f_i(\lambda x + (1-\lambda) y) \leqslant \lambda \sum_{i=1}^{m} \alpha_i f_i(x) + $$
$$+ (1-\lambda) \sum_{i=1}^{m} \alpha_i f_i(y) = \lambda \varphi_1(x) + (1-\lambda) \varphi_1(y),$$
$$\varphi_2(\lambda x + (1-\lambda) y) = f_j(\lambda x + (1-\lambda) y),$$

where j is an integer of the set $[1:m]$. Hence

$$\varphi_2(\lambda x + (1-\lambda) y) \leqslant \lambda f_j(x) + (1-\lambda) f_j(y) \leqslant$$
$$\leqslant \lambda \varphi_2(x) + (1-\lambda) \varphi_2(y)$$

implying in turn the convexity of $\phi_2(x)$. We add the convex function $f_{m+1}(x) \equiv 0$ to the set of the functions $f_i(x)$, and arrive at the convexity of $\phi_3(x)$. ///

<u>THEOREM 1.1.7</u>. Let the function $f(x)$ be convex and let $\phi(z)$ be a monotone increasing convex function of z, where z is scalar. Then the composite function $\phi(f(x))$ is convex in x.

<u>Proof</u>. We take advantage of the inequality (1.1). The monotonicity of ϕ implies that

$$\phi(f(\lambda x + (1-\lambda)y)) \leq \phi(\lambda f(x) + (1-\lambda)f(y)) .$$

By the convexity of ϕ we have

$$\phi(\lambda f(x) + (1-\lambda)f(y)) \leq \lambda\phi(f(x)) + (1-\lambda)\phi(f(y)) ,$$

which proves the convexity of the composite function. ///

2. DIFFERENTIABILITY OF CONVEX FUNCTIONS

1. DIRECTIONAL DIFFERENTIABILITY

THEOREM 1.2.1. If $f(t)$ is a convex function of the scalar argument t, the function

$$\varphi(t) = \frac{f(t) - f(t_0)}{t - t_0}$$

for $t > t_0$ is bounded from below and does not decrease as t increases.

Proof. Let $t_0 < t_1 < t_2$. Then

$$t_1 = t_0 + \lambda(t_2 - t_0), \quad 0 < \lambda = \frac{t_1 - t_0}{t_2 - t_0} < 1.$$

The convexity property of $f(t)$ implies

$$\begin{aligned} f(t_1) = f(t_0 + \lambda(t_2 - t_0)) = f(\lambda t_2 + (1-\lambda)t_0) \leqslant \\ \leqslant \lambda f(t_2) + (1-\lambda) f(t_0) = \\ = \frac{t_1 - t_0}{t_2 - t_0} f(t_2) + \left(1 - \frac{t_1 - t_0}{t_2 - t_0}\right) f(t_0) \end{aligned}$$

yielding in turn

$$\varphi(t_1) = \frac{f(t_1) - f(t_0)}{t_1 - t_0} \leqslant \frac{f(t_2) - f(t_0)}{t_2 - t_0} = \varphi(t_2), \tag{2.1}$$

i.e., ϕ is the monotone nondecreasing function of t for $t > t_0$.

We show that it is bounded as $t \to +t_0$. Let $t_* < t_0 < t$. It is not difficult to obtain an inequality similar to (2.1):

$$\varphi(t_*) = \frac{f(t_0) - f(t_*)}{t_0 - t_*} \leqslant \frac{f(t) - f(t_0)}{t - t_0} = \varphi(t).$$

Thus, the function $\phi(t)$, monotonically decreasing as $t \to +t_0$, is bounded from below and therefore the limit

$$\lim_{t \to +t_0} \frac{f(t) - f(t_0)}{t - t_0}$$

exists, which is called the right derivative of the function $f(t)$ at the point t_0. One can prove similarly that the convex function $f(t)$ has a left derivative at the point t_0. ///

THEOREM 1.2.2. Let the convex function $f(x)$ attain a finite value at a point $x \in E^n$. Then for any direction $q \in E^n$, $\|q\| = 1$ there exists a derivative of f in the direction q:

$$\frac{\partial f(x)}{\partial q} = \lim_{t \to +0} \frac{f(x + tq) - f(x)}{t}.$$

Proof. We consider the line segment joining the point x and $y = x + q$. Also, we define the function of the scalar argument t:

$$\psi(t) = f(ty + (1-t)x) = f(x + tq) .$$

It follows from Theorem 1.1.4 that the function $\psi(t)$ is convex for any $0 \leq t \leq 1$. By Theorem 1.2.2, for the same values of t, $\psi(t)$ has right and left derivatives. Therefore the limit exists:

$$\lim_{t \to +0} \frac{\psi(t) - \psi(0)}{t} = \frac{\partial f(x)}{\partial q},$$

which was to be proved. ///

In our theorem it is of no consequence that the vector of

the direction q be a unit vector; in fact any n-dimensional vector can play this role. In the sequel we shall write the derivative of the function f at a point x in the direction of the vector $q \in E^n$ as

$$D^+f(x, q) = \lim_{t \to +0} \frac{f(x+tq) - f(x)}{t},$$

where $+$ signifies that the derivative is taken on the right. The derivative on the left can be defined similarly in the direction q:

$$D^-f(x, q) = \lim_{t \to -0} \frac{f(x+tq) - f(x)}{t}.$$

It is not hard to see that

$$D^-f(x,q) = -D^+f(x,-q).$$

Using the preceding theorems, we can show that at each point x at which the convex function f is finite, there exist left and right derivatives in the arbitrary direction q, $D^+f(x,q) \geq D^-f(x,q)$.

2. PROPERTIES OF SUBGRADIENTS

DEFINITION 1.2.1. Let the function $f(x)$ be defined everywhere on E^n. To each point $x \in \mathrm{dom}\, f$ we associate the set

$$\partial f(x) = \{z \in E^n: \langle z, y-x\rangle \leqslant f(y) - f(x) \ \forall y \in E^n\}, \tag{2.2}$$

which we call the set of subgradients of the function f at the point x. A concrete element of $\partial f(x)$ is said to be a subgradient of the function $f(x)$ at x.

THEOREM 1.2.3. Let the convex function $f(x)$ attain a finite value at the point x. The vector $z \in \partial f(x)$ iff

$$\langle z,p\rangle \leq D^{+}f(x,p) \tag{2.3}$$

for any vector $p \in E^{n}$.

Proof. Let the subgradient z satisfy (2.2): assuming $y = x + tp$, where $t > 0$. In this case, for any p

$$t\langle z,p\rangle \leq f(x+tp) - f(x) \quad . \tag{2.4}$$

Dividing (2.4) by t, we obtain

$$\langle z, p\rangle \leqslant \frac{f(x+tp)-f(x)}{t}.$$

We let t go to zero and arrive thus at the required inequality (2.3).

Conversely, let (2.3) hold. It is not hard to show that for any vector p the convex function $f(x)$ satisfies

$$D^{+}f(x, p) = \inf_{t>0} \frac{f(x+tp)-f(x)}{t}.$$

Hence, from this condition plus (2.3) we have that (2.4) holds for any p. Letting $y = x + tp$, we arrive at (2.2), which implies in turn that z is a subgradient of the function f at the point x. ///

THEOREM 1.2.4. The set of subgradients $\partial f(x)$ of the convex function f is convex and closed at each point $x \in \text{dom } f$.

Proof. Let $z_1, z_2 \in \partial f(x)$. Then for any $y \in E^{n}$ we have

$$\langle z_1, y-x\rangle \leqslant f(y)-f(x), \tag{2.5}$$
$$\langle z_2, y-x\rangle \leqslant f(y)-f(x). \tag{2.6}$$

Next, multiplying (2.5) by λ, (2.6) by $(1-\lambda)$, where $0 < \lambda < 1$, and summing up, we obtain

$$\langle z_1 + (1-\lambda)z_2, y-x\rangle \leq f(y) - f(x) .$$

Since this inequality holds for any $y \in E^n$, we conclude that

$$\lambda z_1 + (1-\lambda)z_2 \in \partial f(x) , \qquad 0 \leq \lambda \leq 1 ,$$

implying, in turn, the convexity of the set of subgradients.

We prove next the closedness of $\partial f(x)$ at the point x. Let the sequence $z_k \in \partial f(x)$, $\{z_k\} \to z_*$, $z_* \notin f(x)$. Then there exist $\bar{y} \in E^n$ and $\varepsilon > 0$ such that

$$\langle z_*, \bar{y}-x\rangle = f(\bar{y}) - f(x) + \varepsilon . \tag{2.7}$$

For any element, z_k of the sequence $\{z_k\}$ the condition

$$\langle z_k, \bar{y}-x\rangle \leq f(\bar{y}) - f(x) . \tag{2.8}$$

Subtracting (2.7) from (2.8), we obtain

$$\varepsilon \leq \langle z_*-z_k, \bar{y}-x\rangle \leq ||z_* - z_k||\cdot||\bar{y}-x|| .$$

This implies that for any $k > 0$

$$0 < \frac{\varepsilon}{\|\bar{y}-x\|} \leqslant \|z_*-z_k\|,$$

but this contradicts the convergence of the sequence $\{z_k\}$ to z_*. Hence we conclude that $z_* \in \partial f(x)$ and the set of subgradients is therefore closed. ///

From this and the preceding theorems we have that if at x the set of subgradients of the function $f(x)$ is not empty, then

$$D^+f(x, p) = \max_{z \in \partial f(x)} \langle z, p \rangle.$$

The notion of a subgradient is applicable in the case where the function f is defined only on some set X.

DEFINITION 1.2.2. Let the function $f(x)$ be defined on the set X the interior of which is not empty. For the point $x \in X$ we define the set $\partial f(x)$ which is the union of all vectors $z \in E^n$ satisfying

$$\langle z, y-x \rangle \leq f(y) - f(x)$$

for any $y \in X$. The set $\partial f(x)$ is said to be the set of subgradients of the function f at the point x.

For a convex function the set of subgradients is nonempty, bounded, convex at any interior point of the domain of definition.

3. PROPERTIES OF CONVEX DIFFERENTIABLE FUNCTIONS

THEOREM 1.2.5. Let f be a convex function defined on an open set $X \subset E^n$, and let f be differentiable at $\bar{x} \in X^-$. Then

$$\langle f_x(\bar{x}), x-\bar{x} \rangle \leq f(x) - f(\bar{x}) \quad . \tag{2.9}$$

Proof. Since $\bar{x}$ belongs to the open set X, there exists a neighborhood $G_\varepsilon(\bar{x})$ of the point $\bar{x}$, such that $G_\varepsilon(\bar{x})$ belongs entirely to X. Let $x \in X$ and $x \neq \bar{x}$. We connect the points x and $\bar{x}$ by the straight line and put

$$y = \bar{x} + \mu(x-\bar{x}) , \qquad 0 < \mu < 1 \quad . \tag{2.10}$$

If $0 < \mu < \varepsilon/\|x-\bar{x}\|$, then

$$y = (1-\mu)\bar{x} + \mu x \subset G_\varepsilon(\bar{x}) \subset X .$$

The convexity of $f(x)$ on the convex set $G_\varepsilon(\bar{x})$ implies that

$$f((1-\lambda)\bar{x} + \lambda y) \leq (1-\lambda)f(\bar{x}) + \lambda f(y) .$$

Now we make use of the differentiability of f at the point $\bar{x}$ (see Appendix I):

$$f(y)-f(\bar{x}) \geqslant \frac{f((1-\lambda)\bar{x}+\lambda y)-f(\bar{x})}{\lambda} = \\ = \langle f_x(\bar{x}), y-\bar{x}\rangle + \|y-\bar{x}\| \alpha(\bar{x}, \lambda(y-\bar{x})). \tag{2.11}$$

Here the function α satisfies the condition

$$\lim_{\lambda\to 0} \alpha(\bar{x}, \lambda(y-\bar{x})) = 0 .$$

Passing in (2.11) to the limit as $\lambda \to 0$, we obtain

$$\langle f_x(\bar{x}), y-\bar{x}\rangle \leq f(y) - f(\bar{x}) . \tag{2.12}$$

The point y belongs to a convex set; hence, using (2.10) and (1.1), we obtain

$$f(y) - f(\bar{x}) = f(\bar{x} + \mu(x-\bar{x})) - f(\bar{x}) \leq \mu[f(x) - f(\bar{x})] .$$

Substituting this expression in (2.12) and taking into account (2.10), we obtain the required result, (2.9). ///

THEOREM 1.2.6. Let f be a differentiable function on an open convex set X. In this case f is convex on X iff the inequality

$$\langle f_x(x_1), x_2 - x_1\rangle \leq f(x_2) - f(x_1) \tag{2.13}$$

is satisfied for any $x_1, x_2 \in X$.

Proof. Necessity follows from the preceding theorem. We prove sufficiency. If $x_1, x_2 \in X$, $0 \leq \lambda \leq 1$, the convexity of X implies that $x_1 + \lambda(x_2 - x_1) \in X$. From (2.13) we obtain

$$\lambda \langle f_x(x_1+\lambda(x_2-x_1)),\ x_1-x_2\rangle \leqslant f(x_1)-f(x_1+\lambda(x_2-x_1)),$$
$$-(1-\lambda)\langle f_x(x_1+\lambda(x_2-x_1)),\ x_1-x_2\rangle \leqslant$$
$$\leqslant f(x_2)-f(x_1+\lambda(x_2-x_1)) .$$

Multiplying the first of these inequalities by $(1-\lambda)$, the second by λ, and adding them, we obtain the inequality

$$f(\lambda x_2 + (1-\lambda)x_1) \leq \lambda f(x_2) + (1-\lambda)f(x_1)$$

implying in turn the convexity of the function f. ///

If we interchange x_1 and x_2 in (2.13), we obtain the following formula:

$$\langle f_x(x_1),\ x_2-x_1\rangle \leq f(x_2 - f(x_1) \leq \langle f_x(x_2),\ x_2-x_1\rangle . \quad (2.14)$$

THEOREM 1.2.7. Let f be a differentiable function on an open convex set X. In this case f is convex on X iff the inequality

$$0 \leq \langle f_x(x_2) - f_x(x_1),\ x_2 - x_1\rangle \quad (2.15)$$

is satisfied for any $x_1, x_2 \in X$.

Proof. Let $f(x)$ be convex on X and let $x_1, x_2 \in X$. By the preceding Theorem the following inequalities hold:

$$\langle f_x(x_1),\ x_2-x_1\rangle \leqslant f(x_2)-f(x_1),$$
$$\langle f_x(x_2),\ x_1-x_2\rangle \leqslant f(x_1)-f(x_2).$$

Adding these inequalities, we arrive at (2.15).

We prove sufficiency. Let $x_1, x_2 \in X$. Then we have $x_1 + \lambda(x_2 - x_1) \in X$ for any $0 \leq \lambda \leq 1$. By the mean value theorem (see Appendix I) there exists $\bar{\lambda}$, $0 < \bar{\lambda} < 1$ such that

$$f(x_2) - f(x_1) = \langle f_x(x_1 + \bar{\lambda}(x_2 - x_1)),\ x_2 - x_1 \rangle \ . \tag{2.16}$$

By condition (2.15) we have

$$0 \leq \langle f_x(x_1 + \bar{\lambda}(x_2 - x_1)) - f_x(x_1),\ \bar{\lambda}(x_2 - x_1) \rangle \ ,$$

yielding

$$\langle f_x(x_1),\ x_2 - x_1 \rangle \leq \langle f_x(x_1 + \bar{\lambda}(x_2 - x_1)),\ x_2 - x_1 \rangle \ .$$

Taking into account (2.16), we obtain

$$\langle f_x(x_1),\ x_2 - x_1 \rangle \leq f(x_2) - f(x_1)$$

implying, by the preceding Theorem, the convexity of the function $f(x)$. ///

The next two theorems will be proved in a similar manner.

THEOREM 1.2.8. Let the function $f(x)$, strictly convex on an open set $X \subset E^n$, be differentiable at a point $\bar{x} \in X$. Then the inequality

$$\langle f_x(\bar{x}),\ x - \bar{x} \rangle < f(x) - f(\bar{x})$$

holds for any $x \in X$, $x \neq \bar{x}$.

THEOREM 1.2.9. If the function f is differentiable on an open convex set X, then f is strictly convex on X iff the inequality

$$\langle f_x(x_1),\ x_2 - x_1 \rangle < f(x_2) - f(x_1)$$

holds for any $x_1, x_2 \in X$, $x_1 \neq x_2$.

The assertions of these theorems are carried over to concave functions: the signs $\leq$, $<$ being replaced respectively by $\geq$, $>$.

4. PROPERTIES OF TWICE-DIFFERENTIABLE CONVEX FUNCTIONS

THEOREM 1.2.10. Let f be the convex function defined on an open set $X \subset E^n$, and let f be twice differentiable at $x \in X$. Then the matrix $f_{xx}(x)$ is nonnegative definite, i.e., the inequality

$$0 \leq y^T f_{xx}(x)y \qquad (2.17)$$

holds for any $y \in E^n$.

Proof. For any y there exists an integer $\bar{\lambda} \geq 0$ such that for any $0 \leq \lambda \leq \bar{\lambda}$ the points $x + \lambda y \in X$ and, by Theorem 1.2.5,

$$0 \leq f(x+\lambda y) - f(x) - \lambda \langle f_x(x), y \rangle .$$

On the other hand, noting the property of differentiability, we obtain

$$f(x+\lambda y) - f(x) - \lambda \langle f_x(x), y \rangle = \frac{\lambda^2}{2} y^T f_{xx}(x) y + \lambda^2 \|y\|^2 \beta(x, \lambda y),$$

where $\lim_{\lambda \to 0} \beta(x, \lambda y) = 0$.

Hence

$$0 \leq \frac{1}{2} y^T f_{xx}(x)\, y + \|y\|^2 \beta(x, \lambda y) .$$

Letting λ go to zero, we arrive at (2.17).

THEOREM 1.2.11. If the function f is twice differentiable on an open convex set X, then f is convex on X iff the matrix $f_{xx}(x)$ is nonnegative definite on X.

Proof. Necessity follows from the preceding Theorem. Let us

prove sufficiency. If $x_1, x_2 \in X$, then, by Taylor's formula (see Appendix I), the inequality

$$f(x_2) - f(x_1) - \langle f_x(x_1), x_2 - x_1 \rangle = \\ = \frac{1}{2}(x_2 - x_1)^T f_{xx}(x_1 + \tau(x_2 - x_1))(x_2 - x_1)$$

holds for some $0 < \tau < 1$. The right side of this equality is nonnegative, hence the left side also is nonnegative and, by Theorem 1.2.6, the function f is convex on X. ///

The next theorem is also simple to prove.

THEOREM 1.2.12. Let the function f be twice differentiable on an open convex set X. A sufficient condition for f to be strictly convex on X is that the matrix $f_{xx}(x)$ be positive definite on X.

5. PSEUDOCONVEX FUNCTIONS

DEFINITION 1.2.3. Let the function f be defined on some open set in E^n, containing the set X. The function f is said to be pseudoconvex at the point $\bar{x}$ with respect to the set X, if it is differentiable at $\bar{x}$ and

$$0 \leq \langle f_x(\bar{x}), x - \bar{x} \rangle \tag{2.18}$$

for all $x \in X$ implies that $f(\bar{x}) \leq f(x)$.

If the function f is defined everywhere on E^n, is differentiable at the point $\bar{x}$, and the inequality (2.18) for all $x \in E^n$ implies that $f(\bar{x}) \leq f(x)$ for any x, we say that the function f is pseudoconvex at the point $\bar{x}$.

Assume that the function f is convex on a convex set X,

is differentiable at a point $\bar{x} \in X$ at which the function f attains the minimum on X. Then the function f is pseudoconvex at the point $\bar{x}$ with respect to the set X. Indeed, in this case the inequality (2.9) holds; hence the fact that (2.18) is satisfied implies automatically that $f(\bar{x}) \leq f(x)$ for any $x \in X$.

3. NECESSARY AND SUFFICIENT CONDITIONS OF A LOCAL EXTREMUM OF FUNCTIONS OF MANY VARIABLES

1. *BASIC DEFINITIONS*

Let the function f be defined on some set $X \subset E^n$. We say that x_* is a local minimum point of the function f on the set X, if there exists a neighborhood $G(x_*)$ of the point x_*, such that the inequality

$$f(x_*) \leq f(x) \tag{3.1}$$

holds for all x belonging to the intersection of the set X and the set $G(x_*)$. If the strict inequality

$$f(x_*) < f(x) \tag{3.2}$$

holds for all $x \in X \cap G(x_*)$, $x \neq x_*$, we say that x_* is a local isolated or a local strict minimum of the function $f(x)$.

If the inequality (3.1) holds for any $x \in X$, x_* is said to be a minimum or, more precisely, a global minimum of the function f on X. The problem of finding the minimum of the function f on the set X is written as

$$\min_{x \in X} f(x) \quad . \tag{3.3}$$

We denote by

$$X_* = \operatorname*{Arg\,min}_{x \in X} f(x)$$

the set of all the points $x_* \in X$ satisfying (3.1) for any $x \in X$ and call it the set of solutions to the problem (3.3). Usually, in minimization problems it is sufficient to find at least one point $x_* \in X_*$.

In those cases where X coincides with the entire space E^n, (3.3) is said to be the problem of unconstrained minimization of the function f, the point x_* satisfying (3.1) for any $x \in E^n$, is said to be the global minimum of the function $f(x)$. If (3.1) holds for all $x \in G(x_*)$, x_* is said to be an unconstrained local minimum of the function $f(x)$. When (3.2) is satisfied for all x $G(x_*)$, $x \neq x_*$, the point x_* is said to be a local isolated or a local strict minimum of the function $f(x)$.

2. NECESSARY AND SUFFICIENT CONDITIONS FOR A LOCAL MINIMUM

DEFINITION 1.3.1. The point x_* is said to be a stationary point of the differentiable function $f(x)$ if

$$f_x(x_*) = 0 . \tag{3.4}$$

THEOREM 1.3.1 (Necessary Conditions for the Minimum). Let the local minimum of the function f on the set X be attained at the point $x_* \in \operatorname{int} X$. Then:

• 1. if the function f is differentiable at x_*, then x_* is the stationary point of the function f;

• 2. if the function f is twice differentiable at the point

x_*, the matrix $f_{xx}(x_*)$ is nonnegative definite.

Proof. It follows from Definition 1.1.2 that for an arbitrary vector $y \in E^n$ there exists $\bar{\lambda} > 0$ such that for any $0 \leq \lambda < \bar{\lambda}$ the point $x_* + \lambda y \in G(x_*) \subset X$. Using the differentiability of f, we write the condition (3.1) as

$$0 \leq f(x_*+\lambda y) - f(x_*) = \lambda\langle f_x(x_*), y\rangle + \lambda\|y\|\alpha(x_*, \lambda y), \quad (3.5)$$

where the function α is such that $\lim_{\lambda\to 0} \alpha(x_*, \lambda y) = 0$. Taking into account this property, we find from (3.5):

$$\lim_{\lambda\to 0} \frac{f(x_*+\lambda y)-f(x_*)}{\lambda} = \langle f_x(x_*), y\rangle.$$

Hence (3.5) implies

$$0 \leq \langle f_x(x_*), y\rangle .$$

To have this inequality satisfied for an arbitrary y, it is necessary that the condition (3.4) be satisfied.

If the function f is twice differentiable at the point x_*, instead of (3.5) we use the following formula:

$$\begin{aligned} 0 \leq f(x_*+\lambda y) - f(x_*) &= \\ &= \frac{\lambda^2}{2} y^T f_{xx}(x_*) y + \lambda^2\|y\|^2\beta(x_*, \lambda y), \end{aligned} \quad (3.6)$$

where $\lim_{\lambda\to 0} \beta(x_*, \lambda y) = 0$.

Letting λ to zero, we obtain

$$\lim_{\lambda\to 0} \frac{f(x_*+\lambda y)-f(x_*)}{\lambda^2} = \frac{1}{2} y^T f_{xx}(x_*) y \geq 0.$$

By the arbitrariness of the vector $y \in E^n$ the last formula implies that the matrix $f_{xx}(x_*)$ is nonnegative definite, thus completing the proof. ///

THEOREM 1.3.2 (A Sufficient Condition for the Local Minimum). Let the function f defined on X be twice differentiable at the point $x_* \in \text{int } X$, and let the stationarity condition (3.4) be satisfied; also let the matrix $f_{xx}(x_*)$ be positive definite. Then x_* is an unconstrained local isolated minimum of the function f.

Proof. The positive definiteness of the matrix $f_{xx}(x_*)$ implies the existence of $C > 0$ such that $y^T f_{xx}(x_*)y \geq 2C\|y\|^2$ for all $y \in E^n$. We write $\Delta x = x - x_*$ and, then, obtain from the definition of the second derivative if we take into account (3.4) and (3.6) that

$$f(x) - f(x_*) = \frac{1}{2}\Delta x^T f_{xx}(x_*)\Delta x + \|\Delta x\|^2 \beta(x_*, \Delta x) \geqslant$$
$$\geqslant (C + \beta(x_*, \Delta x))\|\Delta x\|^2 .$$

The function $\beta(x_*, \Delta x)$ is such that there exists a neighborhood $G(x_*)$ in which $C + \beta(x_*, \Delta x) > 0$ for all $x \in G(x_*)$. Hence we arrive at the inequality (3.2) holding for any $x \in G(x_*)$, $x \neq x_*$, which was to be proved. ///

3. NECESSARY AND SUFFICIENT CONDITIONS FOR A LOCAL MAXIMUM

It is possible to consider the problem of finding the maximum of the differentiable function f on the set X:

$$\max_{x \in X} f(x) \quad . \tag{3.7}$$

If the maximum in this problem is attained at the interior point x_* of the set X, it is necessary that the condition (3.4) be satisfied at this point; and, if the function f is twice differentiable at x_*, the matrix $f_{xx}(x_*)$ has to be non-positive definite, that is, for any $y \in E^n$ we have the inequality

$$y^T f_{xx}(x_*) y \le 0 .$$

A sufficient condition for the strict local maximum of the function f at the point x_* satisfying (3.4), is that the matrix $f_{xx}(x_*)$ be negative definite.

Thus, if the local maximum or minimum of the differentiable function f is attained at the interior point x_* of the set X, the condition (3.4) is satisfied at this point. Hence in solving extremal problems it is sometimes useful to find stationary points. The problem of finding stationary points of the differentiable function f coincides with the problem of solving the system of equations $f_x(x) = 0$. Hence to solve the problems (3.3) and (3.7) one can use methods for solving the system of equations some of which will be given in Chapter 2.

4. SYLVESTER'S CRITERION

Let $\Delta_1(x), \Delta_2(x), \ldots, \Delta_n(x)$ denote the sequential principal minors of the matrix $f_{xx}(x)$.

THEOREM 1.3.3 (Sylvester's Criterion). In order that the matrix $f_{xx}(x)$ be positive definite, it is necessary and sufficient that the conditions

$$\Delta_1(x) > 0, \quad \Delta_2(x) > 0, \quad \ldots, \quad \Delta_n(x) > 0 \quad (3.8)$$

be satisfied. In order that the matrix $f_{xx}(x)$ be negative definite, it is necessary and sufficient that the conditions

$$\Delta_1(x) < 0, \quad \Delta_2(x) > 0, \; \ldots, \; (-1)^n \Delta_n(x) > 0 \qquad (3.9)$$

be satisfied.

The proof of the criterion can be found in any textbook on Linear Algebra. To each symmetric matrix $f_{xx}(x)$ it is possible to associate the quadratic form

$$\Phi(f) \sim z^T f_{xx}(x) z \;, \qquad z \in E^n \;.$$

The inequalities (3.8) and (3.9) yield necessary and sufficient conditions for the quadratic form $\Phi(x)$ to be positive definite or negative definite. In textbooks on mathematical analysis (see, for instance, G.M. Fikhtengol'ts [1]), only those cases are usually considered where the conditions (3.8) and (3.9) are satisfied; all of other possible cases are "indefinite" since they provide no sufficient conditions for the extremum of the functions. Later on, in Section 5, we shall show that the cases where in (3.8) and (3.9) the inverse strict inequalities hold correspond to the sufficient conditions of the local maximin and minimax.

5. CONVEX FUNCTIONS

THEOREM 1.3.4. Let the function f be convex on E^n. Then each local minimum of the function f is at the same time global minimum on E^n; the set of points of the minimum of the function f is convex. If the function f is strictly convex, X_* consists of a single point.

Proof. Let the condition (3.1) be satisfied for all $x \in G(x_*)$. We take an arbitrary point $y \in E^n$; then for sufficiently small

$0 < \lambda < 1$ we have $x_* + \lambda(y-x_*) \in G(x_*)$. Using the convexity of f, we obtain

$$f(x_*) \le f(x_* + \lambda(y-x_*)) \le \lambda f(y) + (1-\lambda)f(x_*) ,$$

which implies that $f(x_*) \le f(y)$. By the arbitrariness of $y \in E^n$ we conclude that x_* is a point of the global minimum of the function f. ///

Letting $c = f(x_*)$, we see that, by Theorem 1.1.5, the set of minima

$$X_* = \{x \in E^n : f(x) \le c\}$$

is convex.

Assume that the minimum of strictly convex function f is attained at two distinct points $x_1, x_2 \in X$. Then the convexity of X_* and the strict convexity of f imply that

$$c = f(\lambda x_1 + (1-\lambda)x_2) < \lambda f(x_1) + (1-\lambda)f(x_2) = C$$

for any $0 < \lambda < 1$. This contradiction implies in turn the uniqueness of the minimum point. ///

THEOREM 1.3.5. Let f be a convex function on E^n. The condition $0 \in \partial f(x_*)$ holds iff f attains its global minimum on E^n at the point x_*.

Proof. From the definition of the set of subgradients it follows that $0 \in \partial f(x_*)$ iff $f(x) \ge f(x_*)$ for any $x \in E^n$. Then we infer that x_* is the global minimum of $f(x)$.

If the function f is convex on E^n and differentiable at the point x_*, then $f_x(x_*) = 0$ iff x_* is the point of the global minimum of the function f. ///

4. NECESSARY AND SUFFICIENT CONDITIONS FOR A MINIMUM OF FUNCTIONS ON SETS

1. BASIC DEFINITIONS

We assume that the function f is defined on some open set containing the set X. We also assume that the set

$$X_* = \operatorname{Arg}\min_{x \in X} f(x)$$

is not empty. At those points $x \in X$ where the function f is differentiable it is possible to determine the point-set mapping

$$W(x) = \operatorname{Arg}\min_{y \in X} \langle f_x(x),\ y-x \rangle \quad . \tag{4.1}$$

This implies that $W(x) \subset X$. If the function f is differentiable everywhere on X, the condition (4.1) defines the set $W(x) \subset X$ for each point $x \in X$. Thus, W is the point-set mapping of the set X onto itself. The fixed points of the mapping $W(x)$ are such that

$$x \in X \ , \qquad x \in W(x) \quad . \tag{4.2}$$

If $x \notin W(x)$, there exists a point $y \in X$ such that

$$\langle f_x(x),\ y-x) \rangle < 0 \quad .$$

If $x \in W(x)$, we have the inequality

$$0 \le \langle f_x(x),\ y-x \rangle \qquad \forall y \in X \quad , \tag{4.3}$$

the latter being called sometimes "variational inequality"; each point $x \in X$ at which (4.3) is satisfied is said to be the solution of the variational inequality. For x to be a fixed point of the mapping W, it is necessary and sufficient that it be a solution of the variational inequality (4.3).

At each fixed point $x \in E^n$, where $f_x(x) \neq 0$, the set

$$K(x) = \{y \in E^n : \langle f_x(x), y-x \rangle = 0\}$$

is a hyperplane passing through the point x, with normal $f_x(x)$. At the same time, $K(x)$ is the tangent plane to the hypersurface

$$\Gamma(x) = \{y \in E^n : f(y) = f(x)\}$$

at the point x. The set $\Gamma(x)$ is the level set of the function $f(x)$, the gradient $f_x(x)$ is directed along the normal to the level set, inward the set

$$\Gamma_1(x) = \{y \in E^n : f(y) \geq f(x)\} .$$

The hyperplane $K(x)$ generates two half-spaces:

$$0 \leq \langle f_x(x), y-x \rangle, \langle f_x(x), y-x \rangle \leq 0 .$$

Let the point $x \in K$ be such that (4.3) is satisfied at this point. Then the former half-space contains the set X. Hence (4.3) can be interpreted as the condition for the set X to be contained entirely in one of the two closed half-spaces defined by the tangent hyperplane $K(x)$. By Definition 1.1.11, in this situation the tangent hyperplane $K(x)$ is a support to the set X at the point x, the vector $f_x(x)$ is a support to the set X at the point x.

2. FIRST ORDER CONDITIONS FOR A MINIMUM

Theorem 1.4.1. Let the function f be defined on an open set in E^n, containing a convex set X, and let there be a point $\bar{x} \in X_*$ at which the function f is differentiable. Then it is necessary that $\bar{x}$ be a fixed point of the mapping $W(x)$.

Proof. Let x be an arbitrary point of the set X. By Definition 1.1, the convexity of the set X implies that for any $\lambda \in [0,1]$ we have the inclusion $\bar{x} + \lambda(x-\bar{x}) \in X$. Since f is differentiable at the point $\bar{x}$ and $\bar{x} \in X_*$, we have

$$0 \leqslant f(\bar{x}+\lambda(x-\bar{x}))-f(\bar{x}) = \\ = \lambda\langle f_x(\bar{x}),\ x-\bar{x}\rangle + \lambda\|x-\bar{x}\|\alpha(\bar{x},\ \lambda(x-\bar{x})),$$

where

$$\lim_{\lambda\to 0} \alpha(\bar{x},\ \lambda(x-\bar{x})) = 0 .$$

Then we obtain that for any $\lambda \in [0,1]$

$$0 \leq \langle f_x(\bar{x}),\ x-\bar{x}\rangle + \|x-\bar{x}\|\alpha(\bar{x},\ \lambda(x-\bar{x})) .$$

Letting λ to zero, we arrive at the inequality

$$0 \leq \langle f_x(\bar{x}),\ x-\bar{x}\rangle \tag{4.4}$$

holding for any $x \in X$, which implies in turn that $\bar{x} \in W(\bar{x})$. ///

This theorem provides a necessary condition for the minimum of the function f to be global on the set X. As will be shown later (in Section 6.4), this necessary condition is closely connected with a discrete version of Pontryagin's maximum principle in optimal control theory.

The requirement for the set X to be convex, introduced in the Theorem, is essential. It is easy to give examples to show that a violation of this requirement makes the assertion of the Theorem wrong. Nevertheless, a local version of the Theorem is possible which does not require the set X to be convex. The inequality (4.3) will hold then only for the vectors y which belong to the intersection of X and some neighborhood of X_*.

THEOREM 1.4.2. Let the function f be defined on an open set in E^n, containing X, let the set X_* be nonempty and convex, and let there exist a point $\bar{x} \in X_*$ at which the function f is differentiable. Then there exists a neighborhood $G(X_*)$ of the set X_*, such that the inequality (4.4) holds for all $x \in X \cap G(X_*)$.

Definition 1.2.3 of the pseudoconvexity implies the following theorem providing sufficient conditions for the global minimum of the function f on the set X.

THEOREM 1.4.3. Let the function f be pseudoconvex at a point $\bar{x}$ with respect to the set X, and let $\bar{x}$ be a fixed point of the mapping W. Then $\bar{x} \in X_*$.

3. THE CASE OF CONVEX FUNCTIONS

We assume first that the function $f(x)$ is defined and convex on an open set containing the set X, there exists a point $x_* \in X_*$, and the function f is differentiable at the point $x \in X$. Then from (2.14) we have the inequality

$$0 \leq f(x) - f(x_*) \leq \langle f_x(x), x-x_* \rangle , \tag{4.5}$$

which holds for all x belonging to the set X. For $x \in X$,

$\bar{x} \in W(x)$ we have

$$\langle f_x(x), \bar{x}-x \rangle \leq \langle f_x(x), x_* - x \rangle .$$

Hence (4.5) can be expressed in terms

$$0 \leq f(x) - f(x_*) \leq \langle f_x(x), x-\bar{x} \rangle \qquad \forall\, x_* \in X_*, \quad \forall\, \bar{x} \in W(x). \tag{4.6}$$

We shall be using this inequality in the sequel.

Let $x \in X$, $x \in W(x)$. Then

$$0 = \langle f_x(x), \bar{x}-x \rangle \qquad \forall\, \bar{x} \in W(x)$$

and for any $y \in X$ (4.3) is satisfied. From the inequality (4.6) we infer that in this case $x \in X_*$. For any $x \notin X_*$, $x \in X$, by (4.5), we have

$$\langle f_x(x), x_* - x \rangle < 0 .$$

The inequality (4.3) is therefore violated for $y = x_*$.

We arrive at the following result:

In order that the function f, convex and differentiable at the point $x \in X$, to attain at x the minimum on the convex set X, it is necessary and sufficient that the condition (4.3) hold (that is, the point x has to be a fixed point of the mapping $W(x)$ or the gradient $f_x(x)$ has to be a support vector of the set X at the point x).

4. *SECOND ORDER NECESSARY CONDITION FOR A MINIMUM*

It may happen that the condition for a stationary point

$$f_x(\bar{x}) = 0 \tag{4.7}$$

is satisfied at the point $\bar{x} \in X$. If the function f is convex, this implies that $\bar{x} \in X_*$. In the general case the property (4.7) makes the condition (4.4) rather meaningless. Hence one needs to introduce necessary conditions of higher orders.

THEOREM 1.4.4 (Second-Order Necessary Condition for a Minimum). Let the function f be defined on an open set of E^n, containing a convex set X, and let the function f be twice differentiable at the point $\bar{x} \in X_*$ at which (4.7) is satisfied. Then it is necessary that the inequality

$$0 \le (x-\bar{x})^T f_{xx}(\bar{x})(x-\bar{x}) \tag{4.8}$$

hold for any $x \in X$.

Proof. Let x be an arbitrary point of the set X. It follows from the convexity of X that the inclusion

$$\bar{x} + \lambda(x-\bar{x}) \in X$$

holds for any $\lambda \in [0,1]$. Using the condition that the function f is twice differentiable at a point $\bar{x} \in X$, we obtain

$$\begin{aligned} 0 \leqslant f(\bar{x}+\lambda(x-\bar{x})) - f(\bar{x}) = \\ = \lambda^2 (x-\bar{x})^T f_{xx}(\bar{x})(x-\bar{x}) + \lambda^2 \|x-\bar{x}\|^2 \beta(\bar{x}, \lambda(x-\bar{x})). \end{aligned}$$

Here $\lim_{\lambda\to 0} \beta(\bar{x}, \lambda(x-\bar{x})) = 0$. Letting λ go to zero, we arrive at the inequality (4.8). ///

As before, one can define the point-set mapping

$$X_1(\bar{x}) = \operatorname{Arg}\min_{x\in X} (x-\bar{x})^T f_{xx}(\bar{x})(x-\bar{x})$$

at the stationary point $\bar{x}$. To find the set $X_1(\bar{x})$, one needs in this case to solve the problem of minimization of the quadratic form on the set X. It is possible to pose a problem of finding fixed points of this mapping or a problem of solving the quadratic variational inequalities

$$0 \leq (x - \bar{x})^T f_{xx}(\bar{x}) (x - \bar{x}) ,$$

defining thereby the points satisfying the second-order condition for a minimum.

5. PROPERTIES OF MINIMAX PROBLEMS

1. THE STATEMENT OF MINIMAX PROBLEMS

Let $F(x,y)$ be a continuous function defined for all $x \in E^n$ and $y \in E^m$. First, we pose the problems of finding the unconstrained maximin and minimax

$$V_1 = \max_{y \in E^m} \min_{x \in E^n} F(x, y), \tag{5.1}$$

$$V_2 = \min_{x \in E^n} \max_{y \in E^m} F(x, y). \tag{5.2}$$

Using the second problem as an example, we explain the meaning of the expression used. For each fixed $x \in E^n$ we find first the image of the point-set mapping $B(x)$:

$$B(x) = \operatorname{Arg} \max_{y \in E^m} F(x,y) . \tag{5.3}$$

Next we define the function of the maximum

$$\phi(x) = F(x, B(x)) , \tag{5.4}$$

where $F(x, B(x))$ signifies that the value $F(x,\bar{y})$ with $\bar{y} \in B(x)$

is being calculated. It follows from the definition (5.3) that $F(x,\bar{y})$ attains the same value for any $\bar{y} \in B(x)$, which justifies this form of notation.

Next we define the set

$$X_* = \operatorname*{Arg\,min}_{x \in E^n} \phi(x) .$$

Finding X_* and the point-set mapping $B(x)$ is said to be the synthesis of the problem (5.2); at least one pair of points $[x_*,y_*]$, where $x_* \in X_*$, $y_* \in B(x_*)$ is said to be the (global) solution of the minimax problem (5.2). The quantity $V_2 = F(x_*,y_*)$ is said to be the minimax estimate.

By an interior problem we mean the determination of at least one of the elements of the set $B(x)$ for each x. By an exterior problem we mean the determination of at least one point of X_*.

To make our discussion simpler, we assume in this section that all the operations of seeking extrema have solutions. In (5.2) this means that the multivalued mapping B is defined for any x, the set X_* being nonempty.

<u>THEOREM 1.5.1</u>. If the problems (5.1) and (5.2) have solutions, then

$$V_1 \leq V_2 . \tag{5.5}$$

<u>Proof</u>. For any fixed $y \in E^m$ the inequality

$$\min_{x \in E^n} F(x,y) \leq F(x,y)$$

is satisfied. Similarly, for each fixed x

$$F(x,y) \le \max_{y \in E^m} F(x,y) .$$

Hence for any $\bar{y} \in E^m$ and $\bar{x} \in E^n$

$$\min_{x \in E^n} F(x, \bar{y}) \leqslant \max_{y \in E^m} F(\bar{x}, y).$$

By virtue of the arbitrariness of $\bar{x}$ and $\bar{y}$ we obtain

$$\max_{y \in E^m} \min_{x \in E^n} F(x, y) \leqslant \min_{x \in E^n} \max_{y \in E^m} F(x, y).$$

2. CONSTRAINED MINIMAX PROBLEMS

In the problems (5.1) and (5.2) no constraints are imposed on x and y; hence we may say these are problems of seeking an unconstrained maximin and minimax. We consider the case where x and y are restricted by:

$$V_1 = \max_{y \in Y} \min_{x \in X} F(x, y), \tag{5.6}$$

$$V_2 = \min_{x \in X} \max_{y \in Y} F(x, y). \tag{5.7}$$

Solutions to these problems can be found in the same way as was done above (see (5.3) and (5.4)):

$$\begin{gathered} B(x) = \operatorname{Arg} \max_{y \in Y} F(x, y), \quad \varphi(x) = F(x, B(x)), \\ X_* = \operatorname{Arg} \min_{x \in X} \varphi(x). \end{gathered} \tag{5.8}$$

For applications, one question is of great importance: which properties of the function $F(x,y)$ change and which properties remain intact after the operation of maximization is performed over y, i.e., upon the transition of $F(x,y)$ to $\phi(x)$? To answer this question we consider first properties of continuity, convexity, and Lipschitz continuity.

To shorten the notation, let

$$z = [x,y] \in E^{n+m} ,$$

$$Z = X \times Y ,$$

$$F(x,y) = F(z) .$$

We say that the function $F(z)$ satisfies a Lipschitz condition on Z with a constant ℓ, the function $\phi(x)$ satisfies a Lipschitz condition on X with a constant ℓ, if the conditions

$$\begin{aligned} |F(z_1)-F(z_2)| &\leqslant l\|z_1-z_2\|, \\ |\varphi(x_1)-\varphi(x_2)| &\leqslant l\|x_1-x_2\| \end{aligned} \tag{5.9}$$

are respectively satisfied for any $z_1, z_2 \in Z$ and $x_1, x_2 \in X$.

THEOREM 1.5.2. If the function $F(x,y)$ is continuous on the direct product of compact sets $Z = X \times Y$, then the function $\phi(x)$ is continuous as well on X. If the function $F(z)$ satisfies a Lipschitz condition on Z, so does the function $\phi(x)$, with the same constant. If for each $y \in Y$ the function $F(x,y)$ is convex in x on the convex set X, $\phi(x)$ as well is convex on X.

Proof. The function $F(x,y)$, being continuous on the compact set Z, is uniformly continuous on this set. Hence for any $\varepsilon > 0$ there exists $\delta(\varepsilon)$ such that if $\|x_1 - x_2\| < \delta(\varepsilon)$, $x_1, x_2 \in X$, then

$$|F(x_1,y) - F(x_2,y)| < \varepsilon$$

for any $y \in Y$. We consider the expression

$$|\phi(x_1) - \phi(x_2)| = |F(x_1,y_1) - F(x_2,y_2)| ,$$

where $y_1 \in B(x_1)$, $y_2 \in B(x_2)$. We consider two cases. In case one we have

$$0 \leq \phi(x_1) - \phi(x_2) \quad .$$

Taking into account the definition of the set $B(x)$, we obtain

$$0 \leq \phi(x_1) - \phi(x_2) \leq F(x_1,y_1) - F(x_2,y_2) \leq \varepsilon \quad .$$

If the function $F(z)$ satisfies a Lipschitz condition in z, it satisfies as well a Lipschitz condition in x uniformly in y; hence

$$0 \leq \phi(x_1) - \phi(x_2) \leq \ell\|x_1 - x_2\| \quad .$$

In case two

$$0 \leq \phi(x_2) - \phi(x_1) \quad .$$

Similarly we obtain that

$$0 \leq \phi(x_2) - \phi(x_1) \leq F(x_2,y_2) - F(x_1,y_1) \leq \varepsilon \quad ,$$

and if F satisfies a Lipschitz condition, then

$$0 \leq \phi(x_2) - \phi(x_1) \leq \ell\|x_2 - x_1\| \quad .$$

Combining these two cases, we conclude that the inequality

$$|\phi(x_1) - \phi(x_2)| < \varepsilon$$

holds for any $x_1, x_2 \in X$ such that $\|x_1 - x_2\| < \delta(\varepsilon)$, i.e., the function $\phi(x)$ is continuous on the compact set X. If F satisfies a Lipschitz condition, then (5.9) holds.

If F is convex in x, then we have the estimates

$$\varphi(\lambda x_1+(1-\lambda)x_2)=\max_{y\in Y} F(\lambda x_1+(1-\lambda)x_2,\ y)\leqslant$$
$$\leqslant \max_{y\in Y}[\lambda F(x_1,\ y)+(1-\lambda)F(x_2,\ y)]\leqslant$$
$$\leqslant \lambda\varphi(x_1)+(1-\lambda)\varphi(x_2)$$

for any x_1, $x_2 \in X$, $0 \leq \lambda \leq 1$, which proves the convexity of the function ϕ. ///

The property of differentiability of the function F is not preserved in passing to the function ϕ. We have the following theorem.

<u>THEOREM 1.5.3</u> (Danskin-Dem'yanov). Let X be an open set in E^n, let Y be a compact set in E^m, let the functions F and $F_x(x, y)$ be continuous on the set $X \times Y$. Then the function $\phi(x)$ defined by (5.8) has at each point $x \in X$ a derivative in any direction $g \in E^n$,

$$\frac{\partial\varphi(x)}{\partial g}=\max_{y\in B(x)}\langle F_x(x,\ y),\ g\rangle.$$

The proof of this Theorem can be found in Danskin [1], and also in Dem'yanov and Malozemov [1]. In the latter the following theorem is proved which is a generalization of Theorem 1.4.1 to the case of minimization of a maximum function.

<u>THEOREM 1.5.4</u>. Let X be a closed convex set in E^n, let Y be a compact set in E^m, and let the function F be continuous together with $F_x(x,y)$ on the set $X \times Y$. Then, in order that the function $\phi(x)$ attain its minimum on X at a point $x_* \in X$, it is necessary and, in the case of convexity of $\phi(x)$ on X, also sufficient that

$$\inf_{x\in X}\ \max_{y\in B(x_*)} \langle F_x(x_*,y),\ x-x_*\rangle = 0 .$$

To conclude, we give a theorem on existence of solutions of the problems (5.6) and (5.7).

THEOREM 1.5.5. If the function $F(x,y)$ is continuous on the product of compact sets X and Y, then the solutions of the problems (5.6) and (5.7) exist and the inequality

$$V_1 = \max_{y \in Y} \min_{x \in X} F(x, y) \leqslant V_2 = \min_{x \in X} \max_{y \in Y} F(x, y) \tag{5.10}$$

holds.

The proof of this Theorem follows almost word-for-word the proof of Theorem 1.5.1.

3. SADDLE POINTS

We say that the point $z_* = [x_*, y_*]$ is a saddle point of the function $F(x,y)$ in the problems (5.6) and (5.7), if $x_* \in X$, $y_* \in Y$ and the inequalities

$$F(x_*, y) \leq F(x_*, y_*) \leq F(x, y_*) \tag{5.11}$$

hold for any $x \in X$, $y \in Y$.

LEMMA 1.5.1. If the problems (5.6) and (5.7) have solutions, then in order that the equality

$$V_1 = F(x_*, y_*) = V_2 \tag{5.12}$$

hold, it is necessary and sufficient that z_* be the saddle point of the function $F(x,y)$.

Proof. Let $[x_*, y_*]$ be a saddle point of the function $F(x,y)$. Then

$$\max_{y \in Y} F(x_*, y) \leqslant F(x_*, y_*) \leqslant \min_{x \in X} F(x, y_*).$$

The following inequalities are obvious:

$$\min_{x\in X}\max_{y\in Y} F(x, y) \leqslant \max_{y\in Y} F(x_*, y),$$
$$\min_{x\in X} F(x, y_*) \leqslant \max_{y\in Y}\min_{x\in X} F(x, y).$$

Combining the last three groups of inequalities, we obtain that

$$\min_{x\in X}\max_{y\in Y} F(x, y) \leqslant F(x_*, y_*) \leqslant \max_{y\in Y}\min_{x\in X} F(x, y).$$

We compare this inequality with (5.10) and conclude that the equality (5.12) holds.

Let (5.12) hold. We show that the point $[x_*, y_*]$ is saddle. From (5.12) there follows

$$F(x_*, y_*) = \min_{x\in X}\max_{y\in Y} F(x, y) = \max_{y\in Y} F(x_*, y) \geqslant F(x_*, y),$$
$$F(x_*, y_*) = \max_{y\in Y}\min_{x\in X} F(x, y) = \min_{x\in X} F(x, y_*) \leqslant F(x, y_*).$$

Combining these inequalities, we obtain that (5.11) holds, that is, $[x_*, y_*]$ is the saddle point of the function $F(x,y)$. ///

DEFINITION 1.5.1. We say that the function $F(x,y)$ is convex-concave, if $F(x,y)$ is convex in x for all fixed $y \in Y$ and $F(x,y)$ is concave in y for all fixed $x \in X$. The function $F(x,y)$ is said to be strictly convex-concave, if for any fixed y it is strictly convex in x and for any fixed x is strictly concave in y.

THEOREM 1.5.6 (Von Neumann). Let the convex-concave function $F(x,y)$ be continuous on the direct product of convex compact sets X and Y. Then the function $F(x,y)$ has a saddle point.

The proof of this theorem can be found, for instance, in Neumann and Morgenstern [1], and in Davydov [2].

4. THE NOTION OF LOCAL SOLUTIONS

If solving an interior or an exterior problem is treated as that of finding a local extrema, we arrive at the notion of local solutions of a minimax or a maximin problem. Such solutions are obtained if, for example, we attempt to solve the problem (5.2) numerically, using the methods of local unconstrained minimization and maximization.

DEFINITION 1.5.2. Let the points x_* and y_* have respectively the neighborhoods $G(x_*)$ and $G(y_*)$ such that for any vectors $x \in G(x_*)$ the condition

$$g(x) = \operatorname{Arg} \max_{y \in G(y_*)} F(x,y)$$

defines the single-valued function $g(x)$, x_* being the point of the strict minimum of the function $F(x, g(x))$ on $G(x_*)$ and $y_* = g(x_*)$. Then we say that $z_* = [x_*,y_*]$ is a strict local minimax point of the problem (5.2).

If z_* is a strict local minimax point, the strict inequalities

$$F(x_*, g(x_*)) < F(x, g(x)) \quad , \tag{5.13}$$

$$F(x,y) < F(x, g(x)) \tag{5.14}$$

are satisfied for any $x \in G(x_*)$, $x \neq x_*$, $y \in G(y_*)$, $y \neq g(x)$. The inequality (5.14) is satisfied as well for $x = x_*$, $y \neq g(x_*)$.

In a similar manner one can define solutions of the problems (5.1) and (5.2): local or global in x; local or global in y; strict or non-strict in x; or strict or non-strict in y. The definitions given of local solutions are a generalization of well-known notions of local extrema. The minimax problem (5.2) may have several local solutions. However, after they have been found, it is not possible to assert that there is a global solution among those local ones: one needs to consider all global solutions in y, all local solutions in x, and choose among them the solution for which the value of $F(x,y)$ is minimal. To study and find local solutions is crucial for the class of functions defined below, in which the points of the local minimax are at the same time points of the global minimax. For the maximin problems (5.1), local solutions are to be defined in a similar way.

DEFINITION 1.5.3. Let the points x_* and y_* have respectively neighborhoods $G(x_*)$ and $G(y_*)$ such that for any vectors $y \in G(y_*)$ the condition

$$d(y) = \operatorname{Arg} \min_{x \in G(x_*)} F(x,y)$$

defines the single-valued function $d(y)$, y_* being the strict maximum point of the function $F(d(y), y)$ on $G(y_*)$ and $d(y_*) = x_*$. Then $z_* = [x_*, y_*]$ is said to be a strict local maximin point of the problem (5.1). The inequalities (5.13), (5.14) are to be replaced in this case by

$$\begin{aligned} F(d(y), y) &< F(d(y_*), y_*) \quad , \\ F(d(y), y) &< F(x,y) \quad . \end{aligned} \tag{5.15}$$

DEFINITION 1.5.4. The function $F(x,y)$ has a strict local saddle at the point z_*, if z_* is at the same time a strict local minimax and maximin point.

Combining the inequalities (5.13) - (5.15), we obtain that at the saddle point the inequalities

$$\begin{aligned} F(x_*,y) &< F(x_*,y_*) < F(x,y_*) \quad , \\ F(d(y),\ y) &< F(x_*,y_*) < F(x,\ g(x)) \end{aligned} \tag{5.16}$$

are satisfied, where $y \in G(y_*)$, $y \neq y_*$, $x \in G(x_*)$, $x \neq x_*$.

5. NECESSARY AND SUFFICIENT CONDITIONS FOR A LOCAL MINIMAX

Introduce the square symmetric matrix of dimension n^2:

$$\Phi(x,y) = F_{xx}(x,y) - F_{xy}(x,y)\, F_{yy}^{-1}(x,y)\, F_{yx}(x,y) \quad .$$

THEOREM 1.5.7. Let the function $F(z)$ be twice continuously differentiable in some neighborhood $G(z_*)$ of the point z_*. In order that in the problem (5.2) the vector z_* be a strict local minimax point, it is sufficient that the conditions

$$\begin{aligned} F_y(z_*) &= 0_{m1} \ , \\ F_x(z_*) &= 0_{n1} \ , \\ F_{yy}(z_*) &< 0 \ , \\ \Phi(z_*) &> 0 \end{aligned} \tag{5.17}$$

be satisfied. If in the problem (5.2) the vector z_* is a strict local minimax point, it is necessary that $F_y(z_*) = 0_{m1}$, $F_x(z_*) = 0_{n1}$, $F_{yy}(z_*) \leq 0$, and if $F_{yy}(z_*) < 0$, then $\Phi(z_*) \geq 0$.

Proof. To prove sufficiency, one needs to analyze the inequalities (5.13) and (5.14). It follows from $F_y(z_*) = 0_{m1}$, $F_{yy}(z_*) < 0$ that y_* is a strict local maximum point of the function $F(x_*,y)$ in y. Since the matrix $F_{yy}(z_*)$ is negative definite, there exists some neighborhood $G(z_*) = G(x_*) \times G(y_*)$ of the point z_*, where the matrix $F_{yy}(z)$ also is negative definite and for all $x \in G(x_*)$ the implicit equation $F_y(x,y) = 0$ defines uniquely the continuously differentiable function $y = g(x)$, $y_* = g(x_*)$. This function can be obtained as a solution of the Cauchy problem for the following system of quasilinear equations

$$\frac{dg}{dx} F_{yy}(x, g(x)) + F_{xy}(x, g(x)) = 0 \tag{5.18}$$

with n independent variables $[x^1, \ldots, x^n]$ and the initial condition $g(x_*) = y_*$.

We denote by $g(x)$ the solution to the system (5.18), defined on $G(x_*)$ and assuming a value of $G(y_*)$. The values of the vector function $y = g(x)$ are local maximum points of the function $F(x,y)$ in y. Thus, the inequality (5.14) holds. Next, substituting $y = g(x)$ into $F(x,y)$, we obtain the maximum function

$$\phi(x) = F(x, g(x)) .$$

Differentiating ϕ noting (5.18), we obtain

$$\varphi_x(x) = \frac{dF(x, g(x))}{dx} = F_x(x, g(x)),$$

$$\begin{aligned}\varphi_{xx}(x) &= F_{xx}(x, g(x)) + F_{xy}(x, g(x)) \left(\frac{dg}{dx}\right)^T = \\ &= F_{xx}(x, g(x)) - F_{xy}(x, g(x)) F_{yy}^{-1}(x, g(x)) F_{yx}(x, g(x)) = \\ &= \Phi(x, g(x)).\end{aligned}$$

In particular, for $x = x_*$ we have $g(x_*) = y_*$, hence the matrix $\phi_{xx}(x_*)$ is positive definite and x_* is a strict local minimum point of the function $\phi(x)$, and (5.13) holds. Analogous considerations yield the necessary conditions for a local minimax point formulated in the Theorem. ///

Sufficient conditions of the Theorem can be stated as sufficient conditions for the global solution to (5.2). In this case, more rigid requirements ought to be imposed on $F(x,y)$. Arguing that for each $x \in E^n$ the interior problem in (5.2) has a solution and the exterior problem is solvable, we formulate the following sufficient condition for global minimax.

THEOREM 1.5.8. Let the function $F(x,y)$ for each fixed $x \in E^n$ be strictly concave in y and let the function $F(x, B(x))$ be strictly convex in x. Then the point of local minimax in (5.2) is at the same time the global solution of the problem (5.2).

Necessary and sufficient conditions for a maximin in the problem (5.1) will be similar.

THEOREM 1.5.9. Let the function $F(z)$ be twice continuously differentiable in a neighborhood of the point z_*. In order for z_* to be a strict local maximin point of the problem (5.1), it is sufficient that the conditions

$$F_x(z_*) = 0, \quad F_y(z_*) = 0, \quad F_{xx}(z_*) > 0,$$
$$N(z_*) = F_{yx}(z_*) F_{xx}^{-1}(z_*) F_{xy}(z_*) - F_{yy}(z_*) > 0$$

be satisfied.

6. CRITERION FOR MINIMAX DEFINITENESS

If the function $F(x,y)$ is differentiable everywhere in x and y, the conditions of Theorem 1.5.8 can be written in analytical form for any x_1 and x_2 of E_n, $x_1 \neq x_2$ and y_1, y_2 of

E^m, $y_1 \neq y_2$: it is necessary that the inequalities

$$\langle F_x(x_1, g(x_1)), x_2 - x_1\rangle < F(x_2, g(x_2)) - F(x_1, g(x_1)),$$
$$\langle F_y(x, y_2), y_2 - y_1\rangle > F(x, y_2) - F(x, y_1)$$

be satisfied, where

$$g(x_1) = \operatorname{Arg} \max_{y \in E^m} F(x_1, y), \quad g(x_2) = \operatorname{Arg} \max_{y \in E^m} F(x_2, y).$$

If the function $F(x,y)$ is differentiable and strictly convex-concave, the conditions (5.19) will a priori be satisfied. Indeed, for such functions the second inequality holds and the first inequality follows from

$$\langle F_x(x_1, g(x_1)), x_2 - x_1\rangle < F(x_2, g(x_1)) - F(x_1, g(x_1)) \leqslant$$
$$\leqslant \max_{y \in E^m} F(x_2, y) - F(x_1, g(x_1)) = F(x_2, g(x_2)) - F(x_1, g(x_1)).$$

If the function $F(x,y)$ is twice continuously differentiable, a sufficient condition is: for any $x \in E^n$ and $y \in E^m$ the conditions

$$F_{yy}(x,y) < 0, \qquad \Phi(x, g(x)) > 0$$

are satisfied. The latter condition will a priori be satisfied, if a stronger condition holds, the verification of which does not require the knowledge of the function $g(x)$: for any $x \in E^n$, $y \in E^m$

$$F_{yy}(x,y) < 0, \qquad \Phi(x,y) > 0$$

are satisfied.

We write these conditions in a different form and, to do this, introduce the square symmetric matrix

$$R(z) = \left[\begin{array}{c|c} F_{yy}(z) & F_{yx}(z) \\ \hline F_{xy}(z) & F_{xx}(z) \end{array}\right], \tag{5.19}$$

where the matrix R(z) is partitioned into four matrix-squares. We denote the sequential principal minors of the matrix R(z) by $\Delta_1(z)$, $\Delta_2(z)$, ..., $\Delta_{n+m}(z)$. In particular, $\Delta_m(z) = |F_{yy}(z)|$, $\Delta_{m+n} = |R(z)|$.

Now we prove the following

<u>CRITERION</u>. In order that the matrices $\Phi(z)$, $-F_{yy}(z)$ be positive definite, it is necessary and sufficient that the inequalities

$$(-1)^i \Delta_i(z) > 0, \quad i \in [1:m], \tag{5.20}$$

$$(-1)^m \Delta_{m+j}(z) > 0, \quad j \in [1:n], \tag{5.21}$$

hold for the principal minors of the matrix R(z).

The equivalence of the inequalities (5.20) with the negative definiteness of the matrix $F_{yy}(z)$ follows from Sylvester's criterion. In order to prove (5.21), we express the principal minors $\Delta_{m+j}(z)$ of the matrix R(z) in terms of

$$\Delta_{m+j}(z) = \left| \begin{array}{c|c} F_{yy}(z) & F_{yx_j}(z) \\ \hline F_{x_jy}(z) & F_{x_jx_j}(z) \end{array} \right| , \tag{5.22}$$

where F_{yx_j} denotes a m×j-dimensional matrix and $f_{x_jx_j}$ denotes a j×j-matrix. Furthermore, $\dfrac{\partial^2 F}{\partial y^i \partial x^k}$ and $\dfrac{\partial^2 F}{\partial x^i \partial x^k}$ lie on the intersection of the ith row and k^{th} column of these matrices, respectively. Next, we multiply the upper row of the block matrix (5.22) on the left by the j×m-dimensional matrix $F_{x_jy}F_{yy}^{-1}$ and subtract the latter from the lower row. The determinant Δ_{m+j} does not change, as we know; then we obtain

$$\Delta_{m+j}(z) = |F_{yy}(z)| \cdot |F_{x_jx_j}(z) - F_{x_jy}(z)\, F_{yy}^{-1}(z)\, F_{yx_j}(z)| . \tag{5.23}$$

Let $\delta_j(z)$, $j \in [1:n]$, denote the principal minors of the matrix $\Phi(z)$. From (5.23) it follows that $\delta_j(z)$ are connected with the principal minors of the matrix $R(z)$ through

$$(-1)^m \Delta_{m+j}(z) = (-1)^m \Delta_m \delta_j(z) \quad .$$

Thus, if the inequalities (5.20) and (5.21) hold, the principal minors of the matrix $\Phi(z)$ are positive and, therefore, the matrix $\Phi(z)$ is positive definite. The converse is also true, that is, the positive definiteness of $\Phi(z)$ and $-F_{yy}(z)$ imply the conditions (5.20) and (5.21). ///

One particular case is of interest: when y is scalar ($m = 1$). The conditions (5.20) and (5.21) will be in this case:

$$\Delta_i(z) < 0 \quad , \qquad i \in [1:n+1] \quad .$$

According to the results of Section 3, in order that the stationary point z of the function $F(z)$ be an unconstrained local minimum, it is sufficient that the principal minors of the matrix $R(z)$ be strictly positive:

$$\Delta_i(z) > 0 \quad , \qquad i \in [1:n+1] \quad .$$

In order that the stationary point z be an unconstrained local maximum of the function $F(z)$, it is sufficient that the conditions

$$(-1)^i \Delta_i(z) > 0 \quad , \qquad i \in [1:n+1]$$

be satisfied. The remaining, fourth version of the conditions

$$(-1)^i \Delta_i(z) < 0 \quad , \qquad i \in [1:n+1]$$

corresponds to the sufficient maximin conditions in the problem

$$\max_{x \in E^n} \min_{y \in E^1} F(x,y) \quad .$$

The case considered demonstrates a close connection between the criterion proved and Sylvester's criterion. We can elaborate this connection even more, introducing notions similar to those of positive as well as of negative definiteness. For instance, the quadratic form

$$z^T R(z_*) z = x^T F_{xx}(z_*) x + 2y^T F_{yx}(z_*) x + y^T F_{yy}(z_*) y$$

can be called minimax definite at the point z_* if for any $x \neq 0$, $y \neq 0$ we have the inequalities

$$y^T F_{yy}(z_*) y < 0 < \max_{y \in E^m} [x^T F_{xx}(z_*) x + 2x^T F_{xy}(z_*) y + y^T F_{yy}(z_*) y].$$

The maximum of the right side of the inequality in y is attained for $y = F_{yy}^{-1}(z_*) F_{yx}(z_*) x$. Hence this inequality is equivalent to the inequality

$$Y^T F_{yy}(x_*) y < 0 < x^T \Phi(z_*) x \quad .$$

The criterion proved above provides necessary and sufficient conditions for minimax definiteness of the quadratic form. The function $F(z)$ can be called strictly minimax if for all $\tilde{z}$, $z \in E^{n+m}$ the quadratic form $\tilde{z}^T R(z) \tilde{z}$ is minimax definite. For these functions the conditions of Theorem 1.5.8 are satisfied, and their local minimax solution is simultaneously global.

The class of functions thus introduced can be equated with the well-known classes of convex and convex-concave functions.

Indeed, if the local minimum exists in strictly convex functions, then it is unique and it coincides with the global minimum. If the local saddle exists in convex-concave functions, then it is unique and it coincides with the global saddle. If the local minimax exists in strictly minimax functions, then it is unique and it is the global minimax. The class introduced is richer than the class of strictly convex-concave functions (in which $y^T F_{yy}(z)y < 0$, $x^T F_{xx}(z)x > 0$), because the former class contains the latter. It is important to have introduced this class of functions because the problems (5.2) can be solved via iterative local numerical methods.

We consider next two simple examples to illustrate the properties obtained of minimax problems. We assume that x and y are scalar.

<u>EXAMPLE 1</u>. Let

$$F(x,y) = e^{x^2} \sin 2\pi(x-y) .$$

By the periodicity of the function in y, we can restrict ourselves to an interval $0 \leq y \leq 1$. The stationary points are $x = 0$, $y = 0.25$ and $x = 0$, $y = 0.75$. The first pair is a solution of the problem (5.2), in which $V_2 = 1$. Eq. (5.18) has the form

$$\frac{dy}{dx} = 1 + \frac{x}{\pi} \cot 2\pi(y-x) .$$

The straight line passing through the first point is given by

$$y = x + 0.25 . \tag{5.24}$$

It is not difficult to show that the point thus found is the global solution of a minimax problem, and the point $x = 0$ and the function (5.24) yields in the problem a (global) synthesis. The second point $x = 0$, $y = 0.75$ is a local and, simultaneously, global solution of the maximin problem (5.1), in which $V_1 = -1$. No point among those found is saddle.

EXAMPLE 2. For $F(x,y)$ we take the quadratic form:

$$-x^2 + 2kxy - y^2 , \qquad x \in E^1, \quad y \in E^1, \quad k^2 > 1 .$$

In the problem of finding the minimax, the global solution is given by $x = y = 0$, $V_2 = 0$. Eq. (5.18) has a solution given by the linear function $y = kx$. The maximin in the given problem is not attainable on a bounded set.

7. A PARTICULAR CASE

We consider here the simplest version of a minimax problem in which we seek

$$\min_{x \in E^n} \phi(x) , \quad \phi(x) = \max_{i \in [1:c]} f^i(x) , \tag{5.25}$$

where $\{f^i(x)\}$ is a finite set of functions. Theorem 1.5.3 of Danskin and Dem'yanov reads here as follows:

THEOREM 1.5.10. Let the functions $f^i(x)$, $i \in [1:c]$, be continuously differentiable in some neighborhood of the point x. Then the function of the maximum $\phi(x)$ is differentiable at x in any direction q, $||q|| = 1$, with

$$\frac{\partial \varphi(x)}{\partial q} = \max_{i \in B(x)} \langle f_x^i(x), q \rangle,$$
$$B(x) = \{i \in [1:c]: \varphi(x) = f^i(x)\}.$$

If all the functions f_i are continuously differentiable, the necessary condition for the minimum in the problem (5.25) is the following:

THEOREM 1.5.11. In order that x_* be the point of the minimum of the function $\phi(x)$ on E^n, it is necessary, and in the case where $\phi(x)$ is convex, is also sufficient that the inequality

$$\inf_{\|q\|=1} \max_{i \in B(x_*)} \langle f_x^i(x_*), q \rangle \geqslant 0 \tag{5.26}$$

be satisfied, or, which is the same,

$$\inf_{\|q\|=1} \frac{\partial \varphi(x_*)}{\partial q} \geqslant 0. \tag{5.27}$$

The points x_* satisfying (5.26), (5.27), are said to be stationary points of the function of the maximum $\phi(x)$. Let $L_0(x)$ denote the convex hull spanned by the vectors $f_x^i(x)$, where $i \in B(x)$:

$$L_0(x) = \left\{ z \in E^n \colon z = \sum_{i \in B(x)} \alpha_i f_x^i(x) \colon \alpha_i \geqslant 0, \sum_{i \in B(x)} \alpha_i = 1 \right\}.$$

THEOREM 1.5.12. In order that the function $\phi(x)$ attain the minimum at the point x_*, it is necessary, and in the case where $\phi(x)$ is convex is sufficient, that

$$0 \in L_0(x_*) \quad .$$

This condition is a generalization of the condition (3.4) to maximum functions.

6. CONDITIONS FOR A MINIMUM IN NONLINEAR PROGRAMMING PROBLEMS WITHOUT DIFFERENTIABILITY

1. BASIC NOTIONS

The general problem of nonlinear programming consists in finding

$$\min_{x \in X} f(x) \tag{6.1}$$

where the constraint set X is defined by the following condition:

$$X = \{x \in E^n : g(x) = 0,\ h(x) \leq 0\} . \tag{6.2}$$

Here we have introduced two vector-function mappings

$$g(x):\ E^n \to E^e , \qquad h(x):\ E^n \to E^c .$$

We shall say that the function $f(x)$, the vector functions g and h are the functions defining the problem (6.1). The vector function g gives constraints of equality type and h constraints of inequality type. We denote by X_* the set of all solutions to the problem (6.1); X_* is called the solution set. Now we define the set

$$X_0 = \{x \in E^n : g(x) = 0,\ h(x) < 0\} . \tag{6.3}$$

Next we introduce the so-called Lagrange function or simply Lagrangian:

$$L(x,u,v) = f(x) + \langle u, g(x) \rangle + \langle v, h(x) \rangle . \tag{6.4}$$

The vectors $u \in E^e$ and $v \in E^c$ are called Lagrange multipliers or dual vectors.

DEFINITION 1.6.1. Any point $x \in X$ is called a feasible point.

DEFINITION 1.6.2. At the point $[x_*, v_*] \in E^{n+c}$ the complementarity condition is satisfied if

$$v_*^j h^j(x_*) = 0 \qquad j \in [1:c] . \tag{6.5}$$

DEFINITION 1.6.3. At the point $[x_*, v_*] \in E^{n+c}$ the strict complimentarity condition is satisfied if the condition (6.5) holds and for any $j \in [1:c]$ the condition $h^j(x_*) = 0$ implies that $v_*^j > 0$.

DEFINITION 1.6.4. The problem (6.1) is called a convex programming problem if the functions f and d are convex and the vector function $g(x)$ is affine in x.

DEFINITION 1.6.5. The problem (6.1) is called a linear programming problem if the functions defining the problem are affine in x.

DEFINITION 1.6.6. In the problem (6.1) Slater's constraint qualification (CQ) is satisfied if the set X_0 is not empty.

DEFINITION 1.6.7. In the problem (6.1) Karlin's constraint qualification (CQ) is satisfied if there exist no vectors $u \in E^e$, $v \in E_+^c$, $||v|| \neq 0$, for which the inequality

$$0 \leq \langle u, g(x) \rangle + \langle v, h(x) \rangle \tag{6.6}$$

is satisfied for any $x \in E^n$.

If Slater's CQ is satisfied, Karlin's CQ holds. Indeed, at least for $\bar{x} \in X_0$ we then have: $g(\bar{x}) = 0$, $h(\bar{x}) < 0$, and for any $u \in E^e$, $v \in E_+^c$, $||v|| \neq 0$ the inequality

$$\langle u, g(\bar{x}) \rangle + \langle v, h(\bar{x}) \rangle < 0 ,$$

being the converse of the required (6.6).

In Section 5 we shall show that Karlin's CQ is equivalent to Slater's CQ in convex programming problems.

2. SUFFICIENT CONDITIONS FOR A MINIMUM

Here we formulate and prove theorems providing sufficient conditions for a minimum in the problem (6.1). Consider two auxiliary problems of finding the maximin and minimax of the Lagrange function:

$$V_1 = \sup_{u\in E^e} \sup_{v\in E^c_+} \inf_{x\in E^n} L(x, u, v), \tag{6.7}$$

$$V_2 = \inf_{x\in E^n} \sup_{u\in E^e} \sup_{v\in E^c_+} L(x, u, v). \tag{6.8}$$

We say that the point $[x_*,u_*,v_*] \in E^{n+m}$, where $v_* \geq 0$, $m = e + c$, is a saddle point of the function L, if for any $x \in E^n$, $u \in E^e$, $v \in E^c_+$ the inequalities

$$L(x_*,u,v) \leq L(x_*,u_*,v_*) \leq L(x,u_*,v_*) \tag{6.9}$$

are satisfied.

THEOREM 1.6.1. Let $[x_*,u_*,v_*]$ be a saddle point of the Lagrangian L. Then the point x_* is a solution of the problem (6.1) and $[x_*,v_*]$ satisfies the complementarity condition.

Proof. It follows from the left side of (6.9) that for all $u \in E^e$, $v \in E^c_+$ the condition

$$\sum_{i=1}^{e} g^i(x_*)[u^i - u^i_*] + \sum_{j=1}^{c} h^j(x_*)[v^j - v^j_*] \leqslant 0 \tag{6.10}$$

is satisfied. Let $v = v_*$, $u^i = u^i_*$ for all $i \in [1:e]$, except only one component $u^k = u^k_* + 1$. Substituting these vectors into

(6.10), we obtain $g^k(x_*) \le 0$. We may put $u^k = u^k_* - 1$; and then we arrive at the condition $g^k(x_*) \ge 0$. Only one possibility remains: $g^k(x_*) = 0$. By the arbitrariness of k we come to the conclusion that $g(x_*) = 0$. Taking in (6.10) $u = u_*$, $v^j = v^j_*$ for all $j \in [1:c]$ except $v^s = v^s + 1$, we obtain that $h^s(x_*) \le 0$. By the arbitrariness of s, from $[1:c]$ we infer that $h(x_*) \le 0$. Thus, the point $x_* \in X$, that is feasible.

Let us put in (6.10) $u = u_*$, $v = 0$; then we obtain

$$0 \le \langle h(x_*), v_* \rangle \quad . \tag{6.11}$$

At the same time, $h(x_*) \le 0$ and $0 \le v_*$, hence for all $j \in [1:c]$ the inequality $h^j(x_*)v^j_* \le 0$ is satisfied.

Comparing the last inequality with (6.11), we infer that (6.5) holds true. From the right inequality of (6.9), taking into account (6.5), we obtain that for all $x \in E^n$ the condition

$$f(x_*) \le f(x) + \langle u_*, g(x)\rangle + \langle v_*, h(x)\rangle$$

is satisfied. In particular, for any $x \in X$ we have $f(x_*) \le f(x)$. Then we infer that $x_* \in X_*$. ///

If we assume that in (6.7) the interior problem is solvable, we can introduce into consideration the function of dual vectors, letting

$$\gamma(u,v) = \inf_{x \in E^n} L(x,u,v) \quad .$$

Then the exterior problem for (6.7) consists in finding

$$V_1 = \sup_{u \in E^e} \sup_{v \in E^c_+} \gamma(u, v).$$

THEOREM 1.6.2. Let $X_* \neq \emptyset$; then we have the estimates

$$V_1 \leq f(x_*) = V_2 , \qquad x_* \in X_* . \tag{6.12}$$

Proof. First we consider problem (6.8): the interior problem therein is solvable in a simple way:

$$\beta(x) = \sup_{u \in E^e} \sup_{v \in E_+^c} L(x, u, v) = \begin{cases} f(x) & \text{if } x \in X, \\ +\infty & \text{if } x \notin X. \end{cases}$$

Solving the exterior problem, we come to the conclusion that the minimum of the function $\beta(x)$ is attained on the feasible set and coincides with the minimum of the function $f(x)$ on X; hence

$$f(x_*) = \min_{x \in X} f(x) = V_2 .$$

For any $x \in X$, $u \in E^e$, $v \in E_+^c$, we have the inequality

$$f(x) \geq L(x,u,v) .$$

Next we minimize the left and the right sides of the last inequality in $x \in X$; we then obtain

$$f(x) \geqslant f(x_*) = \min_{x \in X} f(x) \geqslant \min_{x \in X} L(x, u, v) \geqslant \inf_{x \in E^n} L(x, u, v) .$$

By the arbitrariness of $u \in E^e$, $v \in E_+^c$, we arrive at the left side of (6.12).

THEOREM 1.6.3. Let the point $[x_*, u_*, v_*] \in E^{n+m}$ in which $x_* \in X$, $v_* \geq 0$, be the solution of the problem (6.7), and let the complementarity condition be satisfied. Then $x_* \in X_*$.

Proof. For $u = u_*$, $v = v_*$ the point x_* must be the solution of the interior problem in (6.7), i.e., the inequality

$$f(x_*)+\langle u_*, g(x_*)\rangle+\langle v_*, h(x_*)\rangle=f(x_*)\leqslant$$
$$\leqslant f(x)+\langle u_*, g(x)\rangle+\langle v_*, h(x)\rangle$$

is satisfied for any $x \in E^n$. If $x \in X$, this inequality implies that

$$f(x_*) \leq L(x,u_*,v_*) \leq f(x) ,$$

whence we infer that $x_* \in X_*$. ///

3. COMPUTATIONAL ASPECTS

The above Theorems suggest possible methods for constructing numerical algorithms for solving the problem (6.1). From Theorem 1.6.1 it follows that when there exists a saddle point of the Lagrangian L it is possible to seek the saddle points of L instead of solving the problem (6.1). We shall consider these methods in Section 4.1. From Theorem 1.6.2 we infer that solving the minimax problem (6.8) yields a solution of the problem (6.1). This approach is quite feasible if we are certain that there exists a bounded solution $[\bar{x},\bar{u},\bar{v}]$ of the minimax problem. Let all $|\bar{u}^i| \leq \tau$, $0 \leq \bar{v}^j \leq \tau$; then we replace the minimax problem (6.8) by the problem

$$P_1=\min_{x\in E^n}\max_{|u^i|\leqslant\tau}\max_{0\leqslant v^j\leqslant\tau} L(x, u, v). \tag{6.13}$$

Solving the interior problem, we obtain

$$P_1=\min_{x\in E^n}\left\{f(x)+\tau\left[\sum_{i=1}^{e}|g^i(x)|+\sum_{j=1}^{c}h^j_+(x)\right]\right\}. \tag{6.14}$$

Thus, having introduced only one hypothesis concerning the existence of bounded solutions to (6.8), we reduced the initial problem (6.1) to that of finding the unconstrained minimum of some auxiliary function of x. In Chapter 3 we shall obtain the same result while studying the method of penalty functions.

Theorem 1.6.3 enables us to seek solutions of the maximin problem (6.7). If its solution is given by the feasible point $(x_* \in X)$, the complementarity condition is satisfied; then x_* is the solution of the initial problem.

We shall use this property in Chapter 4.

4. MODIFIED LAGRANGIANS

All the sufficient conditions given in this section are based on considering auxiliary problems related to the Lagrange function L. Then the following questions arise:

Is it possible to use auxiliary functions of a more general type than the Lagrangian L in order to obtain sufficient conditions for the minimum and, which is more essential, to construct numerical methods for solving the problem (6.1)? Will this technique allow us to guarantee the solvability of auxiliary problems for a wider class of nonlinear programming problems? To which other auxiliary problems, convenient for numerical realization, can we reduce the problem (6.1)?

Currently, intensive research is going on, and it will be possible, apparently, to answer these questions in the future. Theoretical investigations and numerical experiments indicate that it is useful to consider new auxiliary functions usually known as modified or generalized Lagrange functions and to introduce thereby new auxiliary problems different from those of finding saddle points, maximin, or minimax. We cite an unconventional sufficient condition for the minimum in the problem (6.1), which uses a special auxiliary function of the form:

$$H(x,y) = f(x) + \xi(x,y) . \tag{6.15}$$

Here y is a vector whose dimension is not essential at the moment (in particular, y may be scalar), ξ is a continuous function of the arguments. Let

$$x(y) = \operatorname{Arg}\min_{x \in E^n} H(x,y) . \tag{6.16}$$

Assuming that for a vector y the set $x(y)$ is not empty, we proceed to the problem of finding real solutions of the following system:

$$\begin{aligned} g(x) &= 0 , \\ h(x) &\leq 0 , \qquad \text{where} \quad x \in x(y) . \end{aligned} \tag{6.17}$$

The connection between this auxiliary problem and the initial problem (6.1) is more fully unraveled by the following.

<u>THEOREM 1.6.4</u> (Yu. G. Evtushenko [12]). Suppose there exist vectors y_* and x_* satisfying (6.17) and that the function ξ in (6.15) is such that for any $x \in X$ we have the inequality

$$\xi(x,y_*) \leq \xi(x_*,y_*) . \tag{6.18}$$

Then:

- 1. $x_* \in X_* \subset x(y_*)$, (6.19)
- 2. if $\xi(x,y_*)$ is constant for any $x \in X$ then

$$X_* = X \cap x(y_*) . \tag{6.20}$$

<u>Proof</u>. Suppose there exist x_*, y_* satisfying (6.17). Then

$x_* \in X \cap x(y_*)$ and therefore for any $x \in X$

$$H(x_*,y_*) \le H(x,y_*) \quad . \tag{6.21}$$

We use (6.18). From (6.21) we obtain $f(x_*) \le f(x)$ and, taking into account that x is an arbitrary feasible point, we conclude that $x_* \in X_*$.

Let $x' \in X_*$, for this case $f(x') = f(x_*)$ and $x' \in X$. One more time using (6.18), we obtain

$$H(x',y_*) = f(x_*) + \xi(x',y_*) \le H(x_*,y_*) \quad .$$

If the point $[x_*,y_*]$ satisfies (6.17), then the inequality $H(x_*,y_*) \le H(x,y_*)$ holds for all $x \in E^n$ and in particular for x'. Therefore, from the two inequalities

$$H(x',y_*) \le H(x_*,y_*) \le H(x',y_*)$$

we find that $H(x_*,y_*) = H(x',y_*)$, $x' \in x(y_*)$. We have proved (6.19).

To prove (6.20) it suffices to show that

$$x(y_*) \cap X \subset X_* \tag{6.22}$$

since the inverse inclusion has been proved. Let $x' \in x(y_*) \cap X$. Then $H(x_*,y_*) = H(x',y_*)$ hence the function $\xi(x,y_*)$ is constant on X; hence $f(x_*) = f(x')$, thus proving (6.22). ///

Theorem 1.6.4 enables one to reduce the initial problem of nonlinear programming to that of solving a system of nonlinear equalities and inequalities; various numerical methods following from our result here will be described in Section 4.2.

5. SOME AUXILIARY RESULTS

LEMMA 1.6.1. Let (6.1) be a convex programming problem and let the set X_0 defined by (6.3) be empty. Then there exist vectors $u \in E^e$, $v \in E^c_+$, $(u,v) \neq 0$, such that the inequality (6.6) holds for any $x \in E^n$.

Proof. We define two sets

$$\Lambda(x) = \{z,\ y:\ z \in E^e,\ y \in E^c,\ z = g(x),\ y > h(x)\},$$
$$\Lambda = \bigcup_{x \in E^n} \Lambda(x).$$

If the points $[z_1,y_1]$, $[z_2,y_2]$ belong to Λ, then for $0 \leq \lambda \leq 1$ we have

$$(1-\lambda)z_1 + \lambda z_2 = (1-\lambda)g(x_1) + \lambda g(x_2) = g((1-\lambda)x_1 + \lambda x_2),$$
$$(1-\lambda)y_1 + \lambda y_2 > (1-\lambda)h(x_1) + \lambda h(x_2) \geqslant h((1-\lambda)x_1 + \lambda x_2).$$

Hence the set Λ is convex. From the condition of the lemma $X_0 = \emptyset$ it follows that the origin $z = 0$, $y = 0$ does not belong to the set Λ. By the Theorem on separability 1.1.1, there exist $u \in E^e$, $v \in E^c$, $(u,v) \neq 0$, such that from the condition $[a,b] \in \Lambda$ we have

$$0 \leq \langle u,a \rangle + \langle v,b \rangle . \tag{6.23}$$

We can make the components of the vector b as large as possible, hence the vector $v \geq 0$. In (6.23) we put $a = g(x)$, $b = h(x) + \varepsilon d$, $d \in E^e_+$, $\|d\| = 1$, $\varepsilon > 0$, then $[a,b] \subset \Lambda(x) \subset \Lambda$; for any $x \in E^n$ we have

$$0 \leq \langle u,\ g(x) \rangle + \langle v,\ h(x) \rangle + \varepsilon \langle v,d \rangle .$$

Because ε is an arbitrary positive number, we infer that (6.6) is satisfied. ///

<u>LEMMA 1.6.2</u>. For convex programming problems Slater's CQ is equivalent to Karlin's CQ.

<u>Proof</u>. With Definition 1.6.7, we pointed out that when Slater's CQ is satisfied Karlin's CQ holds. We shall show that if Slater's CQ is not satisfied, Karlin's CQ does not hold. By the previous lemma the condition $X_0 = \emptyset$ implies that there exist $u \in E^e$, $v \in E_+^c$, $(u,v) \neq 0$ such that the inequality (6.6) is satisfied. If $\|v\| \neq 0$, Karlin's CQ is violated. We suppose the opposite, that is, $v = 0$. The affine vector function $g(x)$ is representable as $g(x) = Ax + b$, where A is the $e \times n$-matrix. We can assume without loss of generality that the rows of the matrix A are linearly independent, since otherwise the number of restrictions could be reduced without changing the feasible set. Let us write (6.6) in the form

$$0 \leq u^T(Ax+b) = x^T A^T u + \langle b,u\rangle, \qquad \forall x \in E^n, \quad \|u\| \neq 0 . \tag{6.24}$$

We show that (6.24) implies $A^T u = 0$. Otherwise, for x we take

$$x = \begin{cases} -A^T u & \text{if } \langle b, u\rangle \leq 0, \\ -2\dfrac{\langle u, b\rangle A^T u}{u^T A A^T u} & \text{if } \langle b, u\rangle > 0. \end{cases}$$

Substituting these expressions into the right side of (6.24), we obtain that if $\langle b,u\rangle \leq 0$, then

$$x^T A^T u + \langle b,u\rangle = -u^T A A^T u + \langle b,u\rangle < \langle b,u\rangle \leq 0 .$$

If $\langle b,u\rangle > 0$, then $x^T A^T u + \langle b,u\rangle = -\langle b,u\rangle < 0$. But these relations contradict the left-hand inequality in (6.24). Hence

$A^T i = 0$, which contradicts the linear independence of the rows of the matrix A. Therefore, $\|v\| \neq 0$, and Karlin's CQ is not satisfied. ///

6. NECESSARY AND SUFFICIENT CONDITIONS FOR A MINIMUM

The three theorems given in Section 2 permit us to replace the initial problem (6.1) by that of finding saddle points of the Lagrangian, or by that of solving a minimax or a maximin problem. Unfortunately, such reduction of (6.1) to simpler problems is not a universal technique: not in all problems the Lagrangian has finite saddle points and the conditions of Theorem 1.6.3 are satisfied. We cite here some results providing sufficient conditions for the existence of bounded dual vectors. In one of the first works on nonlinear programming, F. John [1] suggested for the analysis of the problem (6.1) that the function

$$L^0(x,u,v,q) = qf(x) + \langle u, g(x)\rangle + \langle v, h(x)\rangle \tag{6.25}$$

should be used.

We say that the point $[x_*,u_*,v_*,q_*] \in E^{n+m+1}$ in which $q_* \in E^1_+$, $u_* \in E^e$, $v_* \in E^c_+$, $(u_*,v_*,q_*) \neq 0$, is a saddle point of the function L^0, if for any $x \in E^n$, $u \in E^e$, $v \in E^c_+$ we have the inequalities

$$L^0(x_*,u,v,q_*) \le L^0(x_*,u_*,v_*,q_*) \le L^0(x,u_*,v_*,q_*) . \tag{6.26}$$

Comparing these inequalities with (6.9), we infer that if $[x_*,u_*,v_*,q_*]$ is a saddle point of the function L^0 and $q_* > 0$, then the totality $\left[x_*, \frac{u*}{q*}, \frac{v*}{q*}\right]$ is the saddle point of the function L. Conversely, if $[x_*,u_*,v_*]$ is a saddle point of the function L, then $[x_*,u_*,v_*,1]$ is a saddle point of the function L^0.

The next theorem is an analog of Theorem 1.6.1 for the function L^0.

THEOREM 1.6.5. Let $[x_*,u_*,v_*,q_*]$ be a saddle point of the function L^0, $(q_*,v_*) \neq 0$. Then $x_* \in X$, (6.5) holds, and either $q_* > 0$, $x_* \in X_*$, or $q_* = 0$ and furthermore in the problem (6.1) Karlin's and Slater's CQ do not hold.

Proof. In the same way as in justifying Theorem 1.6.1, we can show here that $x_* \in X$ and (6.5) holds. If $0 < q_*$, then $x_* \in X_*$. If $q_* = 0$, then $\|v_*\| \neq 0$, from the right inequality of (6.26) we obtain

$$0 \leq \langle u_*, g(x)\rangle + \langle v_*, h(x)\rangle \qquad \forall\ x \in E^n ,$$

indicating that Karlin's and Slater's CQ are violated. ///

THEOREM 1.6.6 (Uzawa-Karlin Saddle Point Theorem). Suppose in the convex programming problem (6.1) the set of solutions in not empty. Then for each $x_* \in X_*$ there exists a saddle point $[x_*,u_*,v_*,q_*]$ of the function L^0, and the complementarity condition is satisfied at the point $[x_*,v_*]$.

Proof. The fact that $x_* \in X_*$ implies that the system

$$f(x) - f(x_*) < 0 , \qquad g(x) = 0, \quad h(x) \leq 0 \qquad (6.27)$$

has no solution. Let us use Lemma 1.6.1. Since the statement "$h(x) \leq 0$ has no solution" implies the statement "$h(x) < 0$ has no solution," the assertion of Lemma 1.6.1 can be formulated as follows: if the system (6.27) has no solution, there exist $u_* \in E^e$, $v_* \in E^c_+$, $q_* \in E^1_+$, not equal to zero at the same time, such that for all $x \in E^n$ the inequality

$$0 \le q_*(f(x) - f(x_*)) + \langle u_*, g(x)\rangle + \langle v_*, h(x)\rangle \qquad (6.28)$$

is satisfied. Putting here $x = x_*$, we obtain that $\langle v_*, h(x_*)\rangle \ge 0$; but since $v_* \ge 0$ and $h(x_*) \le 0$, we infer that (6.5) holds. It follows from (6.28) that for all $x \in E^n$

$$L^0(x_*, u_*, v_*, q_*) = \\ = q_*f(x_*) + \langle u_*, g(x_*)\rangle + \langle v_*, h(x_*)\rangle \le L^0(x, u_*, v_*, q_*),$$

which proves the right inequality in (6.26).

From the condition $h(x_*) \le 0$ we have: $\langle v, h(x_*)\rangle \le 0$ for $v \ge 0$. Noting (6.5), we obtain that for any $u \in E^e$ and $v \in E^c_+$

$$q_*f(x_*) + \langle u, g(x_*)\rangle + \langle v, h(x_*)\rangle \le \\ \le q_*f(x_*) + \langle u_*, g(x_*)\rangle + \langle v_*, h(x_*)\rangle,$$

i.e., the left inequality in (6.26) holds. ///

THEOREM 1.6.7 (Kuhn-Tucker Saddle-Point Theorem). Suppose that in the convex programming problem (6.1) the set of solutions X_* is not empty and either Slater's CQ or Karlin's CQ is satisfied. Then for each $x_* \in X_*$ there exist vectors $\bar{u} \in E^e$, $\bar{v} \in E^c_+$ such that $[x_*, \bar{u}, \bar{v}]$ is a saddle point of the Lagrangian L.

Proof. By the previous theorem the saddle point $[x_*, u_*, v_*, q_*]$ of the function L^0 exists, the condition (6.5) is satisfied at the point $[x_*, v_*]$. If $q_* > 0$, then $\left[x_* \frac{u_*}{q_*}, \frac{v_*}{q_*}\right]$ is the saddle point of the function L.

We prove that $q_* > 0$. Assume the converse, that is, $q_* = 0$.

Then $(u_*, v_*) \neq 0$, and from the right side of (6.26) it follows that for any x the inequality

$$0 \leq \langle u_*, g(x) \rangle + \langle v_*, h(x) \rangle$$

is satisfied, which is impossible for $\|v_*\| \neq 0$, $v_* \geq 0$, since it contradicts Karlin's and Slater's CQ. Arguing in the same way as in proving Lemma 1.6.2, we can show that $\|v_*\| \neq 0$. Hence $q_* \neq 0$, $\bar{u} = \frac{u_*}{q_*}$, $\bar{v} = \frac{v_*}{q_*}$. ///

7. GENERALIZATIONS

The results cited above may be generalized to the case where in place of (6.1) we consider the problem of finding

$$\min_{x \in X \cap U} f(x) \quad , \qquad (6.29)$$

where U is a set in E^n, whose interior is not empty, the set X is defined by the condition (6.2). The set of solutions of the problem (6.29) is denoted as before by X_*, the Lagrangian is defined by the formula (6.4). In the problems (6.7), (6.8) and (6.16), the condition $x \in U$ is used instead of the condition $x \in E^n$. The conditions (6.9), (6.26) defining the saddle points as well as the inequality (6.6) must hold for any $x \in U$. In convex programming problems the set U is required to be convex.

It is easy to see that the theorems and lemmas of this section will still hold for the problem (6.29) if we take into account the additional assertions made. In Theorems 1.6.6 and 1.6.7 and Lemmas 1.6.1 and 1.6.2, it was required that the set U be convex. The representation of a nonlinear programming problem in the form (6.29) will be encountered frequently; it is useful when for various reasons the operation of minimizing the functions on the set U is easily realizable. For example, U is of a simple structure, or the minimum on U can be sought through analytical formulas. The problem (6.29) can be more convenient for theoretical investigations as well, since, imposing special restrictions on U (for example, requiring the set U be compact), we essentially simplify the problem and analysis of the methods for solving it.

7. CONDITIONS FOR A MINIMUM IN NONLINEAR PROGRAMMING PROBLEMS WITH DIFFERENTIABILITY

1. BASIC DEFINITIONS AND PRELIMINARY RESULTS

We consider the problem (6.1), using the Lagrange function (6.4).

DEFINITION 1.7.1. Let the functions defining the problem (6.1) be differentiable at a point x_*; then the point $[x_*, u_*, v_*] \in E^{n+m}$ is said to be a Kuhn-Tucker point if $x_* \in X$, at $[x_*, v_*] \in E^{n+c}$ the complementarity condition (6.5) is satisfied, $v_* \geqslant 0$ and

$$L_x(x_*, u_*, v_*) = f_x(x_*) + g_x(x_*)u_* + h_x(x_*)v_* = 0 . \tag{7.1}$$

DEFINITION 1.7.2. Let the functions defining the problem (6.1) be differentiable at the point x_*. Then we say that the $[x_*, u_*, v_*, q_*] \in E^{n+m+1}$ is a John point if $x_* \in X$, at $[x_*, v_*] \in E^{n+c}$ the complementarity condition (6.5) is satisfied, $q_* \geqslant 0$, $v_* \geqslant 0$, with only some of u_*, v_*, q_* equal to zero, and

$$L_x^0(x_*, u_*, v_*, q_*) = 0 . \tag{7.2}$$

If $[x_*, u_*, v_*, q_*]$ is a F. John point and $q_* = 1$, then $[x_*, u_*, v_*]$ is a Kuhn-Tucker point.

Next we introduce two index sets:

$$\sigma(x) = \{j \in [1:c]:\ h^j(x) = 0\},$$
$$\theta(x, v) = \{j \in [1:c]:\ j \in \sigma(x),\ v^j > 0\}. \tag{7.3}$$

The functions $g^i(x)$, $h^j(x)$ in which $g^i(x) = h^j(x) = 0$, are said to be active constraints at the point x. The index set $\sigma(x)$ defines, thus, the totality of active constraints of the inequality-type at the point x.

DEFINITION 1.7.3. The constraints $g(x) = 0$, $h(x) \leq 0$ satisfy the constraint qualification at the point $x \in X$ if the vector functions $g(x)$, $h(x)$ are differentiable at x and the vectors $g^i_x(x)$, $h^j_x(x)$, $i \in [1:e]$, $j \in \sigma(x)$, are linearly independent.

DEFINITION 1.7.4. The constraints $g(x) = 0$ and $h(x) \leq 0$ satisfy the Arrow-Hurwicz-Uzawa CQ at the point $x \in X$ if the vector function $h(x)$ is differentiable at x, $g(x)$ is continuously differentiable at x, the vectors $g^i_x(x)$ for all $i \in [1:e]$ are linearly independent, and there exists a vector $z \in E^n$ satisfying the relations

$$\begin{aligned} g^T_x(x) z &= 0 , \\ \langle z,\ h^j_x(x)\rangle &< 0 , \qquad \forall\ j \in \sigma(x) . \end{aligned} \tag{7.4}$$

In some cases, one needs to impose special constraints on $g(x)$ and $h(x)$ at the points not belonging to the feasible set. This occurs in the following definition.

DEFINITION 1.7.5. The constraints $g(x) = 0$ and $h(x) \leq 0$ satisfy the strengthened CQ if the vector functions $g(x)$ and $h(x)$

are everywhere differentiable; at each point $x \in X$ the constraint qualification is satisfied and at each point $x \in X$ there exists a vector $z \in E^n$ satisfying the relations

$$g(x) + g_x^T(x)z = 0, \qquad h(x) + h_x^T(x)z \leq 0 .$$

In considering convex programming problems, we can assume without loss of generality that the condition $g(x) = 0$ defines the (n-e)-dimensional linear manifold, since, if this condition were not valid, it would have been possible to reduce the number of constraints of the equality type removing depending conditions; the feasible set X does not change in this situation. Hence in the sequel we will assume that in convex programming problems the rank of the matrix $g_x(x)$ is maximal and equal to e.

For convex programming problems the Arrow-Hurwicz-Uzawa CQ will be satisfied, if the vector function $h(x)$ is differentiable and Slater's CQ holds. Indeed, let $g(\bar{x}) = 0$, $h(\bar{x}) < 0$; then, using the linearity of $g(x)$ and convexity of $h(x)$, we obtain

$$\begin{gathered} g(\bar{x}) = g(x) + g_x^T(x)(\bar{x} - x) = 0, \\ \langle h_x^j(x),\ \bar{x} - x\rangle \leqslant h^j(\bar{x}) - h^j(x) = h^j(\bar{x}) < 0 \quad \forall\ j \in \sigma(x). \end{gathered} \tag{7.5}$$

Letting $z = \bar{x} - x$, we arrive at (7.4)

<u>LEMMA 1.7.1</u> (Motzkin's Lemma on Transposition). Let the matrices A and B be respectively of dimensions $a \times d$, $b \times d$, $b > 0$. Then either the system

$$Az = 0, \qquad Bz < 0, \qquad z \in E^d,$$

or the system

$$A^T z_1 + B^T z_2 = 0 \ ,$$

$$z_1 \in E^a \ , \quad z_2 \in E_+^b \ , \quad \|z_2\| \neq 0$$

has a solution. These two systems cannot have solutions at the same time. The proof can be found in Kuhn and Tucker [2].

LEMMA 1.7.2. Let the functions g, h be differentiable at the point $x \in X$ at which the Arrow-Hurwicz-Uzawa CQ is satisfied, and in a convex programming problem Slater's CQ or Karlin's CQ is satisfied. Then for any $u \in E^e$, $v \in E_+^c$ such that all u^i, v^j, $i \in [1:e]$, $j \in \sigma(x)$, not equal to zero at the same time, we have the conditions

$$g_x(x)\,u + \sum_{j \in \sigma(x)} h_x^j(x)\,v^j \neq 0. \tag{7.6}$$

Proof. If $\sigma(x) = \emptyset$, the required inequality follows from the linear independence of the vectors $g_x^i(x)$, $i \in [1:e]$. If the set $\sigma(x)$ is not empty, we take advantage of the previous Lemma, taking for A the matrix $g_x^T(x)$ and for the rows of the matrix B we take the gradients $h_x^j(x)$, $j \in \sigma(x)$, the system (7.4) is solvable, hence the inequality (7.6) holds for any u, v satisfying the condition of the Lemma. For the convex programming problem the conditions (7.5) are used, which follow from Slater's CQ. ///

2. NECESSARY AND SUFFICIENT CONDITIONS FOR A MINIMUM OF CONVEX PROGRAMMING PROBLEMS

For convex programming problems the differentiability condition in Definitions 1.7.1 and 1.7.2 can be omitted, and instead of

(7.1) and (7.2) we could require the following: let there exist vectors

$$z^0 \in \partial f(x_*) , \qquad z^j \in \partial h^j(x_*) , \qquad j \in [1:c] ,$$

such that respectively

$$z^0 + g_x(x_*) u_* + \sum_{j=1}^{c} z^j v_*^j = 0, \tag{7.7}$$

$$q_* z^0 + g_x(x_*) u_* + \sum_{j=1}^{c} z^j v_*^j = 0. \tag{7.8}$$

<u>THEOREM 1.7.1</u>. Let in the convex programming problem (6.1) the solution set X_* be not empty. Then:

● 1. each Kuhn-Tucker point $[x_*,u_*,v_*]$ is a saddle point of the Lagrange function $L(x,u,v)$, $x_* \in X_*$;

● 2. each F. John point $[x_*,u_*,v_*,q_*]$ is a saddle point of the function $L^0(x,u,v,q)$; for $q_* > 0$ we have: $x_* \in X_*$;

● 3. if Slater's CQ or Karlin's CQ is satisfied, then there exist Kuhn-Tucker points or F. John points $[x_*,u_*,v_*,q_*]$, $q_* > 0$.

<u>Proof</u>. From the convexity of $f(x)$, $h(x)$ and the linearity of $g(x)$ it follows that for arbitrary $x \in E^n$

$$f(x) \geq f(x_*) + \langle z^0, x - x_* \rangle, \tag{7.9}$$

$$h^j(x) \geq h^j(x_*) + \langle z^j, x - x_* \rangle, \tag{7.10}$$

$$g(x) = g(x_*) + g_x(x_*)(x - x_*). \tag{7.11}$$

We multiply (7.10) by v_*^j and sum up over j, multiply (7.11) scalarwise by u_*, and add the inequalities obtained to (7.9). Noting that $v_* \geq 0$ and taking into account the definition of the function L and the condition (7.7), we obtain

$$L(x, u_*, v_*) \geqslant L(x_*, u_*, v_*) + \langle x - x_*, z^0 + + g_x(x_*) u_* + \sum_{j=1}^{c} z^j v_*^j \rangle = L(x_*, u_*, v_*) .$$

Thus, we have proved the right inequality in (6.9). The left inequality has in this case the form

$$\sum_{j=1}^{c} v^j h^j(x_*) \leqslant \sum_{j=1}^{c} v_*^j h^j(x_*).$$

By the constraint qualification the right side of the last formula is zero; and since $x_* \in X$ we obtain that $\langle v, h(x_*)\rangle \leq 0$ for any $v \geq 0$. Hence the inequalities (6.9) are satisfied, and, using Theorem 1.6.1, we infer that $x_* \in X_*$. Assertion 2 can be proved similarly using (7.8).

By Theorem 1.6.7, when Slater's CQ or Karlin's CQ is satisfied, there exists a saddle point $[x_*, u_*, v_*]$ of the function L. From Theorem 1.6.1 if follows that $x_* \in X_*$ and (6.5) holds. By Theorem 1.3.5, (7.7) follows from the right inequality of (6.9). Hence $[x_*, u_*, v_*]$ is a Kuhn-Tucker point. From Theorem 1.6.6 there follows the existence of a saddle point $[x_*, u_*, v_*, q_*]$ of the function L, which is a F. John point; furthermore, by Theorem 1.6.5 $q_* > 0$. ///

3. SUFFICIENT CONDITIONS FOR A MINIMUM IN GENERAL NONLINEAR PROGRAMMING PROBLEMS

For the feasible point x_* and the dual vector v_* we define two convex cones:

$$K_1(x_*, v_*) = \{x \in E^n:\ x^T g_x(x_*) = 0,$$
$$x^T h_x^j(x_*) = 0,\ x^T h_x^s(x_*) \leqslant 0\}, \qquad (7.12)$$
$$K_2(x_*) = \{x \in E^n:\ x^T g_x(x_*) = 0,$$
$$x^T h_x^i(x_*) \leqslant 0,\ x^T f_x(x_*) \leqslant 0\}, \qquad (7.13)$$

where the indices j, s, i assume on all possible values of the sets $\theta(x_*, v_*)$, $\sigma(x_*) \setminus \theta(x_*, v_*)$, $\sigma(x_*)$, respectively (see (7.3)).

LEMMA 1.7.3. Let $[x_*, u_*, v_*]$ be a Kuhn-Tucker point in the problem (6.1). Then the cones $K_1(x_*, v_*)$ and $K_2(x_*)$ coincide.

Proof. We show first that $K_2(x_*) \subset K_1(x_*, v_*)$. Let $\bar{x} \in K_2(x_*)$; then we need to show that

$$\bar{x}^T h_x^j(x_*) = 0 \qquad \forall\, j \in \theta(x_*, v_*)\ . \qquad (7.14)$$

Using (7.1), we obtain

$$\bar{x}^T f_x(x_*) + \bar{x}^T g_x(x_*) u_* + \bar{x}^T h_x(x_*) v_* =$$
$$= \bar{x}^T f_x(x_*) + \sum_{j \in \theta(x_*,\, v_*)} \bar{x}^T h_x^j(x_*) v_*^j = 0.$$

Since each term in this equality is non-positive and all the $v_*^j > 0$ for $j \in \theta(x_*, v_*)$, we infer that (7.14) holds, hence $\bar{x} \in K_1(x_*, v_*)$.

We now prove that $K_1(x_*, v_*) \subset K_2(x_*)$. Let $\bar{x} \in K_1(x_*, v_*)$. Then it follows from (7.1) that

$$\bar{x}^T L_x(x_*, u_*, v_*) = \bar{x}^T f_x(x_*) = 0\ .$$

Hence $\bar{x} \in K_2(x_*)$, which completes proving the lemma. ///

DEFINITION 1.7.6. The matrix of second derivatives $L_{xx}(x_*, u_*, v_*)$ is positive definite on the cone $K_1(x_*, v_*)$ if the quadratic form $x^T L_{xx}(x_*, u_*, v_*) x > 0$ for any nonzero vectors x belonging to the cone $K_1(x_*, v_*)$.

DEFINITION 1.7.7. The matrix of the second derivatives $L_{xx}(x_*, u_*, v_*)$ is uniformly positive definite on the cone $K_1(x_*, v_*)$ with the constant C_1 if $x^T L_{xx}(x_*, u_*, v_*) x \geq C_1 \|x\|^2$ for any vectors x belonging to the cone $K_1(x_*, v_*)$.

We shall state next the lemma of R. Finsler (see Appendix II) specifically for the problem in question.

LEMMA 1.7.4 (R. Finsler). Let the matrix of the second derivatives $L_{xx}(x_*, u_*, v_*)$ of the Lagrange function be positive definite on the cone $K_1(x_*, v_*)$ and let at the point $[x_*, v_*]$ the strict complementarity condition be satisfied. Then there exists a τ_* such that for any $\tau > \tau_*$ the matrix

$$L_{xx}(x_*, u_*, v_*) + \tau \left[g_x(x_*) g_x^T(x_*) + \sum_{j \in \sigma(x_*)} h_x^j(x_*) [h_x^j(x_*)]^T \right]$$

is positive definite.

THEOREM 1.7.2 (McCormick). Let $[x_*, u_*, v_*] \in E^{n+m}$ be a Kuhn-Tucker point in the problem (6.1); let the functions defining the problem be twice differentiable, and let the matrix of the second derivatives $L_{xx}(x_*, u_*, v_*)$ be positive definite on the cone $K_1(x_*, v_*)$. Then x_* is the isolated local minimum of the problem (6.1).

This Theorem is one of the basic results in the theory of nonlinear programming. Subsequent to publication of this Theorem, various modifications and some generalizations have been obtained. Following S. Han and O. Mangasarian [1], we shall formulate and prove a theorem generalizing the result obtained by McCormick.

THEOREM 1.7.3. Let $[x_*, u_*, v_*, q_*] \in E^{n+m+1}$ be a F. John point in the problem (6.1); let the functions defining the problem be twice differentiable at a point x_*, and let the matrix of the second derivatives $L^0_{xx}(x_*, u_*, v_*, q_*)$ be positive definite on the cone $K_1(x_*, v_*)$ or $K_2(x_*)$. Then the point x_* is the isolated local minimum of the problem (6.1).

Proof. Assume the converse, that is x_* is not the local isolated minimum. Then there exists a sequence of feasible points $\{x_k\}$ such that all $x_k \neq x_*$, $\lim_{k\to\infty} x_k = x_*$ and, furthermore, for each point x_k the inequality $f(x_*) \geq f(x_k)$ is satisfied.

Let $y_k = (x_k - x_*) / \| x_k - x_* \|$. It follows from the construction of the points x_k that the conditions

$$0 \geqslant \frac{f(x_k) - f(x_*)}{\| x_k - x_* \|} = y_k^T f_x(x_*) + O(\| x_k - x_* \|),$$
$$0 = \frac{g(x_k) - g(x_*)}{\| x_k - x_* \|} = y_k^T g_x(x_*) + O(\| x_k - x_* \|),$$
$$0 \geqslant \frac{h^j(x_k) - h^j(x_*)}{\| x_k - x_* \|} = y_k^T h_x^j(x_*) + O(\| x_k - x_* \|) \qquad \forall j \in \sigma(x_*)$$

are satisfied. Here $O(z)$ means that $\lim_{z\to 0} O(z) = 0$. Hence there exists a limiting point y of the sequence $\{y_k\}$ such that

$$\| y \| = 1, \quad y^T f_x(x_*) \leqslant 0, \quad y^T g_x(x_*) = 0,$$
$$y^T h_x^j(x_*) \leqslant 0 \quad \forall j \in \sigma(x_*),$$

yielding $y \in K_2(x_*)$.

We shall use the condition for existence of the second derivatives of the functions defining the problem. Then

$$0 \geqslant \frac{f(x_k) - f(x_*)}{\| x_k - x_* \|^2} = \frac{y_k^T f_x(x_*)}{\| x_k - x_* \|} + \frac{1}{2} y_k^T f_{xx}(x_*) y_k + O(\| x_k - x_* \|),$$
$$0 = \frac{g(x_k) - g(x_*)}{\| x_k - x_* \|^2} = \frac{y_k^T g_x(x_*)}{\| x_k - x_* \|} + \frac{1}{2} y_k^T g_{xx}(x_*) y_k' + O(\| x_k - x_* \|),$$
$$0 \geqslant \frac{h^j(x_k) - h^j(x_*)}{\| x_k - x_* \|^2} = \frac{y_k^T h_x^{ji}(x_*)}{\| x_k - x_* \|} + \frac{1}{2} y_k^T h_{xx}^j(x_*) y_k + O(\| x_k - x_* \|).$$

Multiplying the relations obtained respectively by q_*, u_*, v_*^j,

$j \in \sigma(x_*)$, summing them up, and making use of (7.2), we obtain

$$0 \geq \tfrac{1}{2} y_k^T L_{xx}^0(x_*, u_*, v_*, q_*)y + O(\| x_k - x_* \|) \quad .$$

Letting x_k go to x_*, we obtain that $y_k \to y$, having in this case $0 \geq y^T L_{xx}^0 (x_*, u_*, v_*, q_*)y$, which contradicts the positive definiteness of the matrix $L_{xx}^0(x_*, u_*, v_*, q_*)$ on $K_2(x_*)$. ///

If the conditions of Theorem 1.7.2 are satisfied, then for $q_* = 1$ the conditions of Theorem 1.7.3 will be satisfied, which proves the validity of Theorem 1.7.2 (McCormick). Theorem 1.7.3 provides stronger sufficient conditions than Theorem 1.7.2 does; there are problems for which the conditions of Theorem 1.7.3 are assured, but it is not possible to satisfy the conditions of Theorem 1.7.2.

In what follows we shall need the following lemma.

LEMMA 1.7.5. Let the conditions of Theorem 1.7.2 be satisfied at the Kuhn-Tucker point $[x_*, u_*, v_*]$. Then for any vector $[u,v] \in E^m$ such that $u^i > |u_*^i|$, $i \in [1:e]$, $v > v_*$, the point x_* is the strict local minimum of the function

$$L_1(x, u, v) = f(x) + \sum_{i=1}^{e} u^i |g^i(x)| + \sum_{j=1}^{c} v^j h_+^j(x). \tag{7.15}$$

Proof. If the assertion of the Lemma is false, there exists a sequence of points x_k converging to x_*, such that $x_k \neq x_*$ and

$$L_1(x_k, u, v) \leq L_1(x_*, u, v) \quad .$$

This implies that

$$\psi(x_k) = f(x_k) - f(x_*) + \sum_{i=1}^{e} u^i |g^i(x_k)| + \sum_{j=1}^{c} v^j h_+^j(x_k) \leqslant 0.$$

Let

$$y = \lim_{x_k \to x_*} \frac{x_k - x_*}{\|x_k - x_*\|}.$$

The function $\psi(x)$ can be represented as the maximum function by letting

$$|g^i(x)| = \max\,[g^i(x), -g^i(x)] ,$$

$$h^i_+(x) = \max\,[0; h^j(x)] .$$

Then, by Theorem 1.5.10, the derivative of the function $\psi(x)$ in the direction y at the point x_* satisfies the condition

$$\frac{\partial \psi(x_*)}{\partial y} = y^T f_x(x_*) + \sum_{i=1}^{e} u^i\,|y^T g^i_x(x_*)| + \sum_{j \in \sigma(x_*)} v^j \langle y,\ h^j_x(x_*)\rangle_+ \leqslant 0.$$

Using (7.1) we express the gradient $f_x(x_*)$ and put it into the inequality obtained. Then we have

$$\sum_{i=1}^{e} [u^i \cdot |y^T g^i_x(x_*)| - u^i_* \cdot y^T g^i_x(x_*)] + \sum_{j \in \sigma(x_*)} [v^j \langle y,\ h^j_x(x_*)\rangle_+ - v^j_* \langle y,\ h^j_x(x_*)\rangle] \leqslant 0.$$

Since $u^i > |u^i_*|$ and $v^j > v^j_*$, each term in the square brackets is nonnegative and therefore is zero. Hence

$$y^T g^i_x(x_*) = 0, \qquad y^T h^j_x(x_*) = 0, \qquad y^T h^s_x(x_*) \leq 0$$

for all $i \in [1:e]$, $j \in \theta(x_*,v_*)$, $s \in \sigma(x_*) \setminus \theta(x_*,v_*)$. Thus, $y \in K_1(x_*,v_*)$ and $y^T L_{xx}(x_*,u_*,v_*)y > 0$, whence we obtain that for all x_k, beginning with some x_k, the inequality $L(x_k,u_*,v_*) > L(x_*,u_*,v_*)$ is satisfied.

Next we have the inequalities

$$L_1(x_k, u, v) \geqslant L(x_k, u_*, v_*) > L(x_*, u_*, v_*) = \\ = f(x_*) = L_1(x_*, u, v) \geqslant L_1(x_k, u, v).$$

The contradiction obtained proves the Lemma. ///

4. MODIFICATIONS OF McCORMICK'S THEOREM

Later on, while studying numerical methods we shall need other formulations of Theorem 1.7.2. They are the following.

If at $[x_*, v_*]$ the strict complementarity condition is satisfied, then $\sigma(x_*) = \theta(x_*, v_*)$, the cones $K_1(x_*, v_*)$ and $K_2(x_*)$ specified by (7.12) and (7.13) coincide at the Kuhn-Tucker point with the cone

$$K_3(x_*) = \{x \in E^n\colon\ x^T g_x(x_*) = 0, \\ x^T h_x^j(x_*) = 0,\ j \in \sigma(x_*)\}. \tag{7.16}$$

Theorem 1.7.3 can be stated now as follows:

<u>THEOREM 1.7.4</u>. Let $[x_*, u_*, v_*]$ be a Kuhn-Tucker point in the problem (6.1), and let the strict complementarity condition be satisfied at $[x_*, v_*]$; let the functions defining the problem be twice differentiable at x_*, let the matrix $L_{xx}(x_*, u_*, v_*)$ be positive definite on the cone $K_3(x_*)$. Then x_* is the isolated local minimum of the problem (6.1).

Introducing additional variables, we reduce the problem (6.1) to that of finding the minimum in the presence of equality-type constraints only, and for this new problem we state sufficient conditions for the minimum. Let us introduce the vector $p \in E^c$. Also, we express the feasible set in the problem (6.1) in terms of

$$X = \left\{ x \in E^n:\ g(x) = 0,\ h^j(x) + \frac{1}{4}(p^j)^2 = 0,\ j \in [1:c] \right\}.$$

The initial problem (6.1) consists now in minimizing the function f over the two vectors x and p:

$$\min_{x \in X} \min_{p \in E^c} f .$$

Hence it is appropriate to extend the vector x, letting $z = [x, p] \in E^{n+c}$, and to combine the functions defining the constraints:

$$R(z) = \left[g(x),\ h^1(x) + \frac{1}{4}(p^1)^2,\ \ldots,\ h^c(x) + \frac{1}{4}(p^c)^2 \right].$$

Next we represent the problem (6.1) in the form:

$$\min_{z \in Z} f, \qquad Z = \{ z \in E^{n+c}:\ R(z) = 0 \}. \tag{7.17}$$

For this problem we introduce the dual vector $y \in E^m$, $m = e + c$, and write the Lagrange function in the form

$$\begin{aligned} L^1(x, p, y) = f(x) + \sum_{i=1}^{e} y^i g^i(x) + \\ + \sum_{j=1}^{c} y^{e+j} \left[h^j(x) + \frac{1}{4}(p^j)^2 \right]. \end{aligned} \tag{7.18}$$

Also, we write the matrix R_z of dimension $(n+c) \times m$ in terms of

$$R_z(z) = \left[\begin{array}{c|c} g_x(x) & h_x(x) \\ \hline 0_{ce} & \frac{1}{2} D(p) \end{array} \right].$$

If in the problem (6.1) the point $x_* \in X_*$ and the constraints $g(x) = 0$, $h(x) \le 0$ satisfy the CQ, then the point $z_* = [x_*, p_*] \in E^{n+c}$, where

$$p_*^j = 2\sqrt{-h^j(x_*)}\ , \qquad j \in [1:c]\ , \tag{7.19}$$

is feasible, optimal in the problem (7.17), the constraints $R(z) = 0$ satisfy the CQ.

The converse is also true:

if the point $z_* = [x_*\ p_*]$ is feasible, optimal in the problem (7.7), the constraints $R(z) = 0$ satisfy the CQ, then the point x_* is feasible, optimal in the problem (6.1), and the constraints $g(x) = 0$, $h(x) \leqslant 0$ satisfy the constraint qualification at the point x_*.

Let $[x_*, u_*, v_*]$ be the Kuhn-Tucker point in the problem (6.1). Then letting $y_* = [u_*, v_*]$ and using (6.5), (7.18), (7.19) and the condition $x_* \in X_*$, we obtain

$$L_z^1(z_*, y_*) = 0\ , \qquad L_y^1(z_*, y_*) = 0\ . \tag{7.20}$$

In a more elaborate form, these conditions become:

$$\begin{aligned} &L_x^1(z_*, y_*) = f_x(x_*) + R_x(z_*)\, y_* = 0, \\ &L_{p^j}^1(z_*, y_*) = \frac{1}{2} y_*^{e+i} p_*^i = y_*^{e+i}\sqrt{-h^j(x_*)} = 0, \\ &L_y^1(z_*, y_*) = \begin{bmatrix} g(x_*) \\ h(x_*) + \frac{1}{4} D(p_*)\, p_* \end{bmatrix} = 0. \end{aligned} \tag{7.21}$$

Thus, to each Kuhn-Tucker point $[x_*, u_*, v_*]$ for the problem (6.1) there corresponds a stationary point $[z_*, y_*]$ of the Lagrange function $L^1(z,y)$ for the problem (7.17). Conversely, if $[z_*, y_*]$ is the stationary point of the Lagrange function L^1 and $v_*^j = y_*^{e+j} \geq 0$ for $j \in \sigma(x_*)$, then $[x_*, u_*, v_*]$ is the Kuhn-Tucker point in the problem (6.1). Let

$$L^1_{zz}(z_*, y_*) = \left[\begin{array}{c:c} L_{xx}(x_*, u_*, v_*) & 0_{nc} \\ \hdashline 0_{cn} & \frac{1}{2} D(v_*) \end{array}\right]. \tag{7.22}$$

<u>THEOREM 1.7.5</u>. Let the functions defining the problem (6.1) be twice differentiable at the point $x_* \in X$. There exists a vector $y_* \in E^m$ such that conditions (7.20) are satisfied at the point $[z_*, y_*] \in E^{n+m+c}$, where $z_* = [x_*, p_*] \in E^{n+c}$ and the coordinates p_* are defined from (7.19). The matrix $L^1_{zz}(z_*, y_*)$ is positive definite on the subspace

$$K_4(x_*) \;=\; \{z \in E^{n+c} \;:\; z^T R_z(z_*) = 0_{1m}\} \quad . \tag{7.23}$$

Then the point x_* is the isolated local minimum of the problem (6.1).

This Theorem can be proved via the same considerations as in proving Theorem 1.7.3; but it is much simpler to show that the conditions of this Theorem imply the conditions of Theorem 1.7.4. Let $\bar{z} = [\bar{x}, \bar{v}] \in K_4(x_*)$. Then

$$\bar{x}^T g_x(x_*) = 0_{1e}, \quad \bar{x}^T h^j_x(x_*) + \frac{1}{2} \bar{v}^j p^j_* = 0 \quad j \in [1:c].$$

If $j \in \sigma(x_*)$, then, by (7.19), $p^j_* = 0$ and for the vector $z \in K_4(x_*)$ we may take a vector in which only one coordinate $\bar{z}^{n+j} = \bar{v}^j$ is nonzero. From the positive definiteness of $L^1_{zz}(z_*, y_*)$ on $K_4(x_*)$ it then follows that $(\bar{z}^{n+j})^2 v^j_* > 0$ and therefore $v^j_* > 0$ for all $j \in \sigma(x_*)$. Hence at the point $[x_*, v_*]$ the strict complementarity condition is satisfied. From the fact that $p^j{}_* = 0$ for all $j \in \sigma(x_*)$ we infer that each vector $[\bar{x}, \bar{v}]$ belonging to $K_4(x_*)$ is such that the vector $\bar{x} \in K_2(x_*)$. All the conditions of Theorem 1.7.4 are thus satisfied.

This technique for reducing the general problem (6.1) to the problem (7.17), with equality-type constraints only, will be used in the future in describing the numerical methods. We say that the auxiliary vector p introduced above is an artificial vector.

5. NECESSARY CONDITIONS FOR A MINIMUM

For programming problems necessary and sufficient conditions for a minimum are given in Theorem 1.7.1. Now we formulate necessary conditions for a minimum for a general problem of nonlinear programming, (6.29).

THEOREM 1.7.6. Let U be a convex set the interior of which is not empty; let the functions defining the problem (6.29) be given on some open set containing U, and also differentiable on the point x_* belonging to the solutions set X_* of the problem (6.29); and let the vector function g have continuous first partial derivatives at the point x_*. Then:

• 1. it is necessary that $q_* \in E^1_+$, $u_* \in E^e$, $v_* \in E^c_+$ exist, not equal to zero at the same time and such that at the point (x_*, v_*) the complementarity condition (6.5) be satisfied, and for any $x \in \bar{U}$ let the following inequality hold:

$$0 \le \langle x - x_*, L^0_x(x_*, u_*, v_*, q_*)\rangle \quad ; \qquad (7.24)$$

• 2. if the function $L^0(x, u_*, v_*, q_*)$ is pseudoconvex in x with respect to the set U, it is necessary that

$$x_* \in \operatorname*{Arg\,min}_{x \in U} L^0(x, u_*, v_*, q_*) \quad ;$$

• 3. if the set U is open, and the Arrow-Hurwicz-Uzawa CQ is satisfied at the point x_*, then in the problem (6.29) the

Kuhn-Tucker point $[x_*, u_*, v_*]$ exists.

The proof of the theorem can be found, for example, in O. Mangasarian [1]; assertion 1 has been proved also in F.P. Vasil'ev [1], or in N.N. Moiseev et al. [1].

6. PARAMETRIC PROGRAMMING

The functions defining the problem (6.1) may contain a parameter. Parametric programming is in fact the investigation of the dependence between solutions and the parameter. We discuss this subject briefly. We consider the problem of finding

$$\min_{x \in X} \max_{y \in Y(x)} F(x,y) \quad , \tag{7.25}$$

$X = \{x \in E^n : H(x) \le 0\}$, $Y(x) = \{y \in E^m : B(x,y) \le 0\}$, where $H: E^n \to E^k$, $B(x,y): E^n \times E^m \to E^s$. The interior problem in this case is a nonlinear programming problem and consists in finding

$$y(x) = \operatorname{Arg} \max_{y \in Y(x)} F(x,y) \quad . \tag{7.26}$$

We compose the Lagrange function

$$L^2(y, \lambda, x) = F(x, y) - \sum_{i=1}^{s} \lambda^i B^i(x, y).$$

Assume that for any x the set $Y(x)$ is not empty and there exist vector functions $y(x)$, $\lambda(x)$ such that they make a Kuhn-Tucker point:

$$L^2_y(y(x), \lambda(x), x) = F_y(x, y(x)) - \sum_{i=1}^{s} \lambda^i(x) B^i_y(x, y(x)) = 0, \tag{7.27}$$

$$\lambda^i(x) B^i(x, y(x)) = 0, \quad \lambda \geqslant 0, \quad B(x, y(x)) \leqslant 0. \tag{7.28}$$

The exterior problem consists in finding

$$x_* = \operatorname{Arg}\min_{x \in X} \phi(x) , \qquad \phi(x) = F(x, y(x)) .$$

Assuming that the functions $F(x,y)$, $y(x)$ are continuously differentiable, we obtain

$$\frac{d\varphi}{dx} = F_x(x, y(x)) + \frac{\partial y}{\partial x} F_y(x, y(x)).$$

The second term is not, generally speaking, equal to zero, and this makes the solution of the exterior problem more complicated since it requires the matrix $\frac{\partial x}{\partial y}$ be determined. Fortunately, the situation becomes simpler because the following formula holds true:

$$\frac{d\varphi}{dx} = F_x(x, y(x)) - \sum_{i=1}^{s} \lambda^i(x) B_x^i(x, y(x)). \tag{7.29}$$

Indeed, assuming that the functions F, B, $y(x)$ are continuously differentiable in x, we differentiate the first relation in (7.28). We obtain

$$\frac{d\lambda^i(x)}{dx} B^i(x, y(x)) + (B_x^i(x, y(x)) + \\ + \frac{\partial y}{\partial x} B_y^i(x, y(x)))\lambda^i(x) = 0 .$$

Multiplying this equality by $\lambda^i(x)$ and noting the first relation in (7.28), we obtain

$$\frac{\partial y}{\partial x} B_y^i(x, y(x))(\lambda^i(x))^2 = -B_x^i(x, y(x))(\lambda^i(x))^2 .$$

Taking into account (7.27), we obtain

$$\frac{\partial y}{\partial x} F_y(x,\ y(x)) = \sum_{i=1}^{s} \lambda^i(x) \frac{\partial y}{\partial x} B_y^i(x,\ y(x)) =$$
$$= -\sum_{i=1}^{s} B_x^i(x,\ y(x))\, \lambda^i(x)$$

yielding (7.29).

For an exterior problem the Lagrangian has the form

$$L^3(x) = F(x,\ y(x)) + \langle \mu,\ H(x) \rangle .$$

The necessary condition for the minimum is that the conditions

$$F_x(x_*,\ y_*) - \sum_{i=1}^{s} B_x^i(x_*,\ y_*)\lambda_*^i + \sum_{i=1}^{k} \mu_*^i H_x^i(x_*) = 0,$$
$$\mu_*^i H^i(x_*) = 0, \quad \mu_* \geqslant 0, \quad H(x_*) \leqslant 0$$

be satisfied, where $y_* = y(x_*)$, $\lambda_* = \lambda(x_*)$.

Assuming that the functions $\lambda(x)$, $y(x)$ are known, we arrive at the usual problem of nonlinear programming, in which the minimum of the function $F(x,\ y(x))$ is sought.

The next theorem provides sufficient conditions for the functions $y(x)$, $\lambda(x)$ to be differentiable. Let
$\bar{\sigma}(x,\ y(x)) = \{j \in [1:s] : B^j(x,y(x)) = 0\}$.

<u>THEOREM 1.7.7</u>. Let the functions F, B, H be twice continuously differentiable in all the arguments, let at the Kuhn-Tucker point $[y(x),\lambda(x)]$ in the problem (7.26) the sufficient conditions for the maximum following from (McCormick's) Theorem 1.7.2 be satisfied; let at the point $[y(x),\lambda(x)]$ the strict complementarity condition be satisfied, let the gradients $B_x^i(x,\ y(x))$ be linearly independent for $j \in \bar{\sigma}(x,\ y(x))$. Then the functions $y(x)$, $\lambda(x)$ are differentiable, and their derivatives are defined by the following system:

$$\left[\begin{array}{c|c} L^2_{yy}(x, y(x)) & -B_y(x, y(x)) \\ \hline D(\lambda(x)) B^T_y(x, y(x)) & D(B(x, y(x))) \end{array}\right] \begin{bmatrix} \dfrac{dy}{dx} \\ \dfrac{d\lambda}{dx} \end{bmatrix} = -\begin{bmatrix} L^2_{yx}(x, y(x)) \\ D(\lambda) B_x(x, y(x)) \end{bmatrix},$$

the derivative of the function $\phi(x)$ is given by (7.29).

This theorem is a slight modification of Theorem 6 cited in Chapter 2 of A. Fiacco and G. McCormick [1], which contains a proof of Theorem 6 based on the application of the implicit function theorem.

8. NECESSARY CONDITIONS FOR A MINIMUM IN OPTIMAL CONTROL PROBLEMS

1. *THE MAIN PROBLEM OF OPTIMAL CONTROL*

Assume that the behavior of a controllable process is described by the system of ordinary differential equations

$$\frac{dx(t)}{dt} = f(x(t), u(t), t), \qquad 0 \le t \le T, \quad x(0) = x_0, \tag{8.1}$$

where the vector $x(t) \in E^n$, usually known as a state vector; the vector $u(t) \in E^r$ is said to be the vector of controls. All the components of the vector function $f(x,u,t)$ are differentiable in the aggregate of the variables x and u. We consider first the case where the interval T is given and the initial state vector x_0 is fixed. In many applied problems, time plays the role of the independent variable t and the system (8.1) describes a dynamic process; hence it is often referred to as a fixed-time problem. We say that the case where no constraints are imposed on the vector $x(T)$ is a problem with a free right endpoint.

We shall say that a piecewise-continuous function $u = u(t)$ attaining on an interval $0 \leq t \leq T$ arbitrary values in some specified set $U \subset E^r$ is a feasible control.

A given feasible control $u(t)$ uniquely determines the continuous piecewise-differentiable solution $x = x(t)$ of the system (8.1) on the interval $0 \leq t \leq T$. The functional

$$R = b(x(T)) \tag{8.2}$$

is the control performance criterion. The function $b(x)$ is assumed to be everywhere differentiable.

The problem consists in finding the feasible control law $u(t)$ and the corresponding solution $x(t)$ to the system (8.1), such that the function R attains the least possible value.

This problem of optimal control differs from the problem considered in classical Calculus of Variations only in the new requirement $u(t) \in U$. The necessary condition for a minimum for such problems was first stated and proved by L.S. Pontryagin and his colleagues, and is known as the Maximum Principle. Later on, we shall formulate this principle in a form different from the original.

Let us consider the vector function $p(t)$ satisfying the following adjoint system of ordinary differential equations:

$$\frac{dp^i(t)}{dt} = -\sum_{j=1}^{n} \frac{\partial f^j(x(t), u(t), t)}{\partial x^i} p^j(t), \quad i \in [1:n]. \tag{8.3}$$

Using the adopted abridged notation, we can rewrite this system in a compact form:

$$\dot{p}(t) = -f_x(x(t),\ u(t),\ t)\ p(t)\ . \tag{8.3}$$

We require that at the end of the motion the condition

$$p(T) = b_x(x(T)) \tag{8.4}$$

be satisfied. We say that the vector function $p(t)$ thus defined is the adjoint multiplier, or an impulse. We introduce the auxiliary function

$$H(x,u,t,p) = \langle f(x,u,t),\ p\rangle\ ,$$

usually called a Hamiltonian.

Any feasible control $u(t)$ defined on the interval $0 \leqslant t \leqslant T$, which minimizes the functional R is said to be an optimal control and is denoted by $u_*(t)$; we say that the solution to the system (8.1) corresponding to the control $u_*(t)$ is an optimal trajectory and we denote it by $x_*(t)$. If in (8.3) and (8.4) we take $u_*(t)$ and $x_*(t)$ respectively for $u(t)$ and $x(t)$, we obtain the impulse $p_*(t)$. The necessary condition for a minimum in the optimal control problem is as follows.

<u>THEOREM 1.8.1</u>. Let $u_*(t)$ be an optimal control and let $x_*(t)$ be an optimal trajectory of the system (8.1). Then there exists an impulse $p_*(t)$ such that for $0 \leqslant t \leqslant T$ the following conditions are satisfied:

• 1. the minimum of the function $H(x_*(t),\ u,\ t,\ p_*(t))$ with respect to u on the set U is attained at the point $u = u_*(t)$, i.e.,

$$u_*(t) \in \operatorname*{Arg\,min}_{u \in U} H(x_*(t),\ u,\ t,\ p_*(t))\ ; \tag{8.5}$$

●2. if the right sides of the system (8.1) do not depend explicitly on t (the system is autonomous), the function $H(x_*(t),\ u_U(t),\ p_*(t))$ is constant.

This necessary condition for the minimum is said to be the Minimum Principle for optimal control problems. L. S. Pontryagin formulated the Maximum Principle, using in fact the impulse satisfying the system (8.3), but, in contrast to (8.4), the boundary condition was of the form

$$p(T) = -b_x(x(T)) ,$$

hence the condition (8.5) was written differently:

$$u_*(t) \in \operatorname*{Arg\,max}_{u \in U} H(x_*(t),\ u,\ t,\ p_*(t)) \quad .$$

Therefore, the difference is not essential; however, the formulation we have used is more convenient, especially in considering game problems and in establishing the minimum principle satisfying analogous necessary conditions for the minimum for nonlinear programming problems.

Next we make use of the results of Section 4. By Theorem 1.4.1, if for a fixed t the function $H(x_*(t), u, t, p_*(t))$ is differentiable in u at the point $u = u_*(t)$ and the set U is convex, it then follows from (8.5) that the condition

$$u_*(t) \in \operatorname*{Arg\,min}_{u \in U} \langle H_u(x_*(t),\ u_*(t),\ t,\ p_*(t)),\ u - u_*(t)\rangle \tag{8.6}$$

is satisfied.

This necessary condition is usually called a linearized minimum principle.

At present, many different proofs of Theorem 1.8.1 are available. They are published in many books; therefore we shall omit the proof herein. We only refer to the monograph of L. S. Pontryagin, B.G. Boltyanskij, R. V. Gamkrelidze, and E. M. Mishchenko [1], and also to N. N. Moiseev [1] and to R. Gabasov and F. M. Kirillova [2].

We may make an attempt to construct a numerical method for solving optimal control problems using the minimum principle. For each fixed set $x_*(t)$, t, $p_*(t)$ we find from the condition (8.5) the point-set mapping

$$u_* = \beta(x_*, t, p_*) \tag{8.7}$$

on which the Hamiltonian $H(x_*, u, t, p_*)$ reaches the minimum in $u \in U$. Upon substitution of (8.7) into Eqs. (8.1) and (8.3) we obtain the system consisting of $2n$ ordinary differential equations for n-dimensional vector functions $x_*(t)$, $p_*(t)$:

$$\begin{aligned} \dot{x}_* &= f(x_*, \beta(x_*, t, p_*), t) , \\ \dot{p}_* &= -f_x(x_*, \beta(x_*, t, p_*), t) p_* . \end{aligned} \tag{8.8}$$

For this system n conditions are given at the initial time (at the left endpoint) and n conditions at the time (at the right endpoint):

$$x_*(0) = x_0 , \qquad p_*(T) = b_x(x_*(T)) . \tag{8.9}$$

Thus, the solution of the initial problem is formally reduced to that of the boundary value problem for the system of $2n$ differential equations. If one could solve this problem, that is, define the functions $x_*(t)$, $p_*(t)$ satisfying the system (8.8) and the conditions (8.9), one would find the optimal control $u_*(t)$ for the initial problem among the functions representable as (8.7). This technique is quite convenient for analytical solution of the initial problem. However, it is of questionable utility in practical computation.# The reason is that the boundary value problem obtained is usually essentially nonlinear. The mappings β are often discontinuous, which leads to the situation that the right sides of (8.8) become nonsmooth or even non-single-valued. The boundary value problem does not always have a unique solution, and the procedures of solving this problem are frequently divergent. The best that can be done for such a problem is to determine a good initial approximation; only then a real possiblity of solving the boundary value problem develops. These circumstances compel us in the case of optimal control problems to seek different computational methods: we shall describe them in Chapter 6.

2. OPTIMIZATION WITH RESPECT TO CONTROL PARAMETERS

Among optimal control problems one frequently comes across cases where optimization consists not only in choosing the control function $u(t)$ but also in finding the optimal value of a vector of additional control parameters $\xi \in E^s$. The system (8.1) has in this case the form

$$\frac{dx}{dt} = f(x,u,t,\xi) \quad . \tag{8.10}$$

The functional to be minimized also depends on ξ: $R = b(x(T), \xi)$. It is required to find a feasible control function $u(t)$, a vector

Editor's note: Not quite! See J. Stoer and R. Bulirsch, *Introduction to Numerical Analysis*. Berlin Heidelberg New York Tokyo: Springer-Verlag, 1980.

of control parameters ξ and the corresponding solution $x(t)$ to (8.10), such that the functional R assumes the smallest possible value.

The necessary condition for a minimum is given by Theorem 1.8.1, in which instead of $f(x,u,t)$ and $b(x)$ respectively $f(x,u,t,\xi)$ and $b(x,\xi)$ are taken and one additional condition is introduced for a minimum of R in ξ:

$$b_\xi(x(T),\ \xi) + \int_0^T f_\xi(x(t),\ u(t),\ t,\ \xi)\ p(t)\ dt = 0 \ . \tag{8.11}$$

This condition must be satisfied along the optimal trajectory, i.e., for $u(t) = u_*(t)$, $x(t) = x_*(t)$, $p(t) = p_*(t)$, $\xi = \xi_*$, where ξ_* is the optimal vector of control parameters. The condition (8.11) adds to (8.8) and (8.9) scalar relations necessary for defining the components of the vector ξ.

Theorem 1.8.1 and the assertions just listed are fundamental results in optimal control theory. We shall show next how, using there results, various particular problems can be studied.

3. THE PROBLEM WITH FIXED TIME AND FREE RIGHT ENDPOINT

In the main problem formulated above, we shall require in addition that the state vector $x(t)$ at a given finite time T belong to the terminal manifold

$$X = \{x \in E^n : g(x) = 0,\ h(x) \le 0\} \ ,$$

where the differentiable vector functions g and h are the mappings

$$g:\ E^n \to E^e \ , \qquad h:\ E^n \to E^c \ .$$

As well as in the case of nonlinear programming problems, we take for the functional a function analogous to (6.25):

$$R(x, \tilde{u}, v, q) = qb(x) + \sum_{i=1}^{e} \tilde{u}^i g^i(x) + \sum_{j=1}^{c} v^j h^j(x), \tag{8.12}$$

where $q \in E_+^1$, $\tilde{u} \in E^e$, $v \in E_+^c$.

The assertion of Theorem 1.8.1 remains valid; one should only make some changes: instead of (8.4) the following condition is to be used:

$$p(T) = R_x(x(t), \tilde{u}, v, q) \quad . \tag{8.13}$$

Moreover, the complementarity conditions must be satisfied:

$$v^j h^j(x(T)) = 0 \quad , \qquad j \in [1:c] \quad .$$

If the conditions at the right endpoint satisfy the constraint qualification, we may put $q = 1$; the new unknown dual variables $\tilde{u}$, v are "compensated" by the condition $x(T) \in T$. Using the minimum principle, we can reduce this problem to a boundary value problem.

4. LAGRANGE'S PROBLEM. MAYER'S PROBLEM. BOLZA'S PROBLEM

For Lagrange's problem the functional to be minimized will be written as the integral

$$R = \int_0^T B(x(t), u(t), t)\, dt,$$

where B denotes the differentaible function of x and u.

Next, we introduce an additional state variable x^{n+1} through the equation

$$\frac{dx^{n+1}}{dt} = B(x, u, t), \quad x^{n+1}(0) = 0.$$

This problem can be now formulated as the problem of finding the control $u(t)$, ensuring on a fixed interval $[0,T]$ the minimum of the terminal functional

$$R = x^{n+1}(T)$$

in the presence of the relationships

$$\frac{dx}{dt} = f(x, u, t), \quad \frac{dx^{n+1}}{dt} = B(x, u, t).$$

Thus, in the expanded state space E^{n+1} the problem has reduced to the main problem considered above. For the Hamiltonian we take the function

$$H(x, x^{n+1}, u, t, p, p^{n+1}) = \langle f(x, u, t), p\rangle + B(x, u, t)p^{n+1}.$$

Eq. (8.3) the the condition (8.4) are replaced by the following:

$$\frac{dp}{dt} = -f_x p - B_x, \quad p(T) = 0, \quad p^{n+1}(t) \equiv 1.$$

It it is required that at the end of the motion $x \in X$, it then follows from (8.13) that the condition

$$p(T) = \sum_{i=1}^{e} \bar{u}^i g_x^i(x(T)) + \sum_{j \in \sigma(x(T))} v^j h_x^j(x(T))$$

has to be satisfied, where $\sigma(x(T))$ denotes the set of active constraints of the inequality type. The last condition is usually referred to in Variational Calculus as the transversality condition; it has a simple geometric meaning: the vector $p(T) \in E^n$ must be orthogonal to the tangent subspace toward the terminal manifold X.

Mayer's problem with fixed time consists in minimizing a functional of the form

$$R = b(x(T), T) \quad . \tag{8.14}$$

Introducing the additional state variable $x^{n+1} = t$ we reduce the system (8.1) to the autonomous system

$$\frac{dx}{dt} = f(x, u, x^{n+1}), \quad \frac{dx^{n+1}}{dt} = 1, \qquad (8.15)$$
$$R = b(x(T), x^{n+1}(T)).$$

The expanded adjoint multiplier consists of the vector $p(t)$ satisfying the system (8.3), and the scalar p^{n+1} for which

$$\frac{dp^{n+1}(t)}{dt} = -\sum_{i=1}^{n} f_t^i(x, u, t)\, p^i(t) \qquad (8.16)$$

with terminal conditions

$$p(T) = b_x(x(T), T), \quad p^{n+1}(T) = b_T(x(T), T). \qquad (8.17)$$

The Hamiltonian of the expanded system is constant along the optimal trajectory:

$$\langle f(x,u,t), p(t)\rangle + p^{n+1}(t) = \text{const.}$$

The problem has thus been reduced to the basic problem.

In Bolza's problem the functional is given by

$$R = b(x(T), T) + \int_0^T B(x, u, t)\, dt.$$

We also introduce two additional state variables

$$\frac{dx^{n+1}}{dt} = 1, \quad \frac{dx^{n+2}}{dt} = B(x, u, x^{n+1}),$$

then the functional to be minimized has the form

$$R = x^{n+2}(T) + B(x(T), x^{n+1}(T)).$$

In the extended state space E^{n+2} the problem is again reduced to the basic problem, Theorem 1.8.1 is then applicable.

5. THE PROBLEM WITH FREE END TIME T

It is required to find the vector of the controls $u(t)$, the interval $[0,T]$ and the corresponding solution $x(t)$ to the system (8.1), so that the terminal functional (8.14) attains the smallest possible value. We suppose that such an interval exists and is nonzero.

Putting $t = \tau\xi$, we make a substitution of the independent variable in system (8.15). Then the system (8.15) becomes

$$\frac{dx}{d\tau} = \xi f(x, u, x^{n+1}), \quad \frac{dx^{n+1}}{d\tau} = \xi. \tag{8.18}$$

We consider that the new independent variable τ changes on the fixed interval $[0,T_0]$. For the system (8.18) we shall be seeking the control vector $u(\tau)$ and the control parameter ξ, so that for $\tau = T$ the functional $R = b(x(T_0), x^{n+1}(T_0))$ attains the smallest possible value.

We have arrived at the problem with fixed time, the free right endpoint, and one control parameter. It follows from (8.11) that along the optimal trajectory

$$\int_0^{T_0} [\langle f(x(\tau), u(\tau), x^{n+1}(\tau)), p(\tau)\rangle + p^{n+1}(\tau)]\, d\tau = 0 \tag{8.19}$$

is satisfied. We take into account that the integrand is constant. We proceed to the initial variables and, using the condition at the right endpoint (8.17), we obtain

$$\begin{gathered} b_T(x(T), T) + H(x(T), u(T), T, p(T)) = 0, \\ p(T) = b_x(x(T), T). \end{gathered} \tag{8.20}$$

Upon solving the optimization problem (that is of finding $u(t)$ and ξ), the optimal motion time is defined by the formula

$T = T_0\xi$, where T_0 denotes the arbitrary positive number specified before the calculations (one may always assume, for example, that $T_0 = 1$). The assertion of Theorem 1.8.1 concerning the property (8.5) remains valid, the first relationship in (8.20) is formally the additional condition for defining the optimal value of the motion time T.

6. THE MINIMUM TIME PROBLEM

It is required to find the vector $u(t)$, the corresponding solution $x(t)$ to the system (8.1), so that the vector $x(t)$ reaches a given set X in the shortest interval of time.

The interval of variation of the independent variable in this case is unknown. Hence, as in the three previous cases, we introduce the new state variable x^{n+1}, go over to the system (8.15), substitute the independent variable putting $t = \tau\xi$, obtain the system (8.18) for which it is required to find the control $u(\tau)$ and the parameter ξ, on the fixed interval $[0,T_0]$ so that $x(T_0) \in X$ and the functional $x^{n+1}(T_0)$ attain the smallest possible value. Now we introduce the auxiliary function R analogous to (8.12):

$$R = qx^{n+1} + \sum_{i=1}^{e} g^i(x)\tilde{u}^i + \sum_{j=1}^{c} h^j(x)v^j.$$

Here $q \geq 0$, $v \geq 0$. The necessary minimum condition (8.11) leads us to (8.19), yielding in turn

$$H(x(T),\ u(T),\ T,\ p(T)) = -q \leq 0 ,$$

in addition, the complementarity condition $h^j(x)v^j = 0$ appears, as well as the conditions $g(x(T)) = 0$, $h(x(T)) \leq 0$.

Many other problems of optimal control can be reduced to the main problem in a similar way. Theorem 1.8.1 is inapplicable, however, in a quite widely known and important case when mixed constraints are imposed on the state coordinates and controls along a trajectory of the form

$$g(x(t),\ u(t),\ t) = 0\ , \qquad h(x(t),\ u(t),\ t) \leq 0\ .$$

Properties of such problems are investigated in Dubovitskij and Milyutin [1], Smol'yakov [1], and in Anorov [1], wherein the notation for the necessary conditions for a minimum is more complicated and not very suitable for devising numerical methods. We shall not dwell on these conditions, since we do not need them in the future in order to construct numerical methods, which will be based on other ideas originating from Nonlinear Programming.

Chapter 2

CONVERGENCE THEOREMS AND THEIR APPLICATION TO THE INVESTIGATION OF NUMERICAL METHODS

In this chapter we cite the main mathematical results which will be used in the subsequent chapters in justifying numerical methods. We discuss first the methods of stability theory: stability of the first order approximation and the method of Lyapunov functions; then we give theorems on convergence of contraction mappings and point-set mappings, as well as methods for solving systems of non-linear equations and methods for finding the minimax.

1. STABILITY OF THE FIRST ORDER APPROXIMATION

1. BASIC DEFINITIONS

We consider the system of ordinary differential equations

$$\frac{dx}{dt} = f(x,t) \quad , \tag{1.1}$$

where the vector function $f(x,t) \in C^{10}_{xt}(E^n \times I_+)$, i.e., for each $t \geq 0$, f is continuously differentiable over x everywhere on E^n, and for any x is continuous in t on the set I_+. In this case the initial condition $x(0) = x_0$ defines a unique solution of the system (1.1), which we will denote by $x = x(x_0,t)$.

DEFINITION 2.1.1. We say that the system (1.1) is Lagrange stable if for all x_0

• 1. the solution $x(x_0,t)$ exists for all $t \in I_+$;

• 2. the norm $||x(x_0,t)||$ is bounded on I_+.

DEFINITION 2.1.2. The solution $x(x_*,t)$ of the system (1.1) is said to be Lyapunov stable as $t \to \infty$ (or, in short, stable), if for any $\varepsilon > 0$ there exists $\delta = \delta(\varepsilon)$ such that:

• 1. every solution $x(x_0,t)$ of the system (1.1), satisfying the condition

$$||x_0 - x_*|| < \delta \tag{1.2}$$

is defined for $0 \leq t < \infty$;

• 2. for these solutions we have the inequality

$$||x(x_0,t) - x(x_*,t)|| < \varepsilon \qquad \forall\ t \in I_+ .$$

In other words, the solution $x(x_*,t)$ is stable, if all other solutions originating in a sufficiently small neighborhood of the point x_*, remain for any $t \in I_+$ inside the neighborhood constructed around the solution $x(x_*,t)$. The stability implies a continuous, uniform in $t \in I_+$, dependence of the solutions $x(x_0,t)$ of the system (1.1) on the initial point x_0.

DEFINITION 2.1.3. The solution $x(x_*,t)$ $(0 \leq t < \infty)$ is said to be asymptotically stable (as $t \to \infty$), if:

• 1. this solution is Lyapunov stable;

• 2. each solution $x(x_0,t)$ satisfying (1.2), possesses the property

$$\lim_{t\to\infty} ||x(x_0,t) - x(x_*,t)|| = 0 . \tag{1.3}$$

<u>DEFINITION 2.1.4</u>. The solution $x(x_*,t)$ $(0 \leq t < \infty)$ is said to be globally asymptotically stable if this solution is stable and the condition (1.3) holds for the solutions $x(x_0,t)$ of the system (1.1) for all $x_0 \in E^n$.

It is often convenient to reduce the system (1.1) as to make the point $x = 0$ an equilibrium point, i.e., $f(0,t) \equiv 0$ for any $t \in I_+$. In this case, the system (1.1) has a trivial solution $x(0,t) \equiv 0$ which we denote by $x(t) \equiv 0$. We can reformulate Definitions 1.2.3 and 1.2.4 as follows.

<u>DEFINITION 2.1.5</u>. The trivial solution $x(t) \equiv 0$ of the system (1.1) is said to be asymptotically stable if for any $\varepsilon > 0$ there exists $\delta = \delta(\varepsilon)$ such that for any solution to (1.1), satisfying the condition $||x_0|| < \delta$, we have $||x(x_0,t)|| \leq \varepsilon$ for all $t \in I_+$,

$$\lim_{t \to \infty} x(x_0,t) = 0 . \tag{1.4}$$

<u>DEFINITION 2.1.6</u>. The trivial solution $x(t) \equiv 0$ of the system (1.1) is called globally asymptotically stable if:

- 1. it is Lyapunov stable;
- 2. for any $x_0 \in E^n$ the condition (1.4) is satisfied.

<u>DEFINITION 2.1.7</u>. The trivial solution $x(t) \equiv 0$ of the system (1.1) is said to be exponentially stable if there exists a neighborhood of the origin G such that for any $x_0 \in G$ we have the inequality

$$||x(x_0,t)|| \leq N\, e^{-\kappa t} , \tag{1.5}$$

where N, κ are some positive integers not depending on the choice of the point x_0.

The exponential stability of the trivial solution $x(t) \equiv 0$ implies the asymptotic stability of this solution.

Indeed, for an arbitrary $\varepsilon > 0$ we put

$$\|x_0\| < \delta = \frac{\varepsilon}{N} ;$$

then it follows from (1.5) that $\|x(x_0,t)\| \le \varepsilon$ and, furthermore, (1.4) holds.

Methods for assuring asymptotic stability are quite useful for justifying the convergence of numerical methods of optimization. Some numerical methods reduce to finding limit points for a system of ordinary differential equations of the form (1.1), with solutions $x(x_0,t)$ tending as $t \to \infty$ to solutions of the initial optimization problem. We say in this case that the system (1.1) is a numerical method for solving the optimization problem.

2. AUXILIARY LEMMA

LEMMA 2.1.1 (Gronwall). Let the functions $u(t)$ and $v(t)$ be continuous for $t \in [a,\infty)$, let $C > 0$, and let for $t \ge a$ the inequality

$$|u(t)| \le C + \int_a^t |u(s)|\,|v(s)|\,ds$$

be satisfied. Then for $t \ge a$ we have the inequality

$$|u(t)| \le C \exp \int_a^t |v(s)|\,ds .$$

Proof. Let us multiply both sides of the initial inequality by $|v(t)|$:

$$|v(t)||u(t)| \leqslant |v(t)| \left[C + \int_a^t |u(s)||v(s)|\,ds \right].$$

Let $w(t) = \int_a^t |u(s)|\ |v(s)|\ ds$. Then the inequality obtained can be represented as follows:

$$\frac{dw}{dt} \leq |v(t)|[C + w(t)] \quad ,$$

or, noting that $0 < C + w(t)$, we have

$$\frac{w}{C+w(t)} \leq |v(t)| \ .$$

Integrating both sides, we find

$$\ln (C+w(t)) - \ln C \leq \int_a^t |v(s)|\ ds \quad ,$$

which implies

$$C + w(t) \leq C \exp \int_a^t |v(s)|\ ds \ .$$

Taking into account the initial inequality, we obtain

$$|u(t)| \leq C + w(t) \leq C \exp \int_a^t |v(s)|\ ds \ .$$

We have thus arrived at the required inequality. ///

3. THE MAIN THEOREM

We consider a system of the special form

$$\frac{dx}{dt} = Ax + \phi(x,t) \quad , \tag{1.6}$$

where A is a square matrix of the order n.

DEFINITION 2.1.8. The function $v(x,t)$ has an infinitesimal upper limit as $x \to 0$, if for any $\varepsilon > 0$ there exists $\delta = \delta(\varepsilon)$ such that $|v(x,t)| < \varepsilon$ for $||x|| < \delta$ and all $t \in I_+$.

THEOREM 2.1.1 (Lyapunov's Theorem on Stability of the First Order Approximation). Let the matrix A in the system (1.6) be constant and let all the eigenvalues of A have negative real parts. Suppose the function $\|\phi(x,t)\|/\|x\|$ has an infinitesimal upper limit as $x \to 0$. Then the trivial solution $x(t) \equiv 0$ of the system (1.6) is exponentially stable.

Proof. For the system (1.6) we fix an initial point $x_0 \in G$, where G is some neighborhood of the origin. A solution of the system (1.6) satisfies the integral equation:

$$x(x_0,t) = e^{tA}x_0 + \int_0^t e^{(t-s)A} \phi(x(x_0,s), s)\, ds . \qquad (1.7)$$

Here the square matrix $\Phi(t) = e^{tA}$ is definable by the formula

$$\Phi(t) = I_n + \sum_{i=1}^{\infty} \frac{1}{i!} t^i A^i$$

and satisfies the system

$$\frac{d\Phi(t)}{dt} = A\Phi(t) .$$

Differentiating (1.7) over t with these properties taken into account, we obtain that each solution to (1.7) satisfies (1.6). From the fact that the eigenvalues of the matrix A have negative real parts it follows that for $t \geq 0$ there exist $k > 0$, $\eta > 0$ such that

$$||\Phi(t)|| \leq ke^{-\eta t} .$$

Using the last inequality, we obtain from (1.7):

$$\|x(x_0,t)\| \leq k\|x_0\|e^{-\eta t} + k\int_0^t e^{-\eta(t-s)} \|\phi(x(x_0,s), s)\| \, ds \quad .$$

For any $\varepsilon > 0$ there exists δ such that $\|\phi(x,t)\| \leq \frac{\varepsilon\|x\|}{k}$ for each $\|x\| > \delta$ and for any $t \geq 0$. Let $\|x_0\| < \delta$. Then we can find $T > 0$ such that for all $0 < t < T$ the inequality $\|x(x_0,t)\| < \delta$ is satisfied and

$$e^{\eta t}\|x(x_0,t)\| \leq k\|x_0\| + \int_0^t e^{\eta s}\|x(x_0,s)\| \, ds \quad .$$

We use Lemma 2.1.1 and obtain the estimate

$$\|x(x_0,t)\| \leq k\|x_0\|e^{(\varepsilon-\eta)t} \quad , \tag{1.8}$$

which holds for $0 < t < T$. We choose ε so small that $\varepsilon < \eta$. Then it follows from the inequality obtained: $\|x(x_0,t)\| \leq k\|x_0\|$ until $\|x(x_0,t)\| < \delta$. If $\|x_0\| < \frac{\delta}{k}$, then (1.8) holds for all $t \geq 0$, which completes the proof of the Theorem. ///

The analysis of stability of a large class of systems is reduced to the investigation of systems of the form (1.6). Indeed, let the point $a \in E^n$ be an equilibrium for the system (1.1), i.e., $f(a,t) \equiv 0$. Putting $x = a + y$, we interchange the variables. Then we arrive at the system

$$\frac{dy}{dt} = f(a+y, t) \quad .$$

Assume that the vector function $f(x,t)$ is differentiable over x at the point $x = a$; then the system can be represented in the form

$$\frac{dy}{dt} = f(a,t) + f_x^T(a,t)y + \|y\|\gamma(y,t)$$

$$= Ay + \phi(y,t) ,$$

where

$$A = f_x^T(a,t), \qquad \phi(y,t) = \|y\|\gamma(y,t) ,$$

$$\lim_{y\to 0} \|\gamma(y,t)\| = 0 . \tag{1.9}$$

The system (1.1) is thus reduced to (1.6); and if the matrix A is constant and the condition (1.9) holds uniformly in t, we can use Theorem 2.1.1. For the exponential stability of the equilibrium $x(t) \equiv a$ of the system (1.1) it will be sufficient that all eigenvalues of the matrix A have negative real parts. The problem is reduced to the investigation of a linearized system (the system of the first-order approximation, or, as frequently referred to, variational equations): $\dot{y} = Ay$.

The exponential stability of this system implies the exponential stability of the equilibrium point for the system (1.1). The technique described is often used to justify the convergence of numerical methods. As will be shown in Subsection 3.5, these results will also imply the local convergence of the discrete approximation of the system (1.1) for a sufficiently small step of integration.

2. THE METHOD OF LYAPUNOV FUNCTIONS

1. LAGRANGE STABILITY

In this subsection we shall denote by $\{t_k\}$ an infinitely increasing sequence of times t tending to infinity, and by t_k the kth element.

THEOREM 2.2.1. In the system (1.1) let $f(x,t) \in C^{10}_{xt}(X^n \times I_+)$. For a Lagrange stability of the system (1.1) it is sufficient that on $E^n \times I_+$ there exists a function $v(x,t)$ such that:

●1. $w(x) \leq v(x,t)$, where $w(x)$ is a continuous, bounded function;

●2. for each solution $x(x_0,t)$ of the system (1.1) the function $v(x(x_0,t), t)$ is nonincreasing with respect to the variable t.

Proof. For any $t \geq 0$ we have

$$w(x(x_0,t)) \leq v(x(x_0,t),t) \leq v(x_0,t_0) . \tag{2.1}$$

Then the solution $x(x_0,t)$ is bounded. Indeed, if this is not the case, we can find a sequence $\{t_k\}$ converging to $T \leq \infty$, such that

$$\lim_{t_k \to T} ||x(x_0,t_k)|| = \infty$$

and for the infinitely large function w we obtain

$$\lim_{t_k \to T} w(x(x_0,t_k)) ,$$

which contradicts the inequalities (2.1). Thus, the solution $x(x_0,t)$ is unboundedly extendable (to the right) and

$$\sup_{t \in I_+} \|x(x_0, t)\| = \infty \quad . \qquad ///$$

REMARKS. Condition (2.) of the Theorem will a priori be satisfied, if the function $v(x,t)$ is differentiable everywhere over both arguments and its total derivative with respect to the solutions of (1.1) is nonpositive on $E^n \times I_+$, i.e.,

$$\frac{dv(x,t)}{dt} = v_t(x,t) + \langle v_x(x,t),\ f(x,t)\rangle \leq 0 \quad .$$

This condition can be somewhat weakened to requiring only that it hold everywhere outside some bounded set.

2. LYAPUNOV'S THEOREMS ON STABILITY

We consider the case where the process is described by the autonomous system of differential equations

$$\frac{dx}{dt} = f(x) \quad , \tag{2.2}$$

that is, the right sides of the system do not depend on t. We assume that $f(0) = 0$ and $f(x)$ satisfies a Lipschitz condition in some neighborhood G of the point $x = 0$.

We say that the continuous function $v(x)$ is positive definite on G if $v > 0$ everywhere on G, except the point $x = 0$, where $v(x) = 0$. Similarly, if $v < 0$ everywhere on G, except the point $x = 0$, where $v(x) = 0$, we say that the function $v(x)$ is negative definite on G. If the inequality $v \geq 0$ $(v \leq 0)$ holds everywhere on G, we say that the function $v(x)$ on G is nonnegative (nonpositive).

We denote by S_ε, H_ε respectively the spherical surface of an n-dimensional sphere centered at the origin, and its interior:

$$S_\varepsilon = \{x \in E^n : \|x\| = \varepsilon\} ,$$

$$H_\varepsilon = \{x \in E^n : \|x\| < \varepsilon\} .$$

THEOREM 2.2.2 (Lyapunov's Stability Theorem). If there exists a differentiable function $V(x)$ positive definite on G, the total derivative of which, calculated through the system (2.2), is non-positive on G, then the trivial solution $x(t) \equiv 0$ of the system (2.2) is Lyapunov stable.

Proof. Let $\varepsilon > 0$, $H_\varepsilon \subset G$. We write

$$\lambda = \min_{x \in S_\varepsilon} v(x) .$$

Since the function $v(x)$ is continuous and $v(0) = 0$, one can find $\delta \in (0,\varepsilon)$ so small that

$$\sup_{x \in H_\delta} v(x) = \lambda_0 < \lambda .$$

Let $x_0 \in H_\delta$; we consider the solution to (2.2): $x = x(x_0,t)$. If this solution intersected the spherical surface S_ε for some $t = t_1$, we would obtain $v(x(x_0,t_1)) \geq \lambda$. On the other hand,

$$\frac{dv(x)}{dt} = \langle v_x(x), f(x)\rangle \leq 0 .$$

Therefore, $v(x(x_0,t))$ is a non-increasing function of t; hence $v(x(x_0,t)) \leq v(x_0) \leq \lambda_0 < \lambda$ and, therefore, the trajectory (solution) $x(x_0,t)$ does not intersect the surface S_ε for any $t \geq 0$. ///

THEOREM 2.2.3 (Lyapunov's Asymptotic Stability Theorem). If there exists a differentiable function $v(x)$ positive definite on G, whose total derivative in t, calculated through the system (2.2), is negative definite on G, the trivial solution $x(t) \equiv 0$ of the system (2.2) is asymptotically stable.

Proof. By Theorem 2.2.2, for any $R > 0$ there exists $r \in (0,R)$ such that the condition $x_0 \in H_r \subset G$ implies that $x(x_0,t) \in H_R \subset G$ for any $t \geq 0$. We show that for any $\varepsilon \in (0,R)$ there exists T such that $x(x_0,t) \in H_\varepsilon$ for all $t > T$.

For ε one can take $0 < \delta < \min[\varepsilon,r]$ such that $x(x_0,t) \in H_\varepsilon$ for all $t \geq 0$, only if $x_0 \in H_\delta$. Let the trajectory $x(x_0,t)$ not enter H_δ for any $t \geq 0$; then $x(x_0,t) \in H_R \setminus H_\delta$

$$v(x(x_0, t)) - v(x_0) = \int_0^t \frac{dv(x(x_0, t))}{dt}\,dt < tl < 0,$$

where $\ell = \sup_{x \in H_R \setminus H_\delta} \langle v_x(x), f(x) \rangle$. Letting $t \to \infty$, we obtain $\lim_{t\to\infty} v(x(x_0,t)) < 0$. This contradicts the positive definiteness of $v(x)$ on $H_R \subset G$. Hence we can find T such that $x(x_0,T) \in H_\delta$, but then the solution will not intersect S_ε for any $t > T$. Since ε was arbitrarily small, we have $\lim_{t\to\infty} x(x_0,t) = 0$. ///

We shall say in the sequel that the functions $v(x)$ that we have introduced to justify stability are Lyapunov functions, and the method for proving via these functions is the method of Lyapunov functions.

In the Theorems given above the fact that the origin is an equilibrium point for the system (2.2) is not essential. Possibly, $f(a) = 0$; in this case we say that the Lyapunov function $v(x)$ is positive definite if $v(a) = 0$ and $v(x) > 0$ for all $x \neq a$, belonging to some neighborhood of the point a. Lyapunov functions provide sufficient conditions for the trivial solution $x(t) \equiv a$ to be stable.

The method of Lyapunov functions is widely used in investigating various engineering problems in which the system (2.2) is specified and it is required to choose its parameters to ensure a stable equilibrium. In the case of numerical methods of optimization the situation is different: the system (2.2) is to be such that the points of the solution of an initial optimization problem are asymptotically stable equilibrium points. To prove the negative definiteness of a derivative of a Lyapunov function sufficient conditions for the extremum in the initial problem are usually exploited.

3. NON-AUTONOMOUS SYSTEMS

The results obtained in Subsection 1.2 are extendable to a more general case of non-autonomous systems of the form (1.1). To this end, instead of $v(x)$ the function $v(x,t) \subset C^{11}_{xt}(Z)$, where $Z = G \times I_+$, is introduced.

DEFINITION 2.2.1. We say that the function $v(x,t)$ is positive definite on Z if there exists a continuous function $w(t)$ defined on G and such that $0 < w(x) \leq v(x,t)$ for $x \neq 0$ and $w(0) = v(0,t) = 0$. Similarly, the function $v(x,t)$ is negative definite on Z if there exists a continuous function $w(x)$ on G, such that $v(x,t) \leq w(x) < 0$ for $x \neq 0$ and $w(0) = v(0,t) = 0$.

We assume that the system (1.1) is such that $f(0,t) = 0$, $f(x,t) \in C^{11}_{xt}(Z)$. Then the following two theorems due to A.M. Lyapunov hold.

THEOREM 2.2.4. If there exists a differentiable function $v(x,t) \in C^{11}_{xt}(Z)$ positive definite on G, having a non-positive derivative by virtue of the system (1.1), the trivial solution $x(t) \equiv 0$ of the system (1.1) is Lyapunov stable.

THEOREM 2.2.5. If there exists a differentiable function $v(x,t) \in C^{11}_{xt}(Z)$, positive definite on G, admitting an infinitesimal upper limit as $X \to 0$ and having a negative definite derivative in t by virtue of the system (1.1), the trivial solution $x(t) \equiv 0$ of the system (1.1) is asymptotically stable.

The proof is basically the same as that of Theorems 2.2.2 and 2.2.3, and hence it is omitted.

4. ASYMPTOTIC STABILITY THEOREMS

We consider system (1.1), assuming that $f(0,t) = 0$, $f(x,t) \in C^{10}_{xt}(Z)$. Here and below in this subsection, $Z = E^n \times I_+$.

DEFINITION 2.2.2. We say that the function $v(x,t) \in C^{10}_{xt}(Z)$ admits an infinitely large lower limit as $x \to \infty$ if $v(x,t) \to \infty$ uniformly in t as $x \to \infty$, i.e., for any $M > 0$ there exists $R = R(M)$ such that $|v(x,t)| > M$ for all $\|x\| > R$, $t \in I_+$.

DEFINITION 2.2.3. We say that the function $v(x,t) \in C^{10}_{xt}(Z)$ admits a strong infinitesimal upper limit as $x \to 0$ if there exists a function $w(x)$ continuous on E^n, such that $|v(x,t)| \leq w(x)$ for $[x,t] \in Z$ and $w(0) = 0$.

THEOREM 2.2.6 (Barbashin-Krassovskij). Suppose there exists a differentiable function $v(x,t) \in C^{11}_{xt}(Z)$, positive definite on Z, admitting a strong infinitesimal upper limit as $x \to 0$ and an infinitely large lower limit as $x \to 0$, the derivative $\dot{v}(x,t)$

being, by virtue of the system (1.1), negative definite on Z. Then the trivial solution $x(t) \equiv 0$ of the system (1.1) is globally asymptotically stable.

The proof of this Theorem can be found, for example, in B. P. Demidovich [1].

5. APPLICATIONS TO THE CONVERGENCE OF NUMERICAL METHODS

We consider the problem of finding the unconstrained minimum of the differentiable function $f(x)$. To solve the problem, Cauchy suggested a method involving finding the limit points of the following system of differentiable equations:

$$\frac{dx}{dt} = -f(x), \qquad x(0) = x_0 . \tag{2.3}$$

The theorems given above enable us to obtain sufficient conditions for convergence. We assume that there exists at least one solution $x = x_*$ of the minimization problem. We use first the method of Lyapunov functions and introduce the following three functions:

$$v_1(x) = f(x) - f(x_*), \quad v_2(x) = \frac{1}{2}\|f_x(x)\|^2, \quad v_3(x) = \frac{1}{2}\|x - x_*\|^2.$$

If we assume that x_* is an isolated point of the local minimum of the function $f(x)$, there exists a neighborhood $G(x_*)$ in which these functions are positive definite. Differentiating them through the system (2.3), we obtain

$$\dot{v}_1 = -\|f_x(x)\|^2 \leqslant 0, \qquad \dot{v}_2 = -f_x^T(x) f_{xx}(x) f_x(x),$$
$$\dot{v}_3 = \langle f_x(x), x_* - x\rangle.$$

If at the point $x = x_*$ the sufficient conditions of the

minimum given in Theorem 1.3.2 are satisfied, the functions $\dot{v}_1$ and $\dot{v}_2$ will be negative definite at least for $x \in G(x_*)$ and the Lyapunov theorem on asymptotic stability implies the local convergence of the method (2.3) to the point $x = x_*$.

If the function f is strictly convex, then, using Theorem 1.2.5, we obtain that $\dot{v}_3 \le f(x_*) - f(x) \le 0$ for $x \in G(x_*)$. The derivative $\dot{v}_3$ is thus negative definite and, therefore, the method (2.3) converges at least locally to the point $x = x_*$.

Now we invoke the Barbashin-Krassovskij theorem. Assuming that the function $f(x)$ is strictly convex everywhere on E^n, we see that the minimization problem has a unique solution $x = x_*$ and the function $v_3(x)$ admits an infinitesimal upper limit as $x \to x_*$ since $v_3(x_*) = 0$. Furthermore, by Theorem 1.1.2, $v_3(x)$ is an infinitely large function and hence admits an infinitely large lower limit as $\|x\| \to \infty$. The derivative $\dot{v}_3$ is negative definite everywhere on E^n, hence solutions to (2.3) converge to the point $x = x_*$ globally. The method (2.3) is "relaxation" since $f(x(x_0,t))$ is a monotone decreasing function of t. Moreover, in this method, the norm of the gradient of $f(x)$ and the distance between the "current" point $x(x_0,t)$ and the minimum point x_* decrease monotonically.

If we use the theorem on stability of the first-order approximation, we arrive at the system

$$\frac{dy}{dt} = -f_{xx}(x_*)y , \qquad x(x_0,t) = x_* + y(t) .$$

If the matrix $f_{xx}(x_*)$ is positive definite, all the characteristic roots of the matrix $-f_{xx}(x_*)$ are real, strictly negative.

This implies a local exponential convergence of the method (2.3) to the point of local minimum.

As one can see from the example considered, the application of various theorems on convergence, or even of Theorem 2.2.3 but only with different Lyapunov functions, enables one to make a more comprehensive idea about the method in question. Unfortunately, Lyapunov functions can be constructed in this simple manner only in a few cases.

Let the function $f(x)$ be everywhere twice differentiable and let the matrix $f_{xx}(x)$ be everywhere nonsingular. Then we can consider a method involving finding limit points of the following problem:

$$\frac{dx}{dt} = -f_{xx}^{-1} f_x(x) \quad . \tag{2.4}$$

To justify the convergence we use the Lyapunov function $v(x) = \langle f_x(x), f_x(x)\rangle$. Differentiabing v with respect to (2.4), we obtain

$$\frac{dv}{dt} = -2v(x) , \qquad v(0) = \langle f_x(x_0), f_x(x_0)\rangle ,$$

yielding in turn $v(t) = v(0)\, e^{-2t}$. The method converges thus to the stationary points of the function $f(x)$ as $t \to \infty$.

The method (2.4) is usually referred to as the continuous analog of Newton's method which will be described in Section 5.

Now we consider the problem of finding the minimax of (1.5.2). We assume that the function $F(x,y)$ is strictly convex-concave and has a saddle point $[x_*, y_*]$ (see Section 1.5). The simplest numerical method for solving this problem consists in finding limit points of solutions to the following system:

$$\frac{dx}{dt} = -F_x(x,y) , \qquad \frac{dy}{dt} = F_y(x,y) . \tag{2.5}$$

To prove the convergence we use the positive definite functions

$$v_1(x, y) = \frac{1}{2}[\|F_x(x, y)\|^2 + \|F_y(x, y)\|^2],$$
$$v_2(x, y) = \frac{1}{2}[\|x - x_*\|^2 + \|y - y_*\|^2].$$

If the function $F(x,y)$ is twice differentiable on $E^n \times E^m$, and the matrices $F_{xx}(x,y)$ and $-F_{yy}(x,y)$ are everywhere positive definite, then

$$\frac{dv_1}{dt} = -F_x^T(x, y) F_{xx}(x, y) F_x(x, y) + \\ + F_y^T(x, y) F_{yy}(x, y) F_y(x, y) \leqslant 0,$$

the equality sign holds only at the points $x = x_*$, $y = y_*$.

Differentiating v_2 by the system (2.5) and using the strict convexity-concavity condition plus the inequalities (1.5.16), we obtain

$$\dot{v}_2 = \langle F_x(x, y), x_* - x\rangle + \langle F_y(x, y), y - y_*\rangle \leqslant \\ \leqslant F(x_*, y) - F(x, y) + F(x, y) - F(x, y_*) \leqslant 0,$$

the inequality sign holding if and only if $x = x_*$, $y = y_*$. Using the Barbashin-Krassovskij theorem, we conclude that in the case where the function $F(x,y)$ is differentiable and everywhere convex-concave, the method (2.5) converges to a unique solution to the problem (1.5.2) globally.

We consider now a particular case of the problem (1.5.2), where $n = m = 1$, $F(x,y) = xy$. It is easy to verify that the point $[0,0]$ is saddle. The method (2.5) leads to the following system:

$$\frac{dx}{dt} = -y \ , \qquad \frac{dy}{dt} = x \ . \tag{2.6}$$

Differentiating the positive definite function $v(x,y) = x^2 + y^2$ by this system, we obtain that $\dot{v}(x,y) \leq 0$. The system (2.6) is reduced to the equation $\frac{d^2y}{dt^2} + y = 0$, the solution of which

$$y(t) = y_0 \cos t + x_0 \sin t$$

has no limit as $t \to \infty$. This example demonstrates that while using the method of Lyapunov functions one ought to verify thoroughly whether all of the conditions have been satisfied. In the present case the derivative $\dot{v} \equiv 0$, hence the conditions of Theorem 2.2.2, that is, Lyapunov's theorem on stability, are satisfied, but the condition of Theorem 2.2.3, Lyapunov's theorem on asymptotic stability, does not hold.

6. THE NOTION OF CONVERGENCE

From the two examples above one sees that for numerical methods it is essential that solutions of systems of differential equations converge to some set X_*. The fact that the point x_* is an asymptotically stable equilibrium for the system (1.1) implies that the solutions of (1.1) converge locally to x_*. The converse does not, in general, hold. Hence the convergence conditions must be weaker than the asymptotic stability conditions. In numerical methods, the notion of convergence proper can be interpreted more widely than just the tendency of $x(x_0,t)$ to go to x_* as $t \to \infty$. In this connection we state a few definitions.

DEFINITION 2.2.4. The point $\bar{x} \in E^n$ is an ω-limit point of the solution $x(x_0,t)$ to the system (1.1) if there exists a sequence $\{t_k\}$ such that $\lim_{k\to\infty} x(x_0,t_k) = \bar{x}$. The set $\omega(x_0)$ of all ω-limit points of the solution $x(x_0,t)$ is said to be the ω-limit set.

DEFINITION 2.2.5. The method (1.1) converges to the set $X_* \subset E^n$ locally (or on X, or globally) if $\omega(x_0) = X_*$ for any x_0 belonging to some neighborhood $G(X_*)$ (or respectively X or E^n).

The convergence of the method on X to the set X_* implies thus that for any $x_0 \in X$ and any $\varepsilon > 0$ there exists $T(x_0,\varepsilon)$ such that for all $t \geq T(x_0,\varepsilon)$ we have $\mathrm{dis}(x(x_0,t), X_*) < \varepsilon$. Let $R = E^n \times I_+$.

LEMMA 2.2.1. Suppose the system (1.1) admits a differentiable function $v(x,t)$, nonnegative on R and such that its total derivative $\dot{v}(x,t)$, by virtue of the system (1.1), is nonpositive on R; and further, that the solution $x(x_0,t)$ is extendable as $t \to \infty$. Then:

- a. for any sequence $\{t_k\}$ the limit $\lim_{t_k\to\infty} v(x(x_0,t_k), t_k)$ is defined; this limit does not depend on the concrete choice of the subsequence $\{t_k\}$;
- b. there exists a sequence $\{t_i\}$ such that

$$\lim_{t_i\to\infty} \frac{d}{dt} v(x(x_0,t_i), t_i) = 0 .$$

Proof. Since $\dot{v}(x,t) \leq 0$ and $v(x,t) \geq 0$ on R, the limit $\lim_{t\to\infty} v(x(x_0,t), t) = \tilde{v}(x_0)$ is defined. Therefore, assertion (a.) is satisfied for any sequence $\{t_k\}$. We assume that there is no sequence $\{t_i\}$ for which $\lim_{t_i\to\infty} \dot{v}(x(x_0,t_i), t_i) = 0$. Then one can find $\delta > 0$ and $T(\delta) > 0$ such that $\dot{v}(x(x_0,t)), t \to -\delta$ for $t \geq T(\delta)$. This implies that $v(x(x_0,t), t) \to -\infty$; but this contradicts the nonnegativeness of $v(x,t)$ on R. ///

Let us introduce two sets:

$$G_\varepsilon = \{x \in E^n\colon \operatorname{dis}(x, X_*) < \varepsilon\},$$
$$\Gamma_\varepsilon = \{x \in E^n\colon \operatorname{dis}(x, X_*) = \varepsilon\}.$$

Assume that the set X_* is compact.

THEOREM 2.2.7 (Yu. G. Evtushenko [9]). Let solutions to (1.1) be extendable as $t \to \infty$ and let there exist a continuously differentiable function $v(t)$ on E^n, such that $v(x) = 0$ for $x \in X_*$ and $v(x) > 0$ for $x \in X_*$; furthermore, let the derivative $\dot{v}(x)$ by virtue of the system (1.1) satisfy these conditions: for any $\varepsilon > 0$ there exists $t(\varepsilon)$ such that for all $t \geq t(\varepsilon)$ the conditions

$$p(\varepsilon, t) = \sup_{x \in E^n \setminus G_\varepsilon} \dot{v}(x) \leqslant 0, \quad \lim_{t\to\infty} \int_{t(\varepsilon)}^{t} p(\varepsilon, s)\, ds = -\infty \tag{2.7}$$

are satisfied. Then the method (1.1) converges to the set X_* globally.

Proof. We show that for any fixed $\varepsilon > 0$, $x_0 \in E^n$ there exists T such that $x(x_0,t) \in G_\varepsilon$ for all $t > T$. Let $\lambda_\varepsilon = \min_{x\in\Gamma_\varepsilon} v(x)$. Since $v(x) = 0$ everywhere on X_*, it is possible to take $\delta \in (0,\varepsilon)$ so small that

$$0 < \max_{x\in\Gamma_\delta} v(x) = \lambda_\delta < \lambda_\varepsilon .$$

Next we show that for T we can take any value satisfying the following two conditions:

$$T > t(\delta) , \qquad x(x_0,T) \in G_\delta . \tag{2.8}$$

We show that such a value exists. Assume the opposite: for all $t > t(\delta)$, $x(x_0,t) \notin G_\delta$ along the trajectory $x(x_0,t)$ we then have

$$v(x(x_0,t)) - v(x(x_0,t(\delta))) \le \int_{t(\delta)}^{t} p(\delta,s)\, ds .$$

Letting $t \to \infty$ and noting (2.7), we obtain that $v(x(x_0,t)) \to -\infty$, which is impossible since $v(x) \ge 0$; hence we can find T satisfying (2.8). The trajectory $x(x_0,t)$ does not leave the set G_ε for $t \ge T$, since otherwise we could find $T_2 > T_1 > T$ such that

$$x(x_0, T_1) \in \Gamma_\delta, \quad v(x(x_0, T_1)) \leqslant \lambda_\delta,$$
$$x(x_0, T_2) \in \Gamma_\varepsilon, \quad \lambda_\varepsilon \leqslant v(x(x_0, T_2)).$$

At the same time, for $T_1 < t \le T_2$ we have

$$x(x_0,t) \subset E^n \setminus G_\delta .$$

By the first condition of (2.7) on this segment of the trajectory:

$$\sup_{x \in E^n \setminus G_\delta} \dot{v} \le 0 . \tag{2.9}$$

Noting that $\lambda_\delta < \lambda_\varepsilon$, we obtain a contradiction with the inequality (2.9). Hence the method (1.1) converges to the set X_*. Due to the arbitrariness of the initial point x_0 we have proved the global convergence. ///

The assumption concerning the extendability of solutions can be removed when we require instead that the conditions of Theorem

2.2.1 be satisfied, the equality $v(x,t) \equiv w(x) \equiv 0$ hold, and finally that the function $v(x,t)$ admits an infinitesimal upper limit for $\text{dis}\,(x,X_*) \to 0$.

It is interesting to note that the methods do not necessarily converge to the equilibrium points. This problem has been studied in more detail in Yu.G. Evtushenko and V.G. Zhadan [2].

3. THEOREMS ON CONVERGENCE OF ITERATIVE PROCESSES

1. BASIC DEFINITIONS

Let a one-to-one mapping $T: R^n \to R^n$ be defined everywhere on R^n. We pose the problem of finding fixed points of this mapping, i.e., the points belonging to the set

$$X_* = \{x \in R^n : T(x) = x\} .$$

To solve this problem one can use a numerical method involving iterations through the formula

$$x_{k+1} = T(x_k) . \tag{3.1}$$

We shall call this method in the sequel the method of simple iteration. We specify an initial point x_0 and define uniquely the sequence $\{x_k\}$ by (3.1).

<u>DEFINITION 2.3.1</u>. The iterations generated by (3.1) converge locally to the point x_* (or on the set $X \subset R^n$, or globally) if for any initial point x_0 belonging to $G(x_*)$ (or respectively, X, or R^n) there exists a limit of the sequence $\{x_k\}$ coinciding with the point x_*.

Only with a few numerical methods the sequences $\{x_k\}$ are convergent. Nevertheless these methods are widely used, and work effectively. We cite, for example, the penalty function method (see Chapter 3). Methods of this type possess the following essential property: each limit point of the sequence $\{x_k\}$ belongs to the set sought, which is why they are referred to as convergent. A natural analog of Definition 2.2.5 of the method's convergence is

<u>DEFINITION 2.3.2</u>. The method (3.1) converges to the set X_* locally (or on the set X, or globally) if for any x_0 belonging to $G(x_*)$ (or respectively, X, or R^n) the sequence $\{x_k\}$ is defined, has a non-empty set of limit points, and each limit point belongs to X_*.

In investigating numerical methods the following two questions arise:

- 1. Under what conditions is the convergence of the method guaranteed?
- 2. What is the rate of the convergence?

It is worth noting that the answers to these questions are of both theoretical and important practical value: frequently, either the method proper or some stages of the method are constructed on the basis of the convergence guaranteed; the theoretical investigation of the methods is simultaneously the device for creating these methods. Construction of numerical methods for solving optimization problems usually involves mappings under which the set of fixed points coincides with those points at which necessary conditions for the extremum of the given optimization problem are satisfied. Sufficient conditions for the convergence of numerical methods for finding fixed points are equivalent or almost equivalent to the sufficient conditions for the extremum in the optimization problem.

In what follows we shall use two types of estimates of the rate of convergence. Assume that the iterations generated by (3.1) converge to x_*. We define

$$B_p(\{x_k\}) = \begin{cases} 0 & \text{if } x_k = x_* \text{ for all, except a finite number of the subscripts } k; \\ \limsup\limits_{k\to\infty} \dfrac{\|x_k - x_*\|}{\|x_{k-1} - x_*\|^p} & \text{if } x_k \neq x_* \text{ for all except a finite number of the subscripts } k; \\ +\infty & \text{otherwise}. \end{cases}$$

Let $C(T,x_*)$ be the set of all sequences with the limit x_*, generated by the iterative process T. Then the quantities

$$Q_p(x_*) = \sup_{\{x_k\} \subset C(T,\, x_*)} B_p(\{x_k\})$$

are said to be the Q-factor of the process T at the point x_*.

Suppose there are two methods T_1 and T_2 and two Q-factors $Q_p^1(x_*)$, $Q_p^2(x_*)$ are known for them, respectively. If $p \in [1,\infty)$ is such that $Q_p^1(x_*) < Q_p^2(x_*)$, we say that the method T_1 is Q-faster than the method T_2 at the point x_*. It is not difficult to show that such definition is correct, i.e., if for some p the method T_1 is Q-faster than the method T_2 at the point x_*, there exists no other $p' \in [1,\infty)$ at which the method T_2 is Q-faster than T_1. The notion of the "Q-faster" depends on the norm chosen and it may happen that in one norm the process T_1 is Q-faster and in another norm the process T_2 is Q-faster.

Concerning the iterative process (3.1) we say that its rate of convergence is

- linear (the process converges at the rate of geometric pro-progression), if $0 < Q_1(x_*) < 1$;
- superlinear, if $Q_1(x_*) = 0$;
- square, if $0 \leq Q_2(x_*) < \infty$;
- supersquare, if $Q_2(x_*) = 0$.

In some cases it is possible to obtain another numerical characteristic of the convergence rate of iterations, e.g., the number γ, if one succeeds in proving that for any $k = 1,2,\ldots$

$$||x_{k+1} - x_*|| < \gamma||x_k - x_*|| \quad , \tag{3.2}$$

γ being constant for all sequences $\{x_k\}$ over the range of convergence of the iterations (3.1) to the point x_*.

A detailed analysis of Q-factors and definitions of other estimates of convergence rates are given in Ortega and Rheinboldt [1].

2. THE PRINCIPLE OF CONTRACTION MAPPINGS

DEFINITION 2.3.3. We say that the mapping T is a contraction, or contracting (on X) if there exists $0 < q < 1$ such that for any points x, y (belonging to X) the condition

$$\|T(x) - T(y)\| \leq q \|x - y\| \tag{3.3}$$

is satisfied. The q is known as the constant of contraction for the mapping T.

A contraction mapping is continuous. Indeed, for any $\varepsilon > 0$ one may take $\delta = \frac{\varepsilon}{q}$; then from the condition $\|x - y\| < \delta$, by (3.3), it follows that $\|T(x) - T(y)\| < \varepsilon$.

THEOREM 2.3.1 (Banach Fixed Point Principle). Let the mapping T be contracting on R^n. Then:

- 1. the mapping T has only one fixed point x_*;
- 2. the iterations defined by the formula (3.1) converge to x_* globally;
- 3. the convergence rate is estimated by the inequality

$$\|x_k - x_*\| \leq \frac{q^k}{1-q} \|x_1 - x_0\| . \tag{3.4}$$

Proof. It follows from (3.3) that

$$\|x_{k+1} - x_k\| \leq q\|x_k - x_{k-1}\| \leq q^k \|x_1 - x_0\|$$

whence for any $s > 0$ we have

$$\begin{aligned} \|x_{k+s} - x_k\| &\leq \|x_{k+s} - x_{k+s-1}\| + \\ &+ \|x_{k+s-1} - x_{k+s-2}\| + \ldots + \|x_{k+1} - x_k\| \leq \\ &\leq [q^{k+s-1} + q^{k+s-2} + \ldots + q^k] \|x_1 - x_0\| = \\ &= \frac{q^k(1-q^s)}{1-q} \|x_1 - x_0\| \leq \frac{q^k}{1-q} \|x_1 - x_0\| < \varepsilon \end{aligned} \tag{3.5}$$

if $k > N(\varepsilon)$, where $N(\varepsilon)$ is sufficiently large. Therefore, the sequence $\{x_k\}$ is fundamental and, since the space R^n is complete, the limit

$$\lim_{k\to\infty} x_k = x_*$$

exists.

Passing to the limit in the equality (3.2) as $k \to \infty$, with the continuity of $T(x)$ taken into account, we obtain $\lim_{k\to\infty} x_k = T(\lim_{k\to\infty} x_k) = x_*$, or $x_* = T(x_*)$. Therefore, x_* is the fixed point of the mapping T.

Next we prove uniqueness. If there existed, in addition to x_*, another fixed point y, we would have $\|y - x_*\| = \|T(y) - T(x_*)\| \leq q\|y - x_*\|$; but this is possible only when $\|y - x_*\| = 0$, i.e., $y = x_*$. If in the inequality (3.5) we pass to the limit as $s \to \infty$, we shall obtain the estimate (3.4). ///

THEOREM 2.3.2. Let the mapping T contracting on X map the closed set X onto itself ($T: X \to X$). Then the mapping T has only one fixed point $x_* \in X$; the iterations definable by (3.1) converge linearly to x_* on X, and the estimate (3.4) holds.

The proof is almost the same as that of Theorem 2.3.1.

3. MONOTONE MAPPINGS

Let x, y be arbitrary elements of the set $X \subset E^n$.

DEFINITION 2.3.4. We say that the mapping $V: X \to X$ is:

- bounded if the image $V(M)$ of each bounded set $M \subset X$ is bounded;
- potential if there exists a differentiable function $f(x)$ such that $V(x)$ is its gradient; the function f is then called the potential of the mapping $V(x)$;
- monotone if

$$0 \leq \langle V(x) - V(y), x-y \rangle \quad \forall\, x, y \in X ;$$

• uniformly monotone if there exists m (monotonicity constant) such that

$$m\|x-y\|^2 \leq \langle V(x)-V(y),\ x-y\rangle \quad \forall\ x,y \in X\ ; \tag{3.6}$$

• Lipschitz (with constant ℓ) if

$$\|V(x)-V(y)\| \leq \ell\|x-y\| \quad \forall\ x,y \in X\ . \tag{3.7}$$

If the mapping V is simultaneously uniformly monotone and Lipschitz, the conditions (3.6) and (3.7) imply that $m \leq \ell$.

THEOREM 2.3.3. Let the mapping $V\colon E^n \to E^n$ be Lipschitz with constant ℓ and uniformly monotone with monotonicity constant m. Then the mapping $T\colon E^n \to E^n$, defined by

$$T(x) = x - \tau V(x)\ , \tag{3.8}$$

is contracting for $\tau \in \left(0,\ \frac{2m}{\ell^2}\right)$ and

$$\|T(x)-T(y)\| \leq q\|x-y\|\ ,$$

where

$$q = (1-2m\tau+\ell^2\tau^2)^{\frac{1}{2}} < 1\ .$$

Proof. For any x, y we have

$$\begin{aligned}\|T(x)-T(y)\|^2 &= \|x-\tau V(x)-y+\tau V(y)\|^2 = \\ &= \|x-y\|^2-2\tau\langle V(x)-V(y),\ x-y\rangle+\tau^2\|V(y)-V(x)\|^2 \leqslant \\ &\leqslant \|x-y\|^2-2\tau m\|x-y\|^2+\tau^2 l^2\|x-y\|^2 = q^2\|x-y\|^2.\end{aligned}$$ ///

In constructing numerical methods it is expedient to do one's best to ensure the highest rate of convergence. Theorem 2.3.3 allows us to find the best value τ, starting from the minimization of estimation of the relation $\|T(x)-T(y)\|/\|x-y\|$. The minimum value of q with respect to the parameter τ is attained for $\tau = \frac{m}{\ell^2}$ and is

$$q = \left|1 - \left(\frac{m}{\ell}\right)^2\right|^{\frac{1}{2}} < 1 .$$

4. CONVERGENCE OF THE FIRST ORDER APPROXIMATION

We treat now the discrete analog of Theorem 2.1.1 on stability of the first-order approximation. First we prove the following lemma.

LEMMA 2.3.1. Suppose there exists a neighborhood $G(x_*)$ of the fixed point x_* of the mapping T, such that for any $x \in G(x_*)$ the conditions

$$\|T(x) - x_*\| \leq q\|x - x_*\| , \qquad q < 1$$

are satisfied. Then the iterations defined by (3.1) converge linearly locally to the point x_*.

Proof. We have the inequalities

$$\|x_1 - x_*\| = \|T(x_0) - x_*\| \leq q\,\|x_0 - x_*\| .$$

Hence it follows from the condition $x_0 \in G(x_*)$ by induction that all the points x_k belong to $G(x_*)$ and satisfy the condition

$$\|x_k - x_*\| \leq q^k\| x_0 - x_*\| ,$$

hence $\lim_{k\to\infty} x_k = x_*$. ///

THEOREM 2.3.4. Let the mapping $T: R^n \to R^n$ be differentiable at the fixed point x_* and let the spectral radius S of the matrix $T_x(x_*)$ satisfy the condition $S < 1$; then the iterations defined by (3.1) converge linearly locally to x_*,

$$Q_1(x_*) \leq \|T_x^T(x_*)\| . \tag{3.9}$$

Proof. Using the differentiability of the mapping T, we obtain

$$\|T(x) - x_*\| = \|T(x_*) + T_x^T(x_*)(x-x_*) + (x-x_*)T\Phi(x_*, x-x_*) - x_*\| . \tag{3.10}$$

Here the vector function Φ is such that

$$\lim_{x \to x_*} \|\Phi(x_*, x-x_*)\| = 0 . \tag{3.11}$$

Hence for any $\varepsilon > 0$ there exists a neighborhood $G(x_*)$ of the point x_*, such that for all $x \in G(x_*)$ we have

$$\|(x_*, x-x_*)\| \leq \varepsilon .$$

As is well known, for any $\varepsilon > 0$ we can choose a norm of the matrix $T_x^T(x_*)$, such that $\|T_x^T(x_*)\| \leq \varepsilon + S$. Noting that $T(x_*) = x_*$, we obtain from (3.10) that for any $x \in G(x_*)$ we have the inequality $\|T(x) - x_*\| \leq (2\varepsilon + S)\|x-x_*\|$. By the hypothesis of the Theorem, $S < 1$; hence ε can be regarded so small that $2\varepsilon + S < 1$, and the conditions of Lemma 2.3.1 are satisfied implying the local convergence of $\{x_k\}$ to x_*.

From (3.10) it follows that

$$\|T(x) - x_*\| \leq \|x-x_*\| \left[\|T_x^T(x_*)\| + \|\Phi(x_*, x-x_*)\|\right] .$$

Using this inequality for each element of the sequence $\{x_k\}$ generated by (3.1), and noting the property (3.11), we obtain

$$\lim_{x_k \to x_*} \frac{\|x_{k+1} - x_*\|}{\|x_k - x_*\|} \leqslant \lim_{x_k \to x_*} [\|T_x^T(x_*)\| + \|\Phi(x_*, x_k - x_*)\|] = \|T_x^T(x_*)\|,$$

i.e., for the Q-factor the estimate (3.9) holds. ///

We note that a sufficient condition for the iterations (3.1) to converge is that a norm of the matrix $T_x^T(x_*)$ be less than unity, since this will imply that all eigenvalues of the matrix $T_x^T(x_*)$ are less than unity and, therefore, $S < 1$.

It follows from Theorem 2.3.4 that if $T_x(x_*) = 0$ the convergence rate of the iterations (3.1) is superlinear. The fact that the spectral radius of the matrix $T_x(x_*)$ is zero does not guarantee superlinear rate of convergence.

THEOREM 2.3.5. Let the mapping $T: R^n \to R^n$ be differentiable at the fixed point x_*, the matrix $T_x(x_*)$ being symmetric. Then a sufficient condition for the iterations (3.1) to converge locally to the point x_* is that the largest eigenvalue κ of the matrix $T_x(x_*)$ be less than one. The corresponding convergence rate $Q_1(x_*) = \kappa$.

Proof. From the symmetry of the matrix $T_x(x_*)$ it follows that its eigenvalues are real. Let us use Theorem 2.3.4. We assume that the norm of the matrix $T_x(x_*)$ is the operator norm. Then the spectral radius of the matrix $T_x(x_*)$ coincides with its norm as well as with the largest eigenvalue κ. From the condition $|\kappa| < 1$ it follows that the conditions of Theorem 2.3.4 guaranteeing the convergence are satisfied. The symmetry of the matrix $T_x(x_*)$ holds, in particular, if the mapping T is potential. ///

The estimates of the converge rate given by Theorem 2.3.4 and Theorem 2.3.5 are of asymptotic nature. An estimate of the (3.2)-type is given by:

THEOREM 2.3.6. Let the mapping $T: R^n \to R^n$ be differentiable in some neighborhood $G(x_*)$ of the fixed point x_* and let

$$\gamma = \sup_{x \in G(x_*)} ||T_x^T(X)|| < 1 .$$

Then the iterations defined by (3.1) converge locally to x_* and the estimate of thé convergence rate

$$||x_k - x_*|| \leq \gamma ||x_{k-1} - x_*|| = \gamma^k ||x_0 - x_*|| \qquad (3.12)$$

holds.

Proof. From the fact that the spectral radius of the matrix does not exceed any of its norms it follows that the spectral radius of the matrix $T_x^T(x_*)$ is less than unity and therefore the iterations converge locally to the point x_*.

Now we estimate the norm

$$||x_k - x_*|| = ||T(x_{k-1}) - T(x_*)|| .$$

Using the Lagrange formula for mappings (see Appendix I), we obtain

$$\|x_k - x_*\| \leqslant \sup_{0 \leqslant \theta \leqslant 1} \|T_x^T(x_* + \theta(x_{k-1} - x_*))\| \cdot \|x_{k-1} - x_*\| \leqslant$$
$$\leqslant \gamma \|x_{k-1} - x_*\|,$$

yielding (3.12). ///

5. THE CONNECTION BETWEEN THE CONVERGENCE OF DISCRETE PROCESSES AND THE CONVERGENCE OF CONTINUOUS PROCESSES

In Section 2.1 we obtained sufficient conditions for exponential stability of the trivial solution $x(t) \equiv x_*$ for systems of the form

$$\frac{dx}{dt} = F(x) , \qquad F(x_*) = 0 . \qquad (3.13)$$

The analysis involved investigating roots of the equation

$$|F_x^T(x_*) - \lambda I_n| = 0 . \tag{3.14}$$

Integrating (3.13) using the Euler diagram, we arrive at the following iterative process:

$$x_{k+1} = x_k + \alpha F(x_K) = T(x_k) , \qquad x_* = T(x_*) . \tag{3.15}$$

By Theorem 2.3.4, the question whether the process (3.15) has a solution can be answered by analyzing the equation

$$|T_x^T(x_*) - \mu I_n| = |I_n + \alpha F_x^T(x_*) - \mu I_n| = 0 .$$

Transforming it to the form

$$\left|F_x^T(x_*) + \frac{1-\mu}{\alpha} I_n\right| = 0$$

and comparing the result with (3.14), we conclude that the roots of these equations are related in a simple way: $\mu = 1 + \alpha\lambda$.

We express the complex root λ as $\lambda = a + ib$; then

$$\mu = 1 + \alpha a + i\alpha b , \qquad |\mu|^2 = (1 + \alpha a)^2 + \alpha^2 b^2 .$$

The condition $|\mu| < 1$ can be written as

$$\gamma(\alpha) = 2\alpha a + \alpha^2(a^2 + b^2) < 0 .$$

If $\text{Re}\,\lambda = a < 0$, then $\gamma(\alpha) < 0$ for any $0 < \alpha < \alpha_*$, where

$$\alpha_* = -\frac{2\,\text{Re}\,\lambda}{|\lambda|^2} > 0 .$$

Let $\lambda_1, \ldots, \lambda_n$ be roots of Equation (3.14). Then the condition $|\mu| > 1$ holds for any α satisfying the condition

$$0<\alpha<\bar{\alpha}=2\min_{i\in[1:n]}\left[\frac{-\operatorname{Re}\lambda_i}{|\lambda_i|^2}\right].$$

We have arrived at the following theorem.

THEOREM 2.3.7. Let the sufficient conditions of Theorem 2.1.1 on stability of the first-order approximation be satisfied for the system (3.13) at x_*. Then there exists $\bar{\alpha}$ such that for any fixed $0 < \alpha < \bar{\alpha}$ the iterations defined by (3.15) converge linearly locally to the point x_*.

In a particular case where the system (3.13) realizes the Cauchy method (2.3) of unconstrained minimization of the function $f(x)$, the matrix $F_x(x_*) = -f_{xx}(x_*)$. Hence all λ_i are real. If the matrix $f_{xx}(x_*)$ is positive definite, all $\lambda_i < 0$ and

$$\bar{\alpha}=2\min_{i\in[1:n]}\left[-\frac{\lambda_i}{|\lambda_i|^2}\right]=2\min_{i\in[1:n]}\left[\frac{1}{-\lambda_i}\right]=\frac{2}{\eta},$$

where η denotes the largest eigenvalue of the matrix $f_{xx}(x_*)$.

If finding limit points of solutions of the Cauchy problem for the system (3.13) is interpreted as a numerical method for finding an equilibrium point for the system (3.13), then (3.15) is a discrete analog of this method. It follows from this Theorem that the justification of exponential stability of equilibrium for the system (3.13) automatically implies the local linear convergence of the discrete version of the method for a sufficiently small step of integration. This procedure is widely used to justify the convergence.

It is possible to integrate the system (3.13) using higher-order approximation, for instance, the Euler method, the conversion method, the Runge-Kutta method, etc. However, this does not increase the convergence rate but instead makes these methods more cumbersome. This is due to the fact that the application of the high-order schemes does make it possible to "track" more precisely each solution of the system (3.13): although there is no need to do so because of the asymptotic stability of $x = x_*$. Hence in what follows, going over to discrete approximations of numerical methods, we shall use only the simplest Euler polygon.

6. THE APPLICATION TO THE UNCONSTRAINED MINIMIZATION PROBLEM

We pose now the problem of finding the unconstrained minimum of the function $f(x)$. By Theorem 1.3.5, for a convex differentiable function f this problem is equivalent to that of finding a stationary point satisfying the condition $f_x(x_*) = 0$. From Theorem 1.2.7 we obtain that for the convex function f the mapping $f_x(x)$ is monotone. We introduce the mapping

$$T(x) = x - \tau V(x) = x - \tau f_x(x) .$$

We shall seek its fixed points via the method of simple iteration (3.1):

$$x_{k+1} = x_k - \tau f_x(x_k) . \tag{3.16}$$

Assume that there exist ℓ and m such that for any $x, y \in E^n$ the conditions

$$\|f_x(x) - f_x(y)\| \leq \ell\|x - y\| ,$$

$$m\|x - y\|^2 \leq \langle f_x(x) - f_x(y), x-y\rangle$$

are satisfied. The mapping V is therefore a uniformly monotone, Lipschitz potential. It follows from Theorem 2.3.3 that the problem (1.3.4) has a unique solution $x = x_*$, the iterations given by the formula (3.16) converge to x_* globally for any step τ satisfying the condition $\tau \in \left(0, \frac{2m}{\ell^2}\right)$ and the estimate of the convergence rate (3.4) holds, where

$$q = (1 - 2m\tau + \ell^2\tau^2)^{\frac{1}{2}} < 1 .$$

It is possible to justify the local convergence of the iterations (3.16), using no convexity condition for the function f. Instead we require that at the stationary point x_* the sufficient condition for a local minimum should be satisfied (see Theorem 1.3.2) viz. that the matrix $f_{xx}(x_*)$ is positive definite. We consider two characteristic equations

$$|f_{xx}(x_*) - \eta I_n| = 0 ,$$

$$|T_x(x_*) - \kappa I_n| = |-\tau f_{xx}(x_*) + (1-\kappa)I_n| = 0 ,$$

the roots of these equations being related through $\kappa = 1 - \tau\eta$. We arrange eigenvalues of the matrix $f_{xx}(x_*)$ in increasing order: $0 \le \eta_1 \le \eta_2 \le \cdots \le \eta_n$. They are in correspondence with eigenvalues of the matrix $T_x(x_*)$:

$$1 > \kappa_1 = 1 - \tau\eta_1 \ge \kappa_2 \ge \cdots \ge \kappa_n = 1 - \tau\eta_n .$$

By Theorem 2.3.5 a sufficient condition for the iterations (3.16) to converge is that $\kappa_n > -1$. The latter holds for any $0 < \tau < \bar{\tau}$, where $\bar{\tau} = \frac{2}{\eta_n}$. It is appropriate to choose τ from this interval such that the spectral radius $S(\tau)$ of the matrix $T_x(x_*)$ be minimal; hence we seek

$$\min_{0 \leq \tau \leq \bar{\tau}} S(\tau) = \min_{0 \leq \tau \leq \bar{\tau}} \max [|1 - \tau\eta_1|, |1 - \tau\eta_n|].$$

On the interval $(0, \bar{\tau})$ the linear function $1 - \tau\eta_1$ changes from 1 to -1. Hence the minimum of the function $S(\tau)$ will be attained with $1 - \tau\eta_n = -1 + \tau\eta_1$, i.e., for $\tau_* = \dfrac{2}{\eta_1 + \eta_n}$. The spectral radius of the matrix $T_x(x_*)$ is

$$S(\tau_*) = \frac{\eta_n - \eta_1}{\eta_n + \eta_1} < 1.$$

If the function $f(x)$ is differentiable in the neighborhood $G(x_*)$ of the point x_* and for all $x \in G(x_*)$, $z \in E^n$ the inequality

$$\eta_1 ||z||^2 \leq z^T f_{xx}(x) z \leq \eta_n ||z||^2$$

holds, then, by Theorem 2.3.6, the estimate

$$||x_k - x_*|| \leq [S(\tau_*)]^k ||x_0 - x_*||$$

holds for the convergence rate.

7. AN AUXILIARY RESULT

In the sequel, in Chapter 3, we shall need the following lemma.

LEMMA 2.3.2. Let the mapping $T(x): E^n \to E^n$ be such that $T(0) = 0$ and let the inequality $||T(x) - T(z)|| \leq c||x - z||^2$, where c is a number, hold for any $x, z \in E^n$. Then the equation

$$x = T(x + y) \tag{3.17}$$

for each vector y satisfying the condition $||y|| \leq \dfrac{\sqrt{2} - 1}{2c}$, has a unique solution in the sphere $H = \{x \in E^n : ||x|| \leq 2c||y||^2\}$.

Proof. We show that the mapping $T(x+y)$ maps H onto itself. Let $x \in H$. Then

$$\|T(x+y)\| \leqslant c\|x+y\|^2 \leqslant c[\|x\|^2+2\|x\|\cdot\|y\|+\|y\|^2] \leqslant$$
$$\leqslant c\|y\|^2[4c^2\|y\|^2+4c\|y\|+1] \leqslant 2c\|y\|^2.$$

Thus, the point $T(x+y) \in H$. We show next that the mapping $T(x+y)$ is contracting by H. Indeed, if $x, z \in H$, $x \neq z$, then

$$\|x\| \leqslant 2c\|y\|^2, \quad \|z\| \leqslant 2c\|y\|^2,$$
$$\|T(x+y)-T(z+y)\| \leqslant c\|x-z\|^2 \leqslant$$
$$\leqslant c\|x-z\|(\|x\|+\|z\|) \leqslant 4c^2\|x-z\|\|y\|^2 \leqslant$$
$$\leqslant (\sqrt{2}-1)^2\|x-z\| < \|x-z\|.$$

By Theorem 2.3.2 the mapping $T(x+y)$ has only one fixed point corresponding to a solution of Equation (3.17). ///

4. CONVERGENCE OF PROCESSES GENERATED BY MULTIVALUED MAPPINGS

1. PRELIMINARY RESULTS

Let a point-set mapping $W: E^n \to 2^{E^n}$ be defined everywhere on E^n (see Appendix III). We pose the problem of finding the set of all fixed points of the mapping W:

$$X_* = \{x \in E^n : x \in W(x)\} .$$

We assume that this set is not empty. A natural generalization of the method of simple iteration (3.1) to multivalued mappings is the process:

$$x_{k+1} \in W(x_k) . \tag{4.1}$$

In contrast to (3.1), given an initial point x_0 we define nonuniquely the sequence of points $x_1, x_2, \ldots,$ which we denote here as before by $\{x_k\}$.

If for any sequence $\{x_k\}$ obtained from (4.1) the set of limit points is not empty, each convergent subsequence converges to a point belonging to some set Z, we say that the method (4.1) converges to the set Z (globally if this property holds for any $x_0 \in E^n$; or on X if this property holds for all $x_0 \in X$; or is local if this property holds for all x_0 belonging to some neighborhood of the set Z).

We denote the sequence obtained from (4.1) by $\{x_k\}_\Lambda$, where $\Lambda = \{0,1,2,\ldots\}$ is the set of nonnegative integers. We denote by Λ^i an infinite set of nonnegative integers in their natural order, for example, $\Lambda^1 = \{1,5,7,20,\ldots\}$. Also, we denote by $\{x_k\}_{\Lambda^i}$ the subsequence of the given sequence $\{x_k\}$, corresponding to the set Λ^i. The notation $\Lambda^i \subset \Lambda$ means that Λ^i is an infinite subset of Λ. The notation $\{x_k\}_{\Lambda+1}$ is the subsequence obtained from $\{x_k\}_\Lambda$ when 1 has been added to each $k \in \Lambda$.

The justification of the convergence of the processes considered is based on the following lemma similar to Lemma 2.2.1.

LEMMA 2.4.1. Let a continuous function $v(x)$ be defined on the set $X \subset E^n$, and let the sequence of points $\{x_k\}_\Lambda$ of X be such that for any $k \in \Lambda$ the condition $v(x_{k+1}) \le v(x_k)$ is satisfied; also, let a subsequence $\{x_k\}_{\Lambda^1}$, $\Lambda^1 \subset \Lambda$, converging to the point $x_* \in X$ exist. Then

$$\lim_{k\to\infty,\, k\in\Lambda} v(x_k) = \lim_{k\to\infty,\, k\in\Lambda^1} v(x_k) = v(x_*) .$$

Proof. The continuity of $v(x)$ implies

$$\lim_{k\to\infty,\, k\in\Lambda^1} v(x_k) = v(x_*) . \tag{4.2}$$

By the conditions of the Lemma, the sequence $\{v(x_k)\}$ decreases monotonically, hence for any $k \geq 0$ we have $v(x_0) \geq v(x_k) \geq v(x_*)$. We use (4.2) as well as the definition of the limit. For any $\varepsilon > 0$ there exists $N \in \Lambda^1$ such that for any $k > N$ and $K \in \Lambda^1$

$$0 \leq v(x_k) - v(x_*) \leq v(x_N) - v(x_*) < \varepsilon.$$

At the same time for any $i \in \Lambda$, $i \geq N$, the condition $v(x_i) \leq v(x_N)$ is satisfied. Hence for all $x_i \in \{x_k\}_\Lambda$ such that $i > N$, we have

$$|v(x_i) - v(x_*)| < \varepsilon \quad ,$$

proving in turn the Lemma. ///

2. FEJÉR MAPS

DEFINITION 2.4.1. The point-set mapping $W\colon E^n \to 2^{E^n}$, having a non-empty set of fixed points X_*, is called X_*-Fejér if the following two conditions are satisfied:

- •1. $W(y) = y \quad \forall\, y \in X_*$,
- •2. $\|z-y\| < \|x-y\| \ \forall\, y \in X_*, \quad \forall\, x \notin X_*, \quad \forall\, z \in W(x)$.

The simplest example of the X_*-Fejér mapping is the operation of projection onto the compact convex set X_*, considered in Section 1.1 (see Formula (1.1.2)). In this case,

$$W(x) = \operatorname*{Arg\,min}_{y \in X_*} \|x - y\| \ .$$

THEOREM 2.4.1 (I.I. Eremin). Let the point-set mapping $W(x)$ be closed and X_*-Fejér. Then for any fixed point x_0 each sequence $\{x_k\}$ defined from (4.1), converges to the point x_* belonging to X_* and depending only on the choice of x_0.

Proof. Let y be an arbitrary, fixed point of X_*. Let $v(x) = \|x - y\|$. Also, we fix the initial point x_0. The sequence $\{x_k\}$ is bounded since for all k the inequality $\|x_k - y\| \leq \|x_0 - y\|$ holds. Therefore, the set of all limit points of the sequence $\{x_k\}$ is not empty. Let p and q be two limit points. The sequence $\{v(x_k)\}$ decreases monotonically and is bounded below by zero. Hence, according to Lemma 2.4.1,

$$v(p) = v(q) = \|p - y\| = \|q - y\| . \tag{4.3}$$

If $p \in X_*$ or $q \in X_*$, then necessarily $p = q$. Indeed, let $p \in X_*$; then as y we take p, from (4.3) we obtain $\|p - p\| = 0 = \|q - p\|$, i.e., $p = q$.

Let the subsequence $\{x_k\}_{\Lambda^1}$ converge as $k \to \infty$ to a point $p \notin X_*$. We consider the subsequence $\{x_k\}_{\Lambda^1+1}$; its boundedness implies that we can separate a subsequence corresponding to the index set $\Lambda^2 \subset \Lambda^1 + 1$, such that

$$\lim_{k\to\infty,\, k\in\Lambda^2} x_k = q .$$

If we subtract unity from each element of the set Λ^2, we obtain the set $\Lambda^3 \subset \Lambda^1$, with $\lim_{k\to\infty,\, k\in\Lambda^3} x_k = p$. By the closure of the multivalued mapping $W(x)$ we have $q \in W(p)$. Hence, taking an arbitrary vector $y \in X_*$, we obtain $\|q - y\| < \|p - y\|$. But this inequality contradicts (4.3). Whence we conclude that the sequence $\{x_k\}$ converges to a point belonging to the set X_*. ///

3. THE APPLICATION OF THE THEOREM ON FEJÉR MAPS

We consider the problem of finding a point of the set

$$X_* = \{x \in E^n : h^j(x) \leq 0, \ j \in [1:c]\} .$$

We assume that each of the functions $h^j(x)$ is convex and defined everywhere on E^n. It then follows from Theorem 1.1.6 that the function $S(x) = \sum_{j=1}^{c} h_+^j(x)$ is convex as well. For each point $x \in E^n$ we define the mutlivalued mapping

$$W(x) = x - \begin{cases} \frac{\lambda S(x)}{\|H(x)\|^2} H(x) & \text{if} \quad x \notin X_*, \\ 0 & \text{if} \quad x \in X_*. \end{cases} \tag{4.4}$$

Here λ is an integer, $H(x) \in \partial S(x)$ (see Definition 1.2.1).

THEOREM 2.4.2 (I.I. Eremin). Let $X_* \neq \emptyset$, and let the multivalued mapping $W(x)$ be given by (4.4), where $0 < \lambda < 2$. Then each sequence $\{x_k\}$ defined by (4.1) converges to the set X_* globally.

Proof. We show that the mapping W is X_*-Fejér. The justification of the closure follows from Theorem 1.2.4. Let $x \notin X_*$; we show that $\|H(x)\| \neq 0$. We assume the opposite is true: $\|H(x)\| = 0$. Then, by the formula (1.2.2), for any $y \in E^n$ the inequality $S(x) \leq S(y)$ holds, but this contradicts the fact that $x \notin X_*$. Hence $\|H(x)\| \neq 0$ everywhere outside X_*. For arbitrary $y \in X_*$, $x \notin X_*$, $z \in W(x)$ we have

$$B = \|z-y\|^2 = \left\|x - y - \frac{\lambda S(x)}{\|H(x)\|^2} H(x)\right\|^2 =$$
$$= \|x-y\|^2 + \frac{2\lambda S(x)}{\|H(x)\|^2} \langle H(x), y-x\rangle + \frac{\lambda^2 S^2}{\|H(x)\|^2} .$$

Using the inequality (1.2.2), we obtain

$$B \leqslant \|x-y\|^2 + \frac{2\lambda S(x)}{\|H(x)\|^2} (S(y) - S(x)) + \frac{\lambda^2 S^2(x)}{\|H(x)\|^2} =$$
$$= \|x-y\|^2 + \frac{\lambda S^2(x)}{\|H(x)\|^2} (\lambda - 2) < \|x-y\|^2.$$

Therefore, for any $0 < \lambda < 2$ the multivalued mapping $W(x)$ is X_*-Fejér and, by Theorem 2.4.1, the iterations (4.1) converge to X_*. ///

A more detailed discussion of the theory of Fejér maps and their applications to solution of applied problems can be found in I. I. Eremin and V. D. Mazurov [1].

4. NONSTATIONARY PROCESSES

Let the process (4.1) be defined by the nonstationary multivalued mapping $W(x,k)$:

$$x_{k+1} \in W(x_k, k) \quad .$$

We pose the problem of finding the points

$$X_* = \{x \in E^n : x \in W(x,k) \ \ \forall k\} \quad .$$

To illustrate such process, we cite the method of a generalized gradient for finding an unconstrained minimum of the function $f(x)$, in which

$$x_{k+1} = x_k - \alpha_k \gamma_k H(x_k) \ , \qquad H(x) \in \partial f(x) \ . \tag{4.5}$$

Here α_k, γ_k are some positive coefficients; various laws of their variation are feasible -- we shall give three of them:

$$\gamma_k = [d + \|H(x_k)\|]^{-1}, \quad \sum_{k=1}^{\infty} \alpha_k = \infty, \quad \sum_{k=1}^{\infty} \alpha_k^2 < \infty, \tag{4.6}$$

$$\gamma_k = [d + \|H(x_k)\|^2]^{-1}, \quad \sum_{k=1}^{\infty} \alpha_k = \infty, \quad \alpha_k \to 0, \tag{4.7}$$

$$\gamma_k = \text{const}, \qquad \sum_{k=1}^{\infty} \alpha_k = \infty, \quad \alpha_k \to 0, \tag{4.8}$$

where $0 < d$ is arbitrary.

We restrict ourselves to the justification of the method (4.5) with the regulation rule (4.6).

THEOREM 2.4.3. Let the function $f(x)$ be defined everywhere on E^n, be convex, and let the set of its minima X_* be nonempty. Then, using the method of the generalized gradient (4.5) with the regulation (4.6), we can separate from the sequence $\{x_k\}_\Lambda$ a subsequence $\{x_k\}_{\Lambda^1}$ such that

$$\lim_{k\to\infty,\ k\in\Lambda^1} f(x_k)=f(x_*), \qquad x_*\in X_*. \tag{4.9}$$

Proof. Let $x_* \in X_*$, $v_k = \|x_k - x_*\|^2$; then

$$v_{k+1}=v_k-2\alpha_k\gamma_k\langle H(x_k),\ x_k-x_*\rangle+\alpha_k^2\gamma_k^2\|H(x_k)\|^2 ,$$

implying in turn that

$$v_{k+1}=v_0-2\sum_{s=0}^{k}\alpha_s\gamma_s\langle H(x_s),\ x_s-x_*\rangle+\sum_{s=0}^{k}\alpha_s^2\gamma_s^2\|H(x_s)\|^2. \tag{4.10}$$

By the convexity of $f(x)$ on E^n we have $0 \le \langle H(x),\ x-x_*\rangle$. Hence the relations

$$v_{k+1}\leqslant v_0+\sum_{s=0}^{k}\left[\frac{\|H(x_s)\|}{d+\|H(x_s)\|}\right]^2\alpha_s^2\leqslant v_0+\sum_{s=0}^{k}\alpha_s^2$$

follow from (4.10) if we take into account (4.6).

From the boundedness of the right side by a constant not depending on k, it follows that as $k\to\infty$ the sequences $\{v_k\}$, $\{x_k\}$, $\{H(x_k)\}$ are bounded. From (4.10) we obtain also

$$2a\sum_{s=0}^{k}\alpha_s\langle H(x_s),\ x_s-x_*\rangle\leqslant v_0-v_{k+1}+\sum_{s=0}^{k}\alpha_s^2,$$

where the constant a does not depend on k. The right-hand side of the inequality is bounded as $k \to \infty$, by (4.6) $\sum_{s=0}^{\infty} \alpha_s = \infty$; but this is possible only if $\langle H(x_k), x_k - x_* \rangle \to 0$ as $k \to \infty$. From the convexity property

$$f(x_k) \leq f(x_*) + \langle H(x_k), x_k - x_* \rangle$$

we conclude that there exists a subsequence $\{x_k\}_{\Lambda 1}$ such that the condition (4.9) is satisfied. ///

The theorems are similar and the convergence of the method under the regulations (4.7) and (4.8) can be proved similarly. The method of the generalized gradient is extensively studied in the literature; we refer, for example, to Yu.M. Ermol'ev [2] and N.Z. Shor [1].

5. METHODS FOR SOLVING SYSTEMS OF NONLINEAR EQUATIONS

1. *A METHOD OF SIMPLE ITERATION*

Let the mapping $V: R^n \to R^n$. We pose the problem of solving a system of n equations with n unknowns

$$V(x) = 0 . \tag{5.1}$$

Using the formula (3.8), we define the mapping $T(x)$; then the problem considered becomes equivalent to the problem of finding fixed points of the mapping T. The method of simple iteration (3.1) leads us to the following method for solving the system (5.1):

$$x_{k+1} = T(x_k) = x_k - \tau V(x_k) . \tag{5.2}$$

From Theorems 2.3.1 and 2.3.3 we obtain the following result.

THEOREM 2.5.1. Let the mapping $V: R^n \to R^n$ be Lipschitz with constant ℓ and uniformly monotone with monotonicity constant m. Then Equation (5.1) has a unique solution $x = x_*$, and for any $\tau \in \left(0, \frac{2m}{\ell^2}\right)$ the iterations defined by (5.2) converge to x_* globally. We have the following estimate of the convergence rate:

$$\| x_k - x_* \| \leq \frac{\tau q^k}{1-q} \| V(x_0) \| .$$

If the solution $x = x_*$ to the system (5.1) exists and the mapping is differentiable at the point x_*, then, by Theorem 2.3.4, for the local, linear convergence of the iterations (5.2) it is sufficient that the spectral radius of the matrix $T_x^T(x_*) = I_n - \tau V_x^T(x_*)$ be strictly less than one.

Let us consider two equations:

$$|V_x^T(x_*) - \eta I_n| = 0,$$
$$|T_x^T(x_*) - \varkappa I_n| = |-\tau V_x^T(x_*) + (1-\varkappa) I_n| = 0.$$

The roots of these equations are related in a simple way: $\kappa = 1 - \tau\eta$. We write the complex root η as $\eta = a + ib$; then

$$\kappa = 1 - \tau a - i\tau b , \qquad |\kappa|^2 = (1 - \tau a)^2 + \tau^2 b^2 .$$

The condition $|\kappa| < 1$ becomes then

$$(a^2 + b^2)\tau^2 - 2\tau a < 0 . \tag{5.3}$$

If $a = \operatorname{Re} \eta > 0$, (5.3) holds for any $0 < \tau < \bar{\tau}$, where $\bar{\tau} = \frac{2 \operatorname{Re} \eta}{|\eta|^2}$. If $a = \operatorname{Re} \eta < 0$, (5.3) is satisfied for any $\bar{\tau} < \tau < 0$.

Therefore, the spectral radius of the matrix $T_x^T(x_*)$ is less than one iff one of the following conditions is satisfied:

● 1. the eigenvalues $\eta_1, \ldots, \eta_n$ of the matrix $V_x^T(x_*)$ satisfy the conditions $\mathrm{Re}\ \eta_i > 0$, $i \in [1:n]$, and $0 < \tau < \frac{2\,\mathrm{Re}\ \eta_i}{|\eta_i|^2}$, $i \in [1:n]$;

● 2. the eigenvalues $\eta_1, \ldots, \eta_n$ of the matrix $V_x^T(x_*)$ satisfy the conditions $\mathrm{Re}\ \eta_i < 0$, $i \in [1:n]$, and $\frac{2\,\mathrm{Re}\ \eta_i}{|\eta_i|^2} < \tau < 0$, $i \in [1:n]$.

When either condition is satisfied the iterations (5.2) converge locally to the point x_*. If the matrix $V_x(x_*)$ is symmetric, all the roots η_i will be real, and the conditions (1.) and (2.) will be satisfied only if the matrix $V_x(x_*)$ is either positive definite or negative definite.

If the mapping $V(x)$ is potential and the potential is given by the function $f(x)$, the method (5.2) can be regarded as a numerical method for finding local extrema of the function $f(x)$. A case of finding the minimum was considered in Subsection 2.3.6, the method (5.2) coincides in this case with (3.16).

2. AN ANALOG OF THE METHOD OF THE GENERALIZED GRADIENT

As was pointed out in Subsection 2.3.6, methods of minimizing convex functions can be treated as methods for finding fixed points of the corresponding monotone mappings. Hence the method of the generalized gradient described in Section 2.4 is equivalent to the iteration

$$x_{k+1} = x_k - \alpha_k \gamma_k V(x_k) \quad , \tag{5.4}$$

where the nonnegative coefficients α_k, γ_k vary according to one of the rules (4.6) - (4.8). Since the proof of the convergence repeats almost verbatim the proof of Theorem 2.4.3, we restrict ourselves to stating the theorem on convergence.

THEOREM 2.5.2. Let the solution $x = x_*$ to the system (5.1) exist. If the mapping $V(x)$ is bounded and uniformly monotone and the step-length rule (4.6) is used, the iterations defined by (5.4) converge to the point x_* globally.

3. NEWTON'S METHOD

A most widely used method for solving the system (5.1) is Newton's method in which iterations are given by the formulas

$$x_{k+1} = T(x_k), \qquad T(x) = x - [V_x^T(x)]^{-1}V(x), \tag{5.5}$$

i.e., the method of simple iteration for finding fixed points of the mapping $T(x)$. If the mapping V is sufficiently smooth, then

$$T_x(x) = [V_x^T(x)]^{-2} \sum_{i=1}^{n} [V_{xx}^{(i)}(x)] V^{(i)}(x).$$

Therefore, $T_x(x_*) = 0$, which by Theorem 2.3.4 ensures a high rate of convergence $x_k \to x_*$.

Basic properties of the Newton method are reflected in the following theorem.

THEOREM 2.5.3. Let the mapping $V\colon R^n \to R^n$ be differentiable in some neighborhood $G(x_*)$ of the solution x_* of the system (5.1), the derivative $V_x(x)$ being continuous at x_* and the matrix $V_x(x_*)$ being nonsingular. Then the iterations defined by the formulas (5.5) converge locally to the point x_* at a superlinear rate. If, furthermore, there exists a constant ℓ such that

$$\|V_x(x) - V_x(x_*)\| \leq \ell \|x - x_*\| \qquad \forall\, x \in G(x_*) \,, \tag{5.6}$$

the iterations (5.5) possess the local quadratic rate of convergence.

This assertion will obviously follow from Theorem 2.5.4 below.

REMARK. In what follows Theorem 2.5.3 will frequently be applied to various specific problems. To simplify the formulations we shall write instead of the condition (5.6): the derivative $V_x(x)$ satisfies a Lipschitz condition in the neighborhood of the point x_*, i.e.,

$$\|V_x(x) - V_x(y)\| \leq \ell \|x - y\| \qquad \forall\, x,y \in G(x_*) \quad .$$

In the subsequent chapters, the investigation of the method (5.5) is preceded by the justification of the method of simple iteration (5.2) with $\tau = 1$. In justifying the method we prove that the spectral radius of the matrix $I_n - V_x^T(x_*)$ is strictly less than one. Neumann's lemma (see Appendix II) implies then the nonsingularity of the matrix $V_x^T(x_*)$; and to ensure the quadratic convergence rate (5.5) it is required only that either (5.6) be satisfied or the Lipschitz condition be satisfied in the neighborhood of x_*.

4. MODIFICATIONS OF NEWTON'S METHOD

A high rate of convergence of the Newton method explains its wide application to solving various applied problems. At the same time, however, the method has a few shortcomings; the most essential ones are the following:

• 1. time-consuming computations because of the need to compute n^2 partial derivatives;

• 2. the local nature of convergence. If the initial approximation is poor, the method often diverges;

• 3. the need to solve the system of linear equations $V_x^T(x_k)p_k = -V(x_k)$ in order to construct the Newton direction p_k, which requires arithmetic operations of order n^3.

Many works of the recent years have been aimed at elimination of these shortcomings, and as a result, many modifications of the method have been created. We shall discuss some of them.

In a numerical realization of the method, the finite difference approximation of the matrix $V_x(x)$ is usually used instead of the matrix $V_x(x)$. In this case the question arises: How does one choose the size of the step of numerical differentiation and what is the differentiation scheme to be so that the method preserves its high rate of convergence? The theoretical feasibility to use a finite difference approximation follows from the next theorem.

THEOREM 2.5.4. Let the mapping $V: R^n \to R^n$ be differentiable in some neighborhood $G(x_*)$ of the solution x_* to the system (5.1), let the derivative $V_x(x)$ be continuous in x_* and let the matrix $V_x(x_*)$ be nonsingular; define the matrix $W(x)$:

$$\lim_{x \to x_*} \|W(x) - V_x(x)\| = 0 \quad . \tag{5.7}$$

Then the sequence

$$x_{k+1} = \bar{T}(x_k) \ , \qquad \bar{T}(x) = x - [W^T(x)]^{-1}V(x) \ , \tag{5.8}$$

converges locally to the point x_* at a superlinear rate. If, in addition, the inequality (5.6) is satisfied and

$$\|W^T(x) - V_x^T(x)\| \leq c\|V(x)\| , \tag{5.9}$$

the iterations (5.8) possess a local quadratic rate of convergence.

Proof. We show first that the mapping $\bar{T}(x)$ is differentiable at x_* and $\bar{T}_x(x_*) = 0$. This is equivalent to the assertion

$$\lim_{x \to x_*} \frac{\|\bar{T}(x) - \bar{T}(x_*)\|}{\|x - x_*\|} = 0. \tag{5.10}$$

We have the estimates

$$\begin{aligned} \|\bar{T}(x) - \bar{T}(x_*)\| &= \|x - x_* - [W^T(x)]^{-1} V(x)\| \leqslant \\ &\leqslant \|[W^T(x)]^{-1}\| \|W^T(x)(x - x_*) - V(x)\| \leqslant \\ &\leqslant \|[W^T(x)]^{-1}\| (\|W^T(x) - V_x^T(x)\| \|x - x_*\| + \\ &\quad + \|V_x^T(x)(x - x_*) - V(x)\|). \end{aligned} \tag{5.11}$$

From (5.7) and the continuity of the derivative $V_x(x)$ at the point x_* follows

$$\lim_{x \to x_*} \|[W^T(x)]^{-1}\| = \|[V_x^T(x_*)]^{-1}\|,$$

$$\lim_{x \to x_*} \frac{\|V(x) - V(x_*) - V_x^T(x)(x - x_*)\|}{\|x - x_*\|} = 0 ,$$

whence, taken (5.7) and (5.11) into account, we obtain (5.10) and by Theorem 2.3.4 the superlinear rate of convergence $x_k \to x_*$ holds.

Let the inequalities (5.6) and (5.9) be satisfied. Then from (5.11) we have

$$\begin{aligned} \|x_{k+1} - x_*\| &= \|\bar{T}(x_k) - \bar{T}(x_*)\| \leqslant \\ &\leqslant \|[W^T(x_k)]^{-1}\| (c\|V(x_k)\| \|x_k - x_*\| + 2l\|x_k - x_*\|^2) \end{aligned}$$

implying the quadratic rate of convergence, since

$$\lim_{k\to\infty} \frac{\|x_{k+1}-x_*\|}{\|x_k-x_*\|^2} \leq \lim_{k\to\infty} (\|[W^T(x_k)]^{-1}\|)\left(c \lim_{k\to\infty}\left(\frac{\|V(x_k)\|}{\|x_k-x_*\|}\right)+2l\right) \leq$$
$$\leq \|[V_x^T(x_*)]^{-1}\|(c\|V_x^T(x_*)\|+2l).$$

As the simplest finite difference approximation we can take the matrix $W(x)$ in which the ith row is

$$(e^i)^T W(x) = \frac{[V(x+h^{(i)}e^i)-V(x)]^T}{h^{(i)}}. \tag{5.12}$$

Applying Theorem 2.5.4, we can show that if $h_k^{(i)} \to 0$ as $k \to \infty$ $(1 \leq i \leq n)$, the sequence (5.8) involving the approximation (5.12) converges locally to x_* at a superlinear rate. If, in addition, the inequality (5.6) is satisfied and $|h_k^{(i)}| \leq c\|V(x_k)\|$, the rate of convergence is quadratic.

To enlarge the domain of convergence of the method, a special control of the step length α_k is usually carried out on each iteration

$$x_{k+1} = x_k + \alpha_k p_k, \qquad p_k = -[V_x^T(x_k)]^{-1}V(x_k) . \tag{5.13}$$

In the works of V. K. Isaev and V. V. Sonin [1], V. N. Lebedev [1], A. Gleyzal [1], C. Haselgrove [1], and in many others, it has been suggested to reduce in (5.13) the number α_k beginning with $\alpha_k = 1$ until the condition $\|V(x_{k+1})\| \leq \|V(x_k)\|$ is satisfied, and only then go over to the next iteration. This step-length rule, however, does not guarantee the global convergence of the method (5.13). We shall give other step-length rules for the choice of α_k, guaranteeing global convergence and preserving the quadratic rate of convergence near the point x_*:

Let $r > 0$, $\varepsilon \in (0, \min\{1,r\})$, $\lambda \in (0,1)$ be given and let a specific norm $\|\cdot\|$ in R^n be chosen.

RULE 1. As α_k take λ^i, where $i \geq 0$ is the smallest integer satisfying the inequality

$$\|V(x_k + \lambda^i p_k)\|^r \leq (1 - \varepsilon\lambda^i)\,\|V(x_k)\|^r \quad . \tag{5.14}$$

RULE 2. Find an integer $i \geq 0$, as in Rule 1. Choose α_k from the condition

$$\alpha_k = \operatorname{Arg}\min_{0 \leq j \leq i} \|V(x_k + \lambda^j p_k)\| \quad .$$

RULE 3. Choose α_k from the condition

$$\alpha_k = \arg\min_i \|V(x_k + \lambda^i p_k)\| \quad ,$$

where i assumes all possible integer values.

RULE 4. Verify the inequality (5.14) for $i = 0$. If (5.14) is satisfied, put $\alpha_k = 1$; otherwise continue seeking according to Rule 3.

RULE 5. Find α_k from the condition

$$\alpha_k = \arg\min_\alpha \|V(x_k + \alpha p_k)\| \quad .$$

In step-length Rules 1 and 2 the integer i is found by multiplying the initial step $\alpha = 1$ successively by λ until the inequality (5.14) is satisfied. It is interesting to note that for any a priori specified norm in R^n, the quantity V decreases monotonically (from one iteration to the next). This makes it possible in choosing the norm in R^n to take into account the specific nature of the system (5.1).

THEOREM 2.5.5 (O. P. Burdakov [2], [3]). Let the mapping $V: R^n \to R^n$ be differentiable in R^n and let the derivative $V_x(x)$ satisfy the inequalities

$$\|V_x^T(x)-V_x^T(y)\|\leqslant l\|x-y\| \quad \forall x,\ y\in R^n, \tag{5.15}$$

$$\|[V_x^T(x)]^{-1}\|\leqslant m \quad \forall x\in R^n. \tag{5.16}$$

Then, for any values of the parameters $r > 0$, $\lambda \in (0,1)$, $\varepsilon \in (0, \min\{1,r\})$, the modified Newton method (5.13) using step-length Rules 1-5 converges to the unique solution x_* of the system (5.1) from any initial point $x_0 \in R^n$, the rate of convergence being quadratic near the solution.

Proof. We note that by Hadamard's theorem [1] the inequality (5.16) implies the existence and uniqueness of a solution of the system (5.1). We omit the trivial case where $V(x_k) = 0$.

First we consider Rule 1 for $r = 1$. Assuming that $\alpha \in [0,1]$, we use the Newton-Leibniz formula and the inequality (5.15):

$$\|V(x_k+\alpha p_k)\|=\left\|V(x_k)+\alpha\int_0^1 V_x^T(x_k+t\alpha p_k)\,p_k\,dt\right\|=$$

$$=\left\|V(x_k)+\alpha\int_0^1 (V_x^T(x_k+t\alpha p_k)-V_x^T(x_k))\,p_k\,dt-\alpha V(x_k)\right\|\leqslant$$

$$\leqslant(1-\alpha)\|V(x_k)\|+\alpha^2 l\|p_k\|^2.$$

Noting (5.16), we obtain the relation

$$\|V(x_k+\alpha p_k)\| \le (1-\alpha(1-\alpha l m^2\|V(x_k)\|))\,\|V(x_k)\|. \tag{5.17}$$

It is seen that the inequality (5.17) will surely be satisfied for all $\alpha \in [0, \min\{1,\bar{\alpha}_k\}]$, where $\bar{\alpha}_k = \dfrac{1-\varepsilon}{l m^2\|V(x_k)\|}$; then $\alpha_k \ge \min\{1,\lambda\bar{\alpha}_k\}$. It is not hard to see that the sequence $\bar{\alpha}_k$ increases monotonically. Therefore,

$$\|V(x_{k+1})\| \leq q^{k+1} \|V(x_0)\| , \tag{5.18}$$

where $q = 1 - \varepsilon \min \{1, \lambda\bar{\alpha}_0\} < 1$. Whence we conclude that $\|V(x_k)\| \to 0$ as $k \to \infty$. Therefore, the sequence x_k converges to the solution x_* from any initial point. The sequence $\bar{\alpha}_k \to \infty$ therefore, beginning from some iteration, all $\alpha_k = 1$, and by Theorem 2.5.3 the rate of convergence is quadratic. For $r > 1$ the inequality (5.14) is satisfied for all $\alpha \in [0, \min \{1, \lambda\bar{\alpha}_k\}]$. The further considerations are similar to those given above.

Now let $r \in (\varepsilon, 1)$. The inequality (5.14) will be satisfied for all $\alpha \in [0, \min \{1, \bar{\bar{\alpha}}_k\}]$, where

$$\bar{\bar{\alpha}}_k = \left(1 - \frac{\varepsilon}{r}\right) \Big/ lm^2 \|V(x_k)\| .$$

Indeed, from (5.17) we obtain

$$\|V(x_k + \alpha p_k)\| \leqslant \left(1 - \frac{\varepsilon \min \{1, \bar{\bar{\alpha}}_k\}}{r}\right) \|V(x_k)\| \leqslant$$

$$\leqslant \left(1 - \frac{\alpha\varepsilon}{r}\right) \|V(x_k)\| \leqslant (1 - \alpha\varepsilon)^{1/r} \|V(x_k)\|.$$

The further considerations are similar to the case $r = 1$.

Next we consider Rules 2 - 5. Suppose we have x_k and p_k. If we compare the quantities $\|V(x_{k+1})\|$ and $\|\bar{V}(x_{k+1})\|$, obtained through one of Rules 2-5 and Rule 1, respectively, it is not hard to see that $\|V(x_{k+1})\| \leq \|\bar{V}(x_{k+1})\|$. Therefore, the estimate (5.18) remains valid, implying in turn the convergence from any initial point $x_0 \in R^n$. We note next that

$$\|x - x_*\| \leq m\|V(x)\| \qquad \forall\, x \in R^n ,$$

and

$$\|V(x)\| \leq 2\|V_x^T(x_*)\| \ \|x - x_*\|$$

for all x from some neighborhood $G(x_*)$. Applying these estimates to the inequality

$$\|V(x_{k+1})\| \le \|V(x_k + p_k)\| \le \ell m^2 \|V(x_k)\|^2$$

holding near the solution x_*, we obtain the quadratic rate of convergence for Rules 2 - 5. ///

It should be emphasized that we have succeeded in preserving a high rate of convergence of the Newton method owing to the fact that step-length Rules 1-4, beginning from some iteration number yield $\alpha_k = 1$, and in Rule 5 $\alpha_k \to 1$ as $k \to \infty$.

The main results of this Theorem still hold for weaker assumptions. Thus, for example, global convergence can be obtained if one requires the continuity of $V_x(x)$ in R^n rather than the Lipschitz inequality (5.15). One may also get rid of the inequality (5.16). In that case we have the following theorem.

THEOREM 2.5.6 (O.P. Burdakov [3]). Suppose there exists an open convex domain $D \subset R^n$ such that the mapping $V: D \to R^n$ is continuously differentiable in D. Let a point $\bar{x} \in D$ be such that the set

$$L(\bar{x}) = \{x : \|V(x)\| \le \|V(\bar{x})\|,\ x \in D\}$$

is simply connected, compact and the Jacobian $V_x(x)$ is nonsingular in $L(\bar{x})$. We assume that in Rules 1 - 5 all the testing points $x_k + \alpha p_k$ are chosen from D. Then, for any values of the parameters $r > 0$, $\lambda \in (0,1)$, $\varepsilon \in (0, \min\{1,r\})$, the modified Newton method (5.13) converges to a unique solution in $L(\bar{x})$ of the system (5.1) for any $x_0 \in L(\bar{x})$. The rate of convergence is superlinear, and if, in addition, the inequality (5.6) is satisfied the rate of convergence is quadratic as well.

We note that when the set $L(\bar{x})$ is not simply connected, its simply connected components determine domains of attraction of the modification (5.13).

Among step-length Rules 1 - 5 it is impossible to separate the one which is more preferable in all cases. Each rule has its own advantages and disadvantages. In practice, the choice of a specific rule depends on how difficult the computation of the Newton direction p_k is, compared with the computation of the quantity $V(x)$. If these computational difficulties are very different, Rules 3 - 5 are more applicable; if they differ only slightly, Rules 1 and 2 are more applicable.

5. METHODS OF THE QUASI-NEWTON TYPE

The shortcomings (1.) and (3.) of the Newton method, as noted in Section 2.4, are absent in the quasi-Newton methods (they are sometimes referred to as variable metric methods). These methods are based on the notion of sequential approximation, either of V_x^T or of $[V_x^T]^{-1}$. For the approximation, only those values of the mapping V are used which have been already computed on the preceding iterations.

If V_x^T is approximated, the iteration process has the form

$$x_{k+1} = x_k - B_k^{-1}V(x_k) \quad . \tag{5.19}$$

When the matrix $[V_x^T]^{-1}$ is approximated, the computation is done by the formula

$$x_{k+1} = x_k - H_kV(x_k) \quad . \tag{5.20}$$

The $(n\times n)$-matrices B_k and H_k are updated according to the formulas given below. We assume that B_0 and H_0 are a priori specified.

If the mapping V is sufficiently smooth, for small variations $\Delta x_k = x_{k+1} - x_k$ we have

$$V(x_{k+1}) \simeq V(x_k) + V_x^T(x_{k+1})\,\Delta x_k \quad .$$

It is appropriate to require that the matrix B_{k+1} should satisfy the equality

$$V(x_{k+1}) = V(x_k) + B_{k+1}\,\Delta x_k \quad ,$$

which indicates the proximity along the direction Δx_k of the matrix B_{k+1} to $V_x^T(x_{k+1})$. We denote $\Delta V_k = V(x_{k+1}) - V(x_k)$ and rewrite this relation as

$$B_{k+1}\,\Delta x_k = \Delta V_k \quad . \tag{5.21}$$

A similar approach is possible for an approximation of the matrix $[V_x^T(x_{k+1})]^{-1}$. Assume that Δx_k is small; then

$$[V_x^T(x_{k+1})]^{-1}\,\Delta V_k \simeq \Delta x_k \quad .$$

We require that the matrix H_{k+1} satisfy the equality

$$H_{k+1}\Delta V_k = \Delta x_k \quad , \tag{5.22}$$

indicating a sufficient proximity along the vector ΔV_k of the matrix H_{k+1} to the matrix $[V_x^T(x_{k+1})]^{-1}$.

The relations (5.21) and (5.22) are known as quasi-Newton conditions. They do not determine uniquely the matrices B_{k+1} and H_{k+1}. Further hypotheses with regard to these approximations are needed. Techniques for the computation of the matrix B_{k+1} from B_k (as, actually, the matrix H_{k+1} from H_k) are based on quite reasonable requirements. Namely, given B_k, it is necessary to construct a matrix B_{k+1} such that, first, the quasi-Newton relation (5.21) is satisfied and, second, the approximation of B does not change in some (n-1)-dimensional subspace $\pi_k \subset R^n$ not containing Δx_k, i.e.,

$$B_{k+1}p = B_k p \qquad \forall p \in \pi_k . \tag{5.23}$$

Therefore, we have a system of n^2 equations (5.21), (5.23) with respect to n^2 unknown elements of the matrix B_{k+1}. The unique solution is

$$B_{k+1} = B_k + \frac{(\Delta V_k - B_k \Delta x_k) c_k^T}{\langle \Delta x_k, c_k \rangle} , \tag{5.24}$$

where $c_k \in R^n$ is a unique (up to the multiplier) vector orthogonal to π_k. Each concrete technique for choosing the subspace π_k, and therefore the vector c_k, gives a definite quasi-Newton method. Thus, if as π_k one takes an orthogonal complement to Δx_k, i.e., if one requires that the approximation of B does not change along any vector orthogonal to Δx_k, one obtains Broyden's first rank-1 method. We note that in (5.19) one can get rid of the inversion of the matrix B_k and update the matrix B_k^{-1} instead of B_k. To do this, we apply the Sherman-Morrison formula (see Appendix II) yielding

$$B_{k+1}^{-1} = B_k^{-1} + \frac{(\Delta x_k - B_k^{-1}\Delta V_k) c_k^T B_k^{-1}}{\langle B_k^{-1} \Delta V_k, c_k \rangle} . \tag{5.25}$$

Now we concentrate on the updating formulas for H_k. The matrix H_{k+1} must satisfy the quasi-Newton relation (5.22) coinciding in this case with the matrix H_k in the (n-1)-dimensional subspace $\tau_k \subset R^n$ not containing ΔV_k, i.e.,

$$H_{k+1}p = H_k p \qquad \forall\ p \in \tau_k ,$$

which defines uniquely the updating formula

$$H_{k+1} = H_k + \frac{(\Delta x_k - H_k \Delta V_k) d_k^T}{\langle \Delta V_k, d_k \rangle}, \tag{5.26}$$

in which the vector $d_k \in R^n$ is orthogonal to the subspace τ_k. Specifying the form of the subspace τ_k (therefore, of the vector d_k as well), one can obtain a concrete Newton method. If, for example, as τ_k one takes the orthogonal complement to ΔV_k, then $d_k = \Delta V_k$ and one obtains Broyden's second rank-1 method.

There is a certain relationship between the formulas (5.24) and (5.26). Thus, if the matrices B_k and H_k are nonsingular and $H_k = B_k^{-1}$, then for $c_k = B_k^T d_k$, by (5.25), we obtain $H_{k+1} = B_{k+1}^{-1}$.

The formulas (5.24) and (5.26) define a whole class of quasi-Newton methods. We cite here the best known:

Broyden's first method $(c_k = \Delta x_k)$:

$$H_{k+1} = H_k + \frac{(\Delta x_k - H_k \Delta V_k) \Delta x_k^T H_k}{\langle \Delta V_k, H_k^T \Delta x_k \rangle}; \tag{5.27}$$

Broyden's second method $(d_k = \Delta V_k)$:

$$H_{k+1} = H_k + \frac{(\Delta x_k - H_k \Delta V_k) \Delta V_k^T}{\langle \Delta V_k, \Delta V_k \rangle}; \tag{5.28}$$

Pearson's method $(c_k = \Delta V_k)$:

$$H_{k+1} = H_k + \frac{(\Delta x_k - H_k \Delta V_k) \Delta V_k^T H_k}{\langle V_k, H_k \Delta V_k \rangle}; \tag{5.29}$$

McCormick's method $(d_k = \Delta x_k)$:

$$H_{k+1} = H_k + \frac{(\Delta x_k - H_k \Delta V_k)\Delta x_k^T}{\langle \Delta V_k, \Delta x_k \rangle}; \tag{5.30}$$

the symmetric rank-one method $(d_k = \Delta x_k - H_k \Delta V_k)$:

$$H_{k+1} = H_k + \frac{(\Delta x_k - H_k \Delta V_k)(\Delta x_k - H_k \Delta V_k)^T}{\langle \Delta V_k, \Delta x_k - H_k \Delta V_k \rangle}. \tag{5.31}$$

In Thomas's method the computations are made by the formulas (5.20) and (5.26), in which

$$W_{k+1} = (1 + \|\Delta x_k\|)\left(\|\Delta x_k\| I_n + W_k - \frac{d_k d_k^T}{\langle \Delta x_k, d_k \rangle}\right), \tag{5.32}$$

where $d_k = \left(W_k + \frac{\|\Delta x_k\|}{2} I_n\right)\Delta x_k$, $\|\cdot\|$ is a Euclidean norm and $W_0 = I_n$.

The matrix V_x is symmetric for some equations of the system (5.1). Hence it is useful to consider quasi-Newton methods which account for this characteristic. For these methods the quasi-Newton relation (5.21) is satisfied, and if the matrix B_k is symmetric, so is the matrix B_{k+1}. We cite here the arguments used first by M. Powell [2] in obtaining a symmetric version of Broyden's first method and later by J. Dennis [1] in deriving formulas of a more general kind.

Let the matrix B_k be symmetric. Then the approximation

$$\bar{B}_{k,1} = B_k + \frac{(\Delta V_k - B_k \Delta x_k)c_k^T}{\langle \Delta x_k, c_k \rangle}$$

obtained by the formula (5.24) is not symmetric although it does satisfy the relation (5.21). The symmetry of this approximation can be obtained letting

$$\bar{B}_{k,2} = \frac{\bar{B}_{k,1} + \bar{B}_{k,1}^T}{2} .$$

and, since the matrix $\bar{B}_{k,2}$ does not satisfy the relation (5.21), we shall continue the process described. In this case, we obtain a sequence of matrices $\{\bar{B}_{k,i}\}_{i=0}^{\infty}$ such that

$$\begin{aligned} \bar{B}_{k,2i+1} &= \bar{B}_{k,2i} + \frac{(\Delta V_k - \bar{B}_{k,2i}\,\Delta x_k)\,c_k^T}{\langle \Delta x_k, c_k\rangle}, \\ \bar{B}_{k,2i+2} &= \frac{\bar{B}_{k,2i+1} + \bar{B}_{k,2i+1}^T}{2}, \end{aligned} \qquad i = 0, 1, 2, \ldots,$$

where $\bar{B}_{k,0} = B_k$. It turns out that $\{\bar{B}_{k,i}\}_{i=0}^{\infty}$ has the limit

$$B_{k+1} = B_k + \frac{(\Delta V_k - B_k\,\Delta x_k)\,c_k^T + c_k\,(\Delta V_k - B_k\,\Delta x_k)^T}{\langle c_k, \Delta x_k\rangle} - \frac{\langle \Delta V_k - B_k\,\Delta x_k, \Delta x_k\rangle}{\langle c_k, \Delta x_k\rangle^2}\, c_k c_k^T . \tag{5.33}$$

It is not hard to see that the matrix B_{k+1} is symmetric and satisfies the quasi-Newton relation (5.21). In the relation (5.19) we can get rid of the necessity of inverting the matrix B_k. To do this, we apply the Sherman-Morrison formula to (5.33) and obtain

$$\begin{aligned} B_{k+1}^{-1} = B_k^{-1} &+ [\langle c_k, B_k^{-1}\,\Delta V_k\rangle\,(B_k^{-1}c_k\,(\Delta x_k - B_k^{-1}\,\Delta V_k)^T + \\ &+ (\Delta x_k - B_k^{-1}\Delta V_k)c_k^T B_k^{-1}) - \langle \Delta V_k, \Delta x_k - B_k^{-1}\,\Delta V_k\rangle B_k^{-1}c_k c_k^T B_k^{-1} + \\ &+ \langle c_k, B_k^{-1}c_k\rangle\,(\Delta x_k - B_k^{-1}\,\Delta V_k)\,(\Delta x_k - B_k^{-1}\,\Delta V_k)^T] \times \\ &\times [\langle c_k, B_k^{-1}\,\Delta V_k\rangle^2 + \langle c_k, B_k^{-1}\,\Delta V_k\rangle\,\langle \Delta V_k, \Delta x_k - B_k^{-1}\,\Delta V_k\rangle]^{-1}. \end{aligned} \tag{5.34}$$

Suppose we have a symmetric approximation H_k of the matrix $[V_x^T(x_k)]^{-1}$. The sequence of the matrices $\{H_{x,i}\}_{i=0}^{\infty}$ obtained as a result of alternation of a quasi-Newton updating by the formula (5.26) and symmetrization, converges to the matrix

$$H_{k+1}=H_k+\frac{(\Delta x_k-H_k\Delta V_k)d_k^T+d_k(\Delta x_k-H_k\Delta V_k)^T}{\langle d_k,\ \Delta V_k\rangle}-\\ -\frac{\langle\Delta x_k-H_k\Delta V_k,\ \Delta V_k\rangle}{\langle d_k,\ \Delta V_k\rangle^2}d_kd_k^T. \tag{5.35}$$

It is seen that the matrix H_{k+1} is symmetric and satisfies the quasi-Newton relation (5.22). We note that there is no such obvious connection between the formulas (5.33) and (5.35) as the one between (5.24) and (5.26) for $c_k = B_k^T d_k$.

The relations (5.33) and (5.35) determine two intersecting (but not coinciding) classes of quasi-Newton methods. The following concrete methods correspond to the concrete values of the vectors c_k and d_k:

the Davidon-Fletcher-Powell (DFP) method ($c_k = \Delta V_k$):

$$H_{k+1}=H_k+\frac{\Delta x_k\Delta x_k^T}{\langle\Delta x_k,\ \Delta V_k\rangle}-\frac{H_k\Delta V_k\Delta V_k^TH_k}{\langle\Delta V_k,\ H_k\Delta V_k\rangle}; \tag{5.36}$$

the Broyden-Fletcher-Goldfarb-Shanno (BFGS) method ($d_k = \Delta x_k$):

$$H_{k+1}=H_k+\frac{(\Delta x_k-H_k\Delta V_k)\Delta x_k^T+\Delta x_k(\Delta x_k-H_k\Delta V_k)^T}{\langle\Delta x_k,\ \Delta V_k\rangle}-\\ -\frac{\langle\Delta x_k-H_k\Delta V_k,\ \Delta V_k\rangle}{\langle\Delta x_k,\ \Delta V_k\rangle^2}\Delta x_k\Delta x_k^T; \tag{5.37}$$

Powell symmetric Broyden (PSB) method ($c_k = x_k$):

$$B_{k+1}=B_k+\frac{(\Delta V_k-B_k\Delta x_k)\Delta x_k^T+\Delta x_k(\Delta V_k-B_k\Delta x_k)^T}{\langle\Delta x_k,\ \Delta x_k\rangle}-\\ -\frac{\langle\Delta V_k-B_k\Delta x_k,\ \Delta x_k\rangle}{\langle\Delta x_k,\ \Delta x_k\rangle^2}\Delta x_k\Delta x_k^T. \tag{5.38}$$

The last method can be described by an expression for B_{k+1}^{-1} by the formula (5.34), in which one needs to take $c_k = x_k$ (we omit it here because it is too cumbersome).

We proceed to describe properties of quasi-Newton methods. D. Gay in [1] proves the convergence of the methods (5.24) and (5.26) to a root of the system of linear equations for any initial point $x_0 \in R^n$ in a number of steps not exceedings $2n$. In [1], D. Gay also proves the local $2n$-step quadratic convergence rate of the Broyden first and second methods for a nonlinear case. The local superlinear convergence to a solution of the system of nonlinear equations (5.1) has been proved for most of the methods cited. The next theorem has a great potential for the investigation of rates of convergence.

THEOREM 2.5.7 (J. Dennis and J. Moré [1]). Let the mapping $V\colon R^n \to R^n$ be continuously differentiable on an open convex set D. Let there exist a point $x_* \in d$ such that $V(x_*) = 0$ and let the matrix $V_x(x_*)$ be nonsingular. Assume that for some $x_0 \in D$ the sequence x_k constructed in accord with (5.19), is contained in D, $x_k \neq x_*$ for all $k \geq 0$ and converges to x_*. In this case the sequence x_k converges to x_* superlinearly iff

$$\lim_{k \to \infty} \frac{\| [B_k - V_x^T(x_*)](x_{k+1} - x_k) \|}{\| x_{k+1} - x_k \|} = 0. \tag{5.39}$$

Proof. Let the relation (5.39) be satisfied. We consider

$$\begin{aligned}[B_k - V_x^T(x_*)](x_{k+1} - x_k) &= -V(x_k) - V_x^T(x_*)(x_{k+1} - x_k) = \\ &= V(x_{k+1}) - V(x_k) - V_x^T(x_*)(x_{k+1} - x_k) - V(x_{k+1}).\end{aligned} \tag{5.40}$$

From the continuity of the Jacobian V_x in x_* and the relation (5.39) it follows that

$$\lim_{k \to \infty} \frac{\| V(x_{k+1}) \|}{\| x_{k+1} - x_k \|} = 0, \tag{5.41}$$

and since $V(x_*) = 0$ and the matrix $V_x(x_*)$ is nonsingular, there exists $\beta > 0$ such that

$$\|V(x_{k+1})\| = \|V(x_{k+1}) - V(x_*)\| \geq \beta \|x_{k+1} - x_*\|.$$

Therefore,

$$\frac{\|V(x_{k+1})\|}{\|x_{k+1} - x_k\|} \geq \frac{\beta \|x_{k+1} - x_*\|}{\|x_{k+1} - x_*\| + \|x_k - x_*\|} = \beta \frac{\rho_k}{1+\rho_k},$$

where $\rho_k = \|x_{k+1} - x_*\| / \|x_k - x_*\|$. Thus, from (5.39) we obtain the convergence $\frac{\rho_k}{1+\rho_k} \to 0$ as $k \to \infty$, and therefore the convergence $\rho_k \to 0$, which was to be proved.

We assume now that the sequence $x_k \to X_*$ at a superlinear rate. Then

$$\lim_{k \to \infty} \frac{\|x_{k+1} - x_k\|}{\|x_k - x_*\|} = 1, \tag{5.42}$$

since

$$\left| \frac{\|x_{k+1} - x_k\|}{\|x_k - x_*\|} - \frac{\|x_k - x_*\|}{\|x_k - x_*\|} \right| \leq \frac{\|x_{k+1} - x_*\|}{\|x_k - x_*\|}.$$

Applying (5.42) to the relation

$$\frac{\|V(x_{k+1})\|}{\|x_{k+1} - x_k\|} = \frac{\|V(x_{k+1}) - V(x_*)\|}{\|x_k - x_*\|} \cdot \frac{\|x_k - x_*\|}{\|x_{k+1} - x_*\|}$$

and using the continuity of V_x in x_*, we obtain (5.41). The latter implies (5.39) if we take into account (5.40). ///

As follows from the Theorem, for the superlinear convergence $x_k \to x_*$ it is not at all necessary that $B_k \to V_x^T(x_*)$. A sufficient condition is the convergence of the sequence B_k to $V_x^T(x_*)$ only along the displacement Δx_k, which is characteristic of quasi-Newton methods.

The matrix B_k may, generally speaking, never become equal to $V_x^T(x_*)$ even in the linear case $(V_x^T(x) \equiv A)$, where $\Delta V = A\Delta x$. Thus, after the first iteration we have $B_1\Delta x_0 = \Delta V_0$, and after the next iteration $B_2\Delta x_1 = \Delta V_1$, and not at all necessarily $B_2\Delta x_0 = \Delta V_0$. In other words, on each iteration the current information is embedded in the matrix B_{k+1}; however, in this case the old information is lost in part. This creates necessity for a method which, along with the equality (5.21) would preserve the relation

$$B_k\Delta x_i = \Delta V_i , \quad k-n+1 \leq i < k, \quad i \geq 0,$$

i.e., satisfy for all k the relation

$$B_{k+1}\Delta x_i = \Delta V_i , \quad k-n+1 \leq i \leq k, \quad i \geq 0. \tag{5.43}$$

For methods of the form (5.20) this means that

$$H_{k+1}\Delta V_i = \Delta x_i , \quad k-n+1 \leq i \leq k, \quad i \geq 0. \tag{5.44}$$

This is a property of the sequential (n+1)-point secant method in which $B_{k+1} = [\Delta V]_k[\Delta x]_k^{-1}$ (or $H_{k+1} = [\Delta x]_k[\Delta V]_k^{-1}$), where the $(n\times n)$-matrices

$$[\Delta x]_k = [\Delta x_k, \Delta x_{k-1}, \ldots, \Delta x_{k-n+1}],$$
$$[\Delta V]_k = [\Delta V_k, \Delta V_{k-1}, \ldots, \Delta V_{k-n+1}].$$

In the linear case $(v_x^T(x) \equiv A)$, as is not hard to see, $B_n^T = A$ (or $H_n^T = A^{-1}$). In the nonlinear case, if the matrices $[\Delta x]_k$ and $[\Delta V]_k$ are nonsingular, the point x_{k+1} will be the unique root of the unique linear mapping having the values $\{V(x_{k-i})\}_{i=0}^n$ at the points $\{x_{k-i}\}_{i=0}^n$. This fact points out that the sequential (n+1)-point secant method is a generalization of the standard (one-dimensional) secant method to the case of n variables.

To avoid in the secant method the inversion of the matrix $[\Delta x]_k$ (or $[\Delta V]_k$), it is possible to represent this matrix in quasi-Newton form. To do this, in the formula (5.25) the vector c_k must be chosen such that

$$\langle c_k, \Delta x_i \rangle = 0 , \qquad k-n+1 \le i < k, \quad i \ge 0 ,$$

$$\langle c_k, \Delta x_k \rangle \ne 0 . \tag{5.45}$$

This choice of c_k guarantees that the relation (5.43) is satisfied. Similarly, another representation of the secant method in the form (5.26) is related with

$$\langle d_k, \Delta V_i \rangle = 0 , \qquad k-n+1 \le i < k, \quad i \ge 0 ,$$

$$\langle d_k, \Delta V_k \rangle \ne 0 , \tag{5.46}$$

which guarantees (5.44).

If the Jacobian $V_x(x)$ of the system (5.1) is symmetric, a symmetric version of the secant method (see [1]) reflects this specific feature most fully. In this case computations are made through (5.19), (5.34), (5.45).

<u>THEOREM 2.5.8</u> (L. Bittner [1], L. Tornheim [1]). Let the mapping V satisfy the conditions of Theorem 2.5.7. We assume that $\|x_0 - x_*\|$ is sufficiently small and for appropriate $\sigma > 0$, $r > 0$ the iterations of the sequential $(n+1)$-point secant method

$$x_{k+1} = x_k - [\Delta x]_k [\Delta V]_k^{-1} V(x_k)$$

are such that $\|[\Delta x]_k\| \leq r$, and the matrices $[\Delta x]_k$ are uniformly nonsingular, i.e.,

$$\left|\det\left(\frac{\Delta x_k}{\|\Delta x_k\|}, \frac{\Delta x_{k-1}}{\|\Delta x_{k-1}\|}, \ldots, \frac{\Delta x_{k-n+1}}{\|\Delta x_{k-n+1}\|}\right)\right| \geqslant \sigma > 0. \tag{5.47}$$

Then $x_k \to x_*$ at a superlinear rate. If, in addition, the Lipschitz inequality (5.6) is satisfied, the order of convergence is not less than the unique positive root of the equation $t^{n+1} - t^n - 1 = 0$.

Proof. We omit the proof; it can be found, for example, in Ortega and Rheinboldt [1].

Together with the high rate of convergence and the need to compute on each iteration only one value of the mapping V, the sequential (n+1)-point secant method has the essential disadvantage that the matrices $[\Delta x]_k$ be required to be uniformly nonsingular. The sequence x_k, which does not satisfy the inequality (5.47), can become the source of instability of the method. Research of the last several years in this area has been oriented mainly to construction of versions of the secant method stable with respect to the linear dependence of the vectors $\{\Delta x_{k-i}\}_{i=0}^{n}$. In this connection we ought to mention works of B. Gragg and G. Steward [1], D. Gay and R. Schnabel [1], J. Martinez [1], and O. P. Burdakov [1].

The minimization of the function f(x) manifests some new properties of methods of quasi-Newton type, of the form

$$x_{k+1} = x_k - \alpha_k H_k f_x(x_k) \quad . \tag{5.48}$$

It turns out that if the function f(x) is quadratic, that is,

$$f(x) = \frac{\langle Ax, x\rangle}{2} + \langle b, x\rangle + d \quad ,$$

where the matrix A is symmetric and strictly positive definite, and the length of the step α_k is chosen from the condition for exact minimum

$$\alpha_k = \arg\min_{\alpha} f(x_k - \alpha H_k f_x(x_k)) \ ,$$

some quasi-Newton methods -- such as, for example, the Davidon-Fletcher-Powell method (5.36), the Broyden-Fletcher-Goldfarb-Shanno method (5.37), the symmetric method of the first rank (5.31) -- generate a sequence of conjugate directions. In this case

$$\langle A\Delta x_i, \Delta x_j\rangle = 0 \qquad \forall\ i \neq j \quad ,$$

and, as a consequence, there is convergence to the point of the minimum of a quadratic function in a number of steps not exceeding n. If the method completes n steps, then $H_n = A^{-1}$. In the case of minimization of a non-quadratic function this property leads us to the n-step quadratic rate of convergence of the form

$$\|x_{k+n} - x_*\| \leq c\|x_k - x_*\|^2 \quad . \tag{5.49}$$

If for all $k \geq n$ the directions $\{\Delta_{k-i}\}_{i=0}^{n-1}$ are uniformly linearly independent, one can prove a higher rate of convergence than (5.49), of the order exceeding the unique positive root of the equation $t^{n+1} - t^n - 1 = 0$. This estimation of the convergence rate of quasi-Newton methods generating conjugate directions coincides with the estimation described in Theorem 2.5.8 for the sequential $(n+1)$-point secant method.

5. NUMERICAL METHODS FOR FINDING A MINIMAX

Several factors contributed to the appearance of numerical methods for finding the minimax: first, many practical problems of making decisions in conflict situations reduce to the problem of finding the minimax, and numerical methods of solving these problems are needed in a wide class of applications. Second, these methods are useful in solving problems of nonlinear programming and optimal control, which we will discuss here and again later on.

To begin, we discuss the simplest case where a local minimax is being sought (see Definition 1.5.2), and then we discuss in brief methods of finding global solutions.

1. METHODS FOR FINDING A LOCAL MINIMAX

We assume that in the problem (1.5.2) the function $F(x,y)$ is defined and differentiable on $E^n \times E^m$. We seek the limit (as $t \to \infty$) points of the solution of the following Cauchy problem:

$$\frac{dx}{dt} = -\varepsilon F_x(x,y) , \qquad \frac{dy}{dt} = F_y(x,y) ,$$
$$x(0) = x_0 , \qquad y(0) = y_0 . \tag{6.1}$$

The discrete version of this system has the form

$$x_{k+1} = x_k - \alpha\varepsilon F_x(x_k,y_k) ,$$
$$y_{k+1} = y_k + \alpha F_y(x_k,y_k) . \tag{6.2}$$

Here $0 < \varepsilon << 1$ is a small parameter, and the integration step is $\alpha > 0$. We shall show that the following theorem holds.

<u>THEOREM 2.6.1</u> (N.I. Grachev, Yu.G. Evtushenko [4]). Let the

function $F(x,y)$ be twice continuously differentiable in the neighborhood of the point $z_* = [x_*, y_*]$, where sufficient conditions for the local minimax, given in Theorem 1.5.7, are satisfied. Then there exist $\bar{\varepsilon} > 0$, $\bar{\alpha} > 0$ such that for any fixed $0 < \varepsilon < \bar{\varepsilon}$ and $0 < \alpha < \bar{\alpha}$ the solutions $x(\varepsilon, x_0, y_0, t)$, $y(\varepsilon, x_0, y_0, t)$ of the system (6.1) and the iterations $x_k(\varepsilon, x_0, y_0)$, $y_k(\varepsilon, x_0, y_0)$ in the scheme (6.2) converge locally to the point z_*.

The Theorem provides relatively simple methods for solving the problem (1.5.2). The presence of slowly- and rapidly-varying variables in (6.1) and (6.2) makes the computations complicated; however, in some problems involving large dimensions, such an approach is quite effective. We change in (6.1) the independent variable $\tau = \varepsilon t$ and obtain

$$\frac{dx}{d\tau} = -F_x(x,y) , \qquad \varepsilon \frac{dy}{d\tau} = F_y(x,y) . \tag{6.3}$$

Systems of the form (6.3) have been studied in the theory of singular perturbations of ordinary differential equations: A. N. Tikhonov [1], A.A. Dorodnitsyn [1], E.F. Mishchenko and L.S. Pontryagin [1]. A detailed bibliography is given in A.B. Vasil'eva and V.F. Butuzov [1], and in E.F. Mishchenko and N.Kh. Rozov [1].

Following the lines of the methods of singular perturbations, one can consider the so-called degenerate system obtained from (6.3), for $\varepsilon = 0$:

$$\frac{dx}{d\tau} = -F_x(x,y) , \qquad F_y(x,y) = 0 . \tag{6.4}$$

If the conditions of Theorem 1.5.8 are satisfied and the function F is sufficiently smooth, the second equation defines the unique function $y = g(x)$, with

$$g(x) = \operatorname{Arg}\max_{y \in E^m} F(x,y) . \tag{6.5}$$

Substituting this expression into $F(x,y)$, we obtain the following maximum function:

$$\phi(x) = F(x, g(x)) . \tag{6.6}$$

Substituting $y = g(x)$ into the first equation of (6.4), we obtain the system

$$\frac{dx}{d\tau} = -\frac{d\phi}{dx} = -F_x(x, g(x)) , \tag{6.7}$$

which coincides with the Cauchy method (see (2.3)), which has been applied to the minimization of the function $\phi(x)$. According to the results of Section 2 for the local convergence of the method (6.7) it is sufficient that the matrix

$$\phi_{xx}(x) = F_{xx}(x,g(x)) - F_{xy}(x,g(x))F_{yy}^{-1}(x,g(x))F_{yx}(x,g(x)) \tag{6.8}$$

be positive definite at the point $x = x_*$.

In singular perturbation theory it has been proved that if ε is sufficiently small and certain assumptions are satisfied, a solution of the system (6.4) approximates the solution (6.3). We go backward: instead of solving the degenerate system (6.4), which may be too complicated for large values of m, we integrate the system (6.1) equivalent to the singularly perturbed system (6.3).

Now we turn to proving the Theorem. We write the first variational equation for the system (6.1):

$$\delta\dot{z} = M(\varepsilon, z_*)\delta z ;$$

here

$$M(\varepsilon, z_*) = \left[\begin{array}{c|c} -\varepsilon F_{xx}(z_*) & -\varepsilon F_{xy}(z_*) \\ \hline F_{yx}(z_*) & F_{yy}(z_*) \end{array}\right], \quad z_* = [x_*, y_*],$$
$$\delta z = [\delta x, \delta y] \in E^{n+m}, \quad \delta x = x(t) - x_*, \quad \delta y = y(t) - y_*.$$

The continuous, bounded (as a function of ε) matrix $M(\varepsilon, z_*)$ satisfies the following condition: the roots of the characteristic equation

$$|M(\varepsilon, z_*) - \lambda I_{n+m}| = 0 \qquad (6.9)$$

as $\varepsilon \to 0$ split into two groups:

the first group of m roots are close to those of the equation

$$|F_{yy}(z_*) - \lambda I_m| = 0 \ ;$$

the second group of n "small" roots of order $\varepsilon\mu$, where μ are close to those of the equation

$$\left|\begin{array}{c|c} F_{xx}(z_*) + \mu I_n & F_{xy}(z_*) \\ \hline F_{yx}(z_*) & F_{yy}(z_*) \end{array}\right| = 0.$$

Multiplying the lower row-matrix on the left by $F_{xy}(z_*)F_{yy}^{-1}(z_*)$ and subtracting this matrix from the upper matrix, we obtain that μ is the root of the following equation:

$$|\Phi(z_*) + \mu I_n| = 0,$$
$$\Phi(z) = F_{xx}(z) - F_{xy}(z) F_{yy}^{-1}(z) F_{yx}(z). \qquad (6.10)$$

By the sufficient condition of the local minimax the matrix $\Phi(z_*)$ is positive definite, hence all the roots μ are real and strictly negative. Therefore, for sufficiently small values all the roots of Equation (6.9) have strictly negative real parts. By the theorem on stability of the first-order approximation, the stationary point z_* is an asymptotically stable equilibrium for system (6.1). This implies the local convergence of the method (6.1) to z_* as well as the convergence of the difference version of (6.2) for sufficiently small values of α (see Section 3). ///

The methods (6.1) and (6.2) can be used to find local maximins in the problem (1.5.1). In this case, however, one should take $\varepsilon \gg 1$, i.e., ε is regarded as a large parameter. When the sufficient conditions for the local maximin given in Theorem 1.5.9 are satisfied, the solutions of (6.1) and (6.2) converge locally to the local maximin if ε is sufficiently large and α is sufficiently small. A new small parameter ε in (6.1) leads us to the situation in which the variables x change considerably at a slower rate than the variables y do. This makes the integration of (6.1) more complicated. In the case where $F(x,y)$ is a strictly convex-concave function, in the problem (1.5.1) there is a saddle point and we can put $\varepsilon = 1$, simplifying thereby the calculations; the method (6.1) becomes thus the method (2.5) for finding saddle points. The method (6.1) can be used, because it is so simple, to solve elementary game problems. (We shall dwell upon this issue in more detail later on, in Section 8 of Chapter 6). A theorem on the global convergence of (6.1) and (6.2) has been proved under the assumption that somewhat stronger requirements than the conditions stated in Theorem 1.5.8 are satisfied (see Yu.G. Evtushenko [12]).

A "rapid" motion along y can be made, using the Newton method:

$$\dot{x} = -\varepsilon F_x(x,y) , \qquad \dot{y} = -F_{yy}^{-1}(x,y)F_y(x,y) .$$

It is not hard to show that eigenvalues of the matrix in the variation equation consist of n numbers close to the roots of (6.10), plus m numbers close to -1. This implies the local convergence of the method under usual assumptions.

2. REDUCTION TO THE PROBLEM OF FINDING SADDLE POINTS

In the problem (1.5.2) we replace the vector y by a new m-dimensional vector p:

$$p = y - g(x) \quad , \tag{6.11}$$

where $g(x)$ is defined by (6.5).

We consider next a new problem of finding an unconstrained minimax:

$$\min_{x \in E^n} \max_{p \in E^m} B(x, p), \quad B(x, p) = F(x, p + g(x)). \tag{6.12}$$

A remarkable property of the transformation made above is that from the existence of the synthesis of the problem (1.52) it follows that the problem (6.12) has a saddle point. Indeed, Theorem (1.5.1) implies the inequality

$$\max_{p \in E^m} \min_{x \in E^n} B(x, p) \leqslant \min_{x \in E^n} \max_{p \in E^m} B(x, p).$$

On the other hand,

$$\max_{p \in E^m} \min_{x \in E^n} B(x, p) \geqslant \min_{x \in E^n} B(x, 0) =$$
$$= \min_{x \in E^n} \max_{y \in E^m} F(x, y) = \min_{x \in E^n} \max_{p \in E^m} B(x, p).$$

Comparing this inequality with the preceding inequality, we obtain

$$\max_{p \in E^m} \min_{x \in E^n} B(x, p) = \min_{x \in E^n} \max_{p \in E^m} B(x, p) \ ,$$

i.e., the problem (6.12) has a saddle point.

Differentiating $B(x,p)$ over x and p, we obtain the following formulas:

$$\frac{\partial B}{\partial x}=F_x+\frac{dg}{dx}F_y, \quad \frac{\partial B}{\partial p}=F_y, \quad \frac{\partial^2 B}{\partial x\,\partial p}=F_{xy}+\frac{dg}{dx}F_{yy},$$
$$\frac{\partial^2 B}{\partial x^2}=F_{xx}+\frac{dg}{dx}\left(F_{yx}+\frac{\partial^2 B}{\partial p\,\partial x}\right)+\sum_{i=1}^{m}\frac{d^2 g^i}{dx^2}F_{y^i}, \quad \frac{\partial^2 B}{\partial p^2}=F_{yy}.$$

Here, for the sake of brevity, we omit the arguments of the functions. Letting $p_* = y_* - g(x_*)$, we then obtain

$$B_x(x_*, 0)=F_x(z_*)=0, \quad B_p(x_*, 0)=F_y(z_*)=0,$$
$$B_{pp}(x_*, 0)=F_{yy}(z_*)<0, \quad B_{xp}(x_*, 0)=0,$$
$$B_{xx}(x_*, 0)=\Phi(z_*)>0.$$

These formulas show that the transformation (6.11) has the following property: the stationary points $[x,y]$ of the function $F(x,y)$ turn into the stationary points $[x,p]$ of the function $B(x,p)$; if at the point $[x_*,y_*]$ sufficient conditions for a strict local minimax are satisfied, the point $[x_*,y_*]$ becomes the point $[x_*,p_*]$, where the sufficient conditions for a strict local saddle of the function $B(x,p)$ are satisfied, and $F(x_*,y_*) = B(x_*,0)$, which allows us to use the well-known iterative numerical methods for finding saddle points in order to find the local minimax and maximin. We shall elucidate our point using two simple methods by way of example.

To find saddle points of the function $B(x,p)$, we use the gradient method (2.5). We compose a system of $n+m$ ordinary differential equations

$$\frac{dx}{dt} = -B_x(x,p), \qquad \frac{dp}{dt} = B_p(x,p).$$

Next, in this system we go from the variables x, p to the initial variables x, y. By doing this we transform the method for finding a saddle point into a method for finding a local minimax. We obtain

$$\dot{x} = -F_x - g_x F_y , \qquad \dot{y} = F_y + g_x^T x . \qquad (6.13)$$

Now we use another method. For the variable p we use the Newton method, and for x, the gradient method:

$$\dot{x} = -B_x(x,p) , \qquad \dot{p} = -B_{pp}^{-1}(x,p)\, B_p(x,p) .$$

In terms of the initial variables the method has the form

$$\dot{x} = -F_x - g_x F_y , \qquad \dot{y} = -F_{yy}^{-1} F_y + g_x^T x . \qquad (6.14)$$

The practical application of the methods obtained becomes difficult because the expression

$$g_x^T = -F_{yy}^{-1}(x,g(x))F_{yx}(x,g(x)) \qquad (6.15)$$

contains the function $g(x)$, which can be computed only by solving a minimization problem. However, simplified versions of the methods obtained are feasible: in (6.13) and (6.14) we can put

$$g_x^T = -F_{yy}^{-1}(x,y)F_{yx}(x,y) . \qquad (6.16)$$

In what follows we shall prove the convergence of these methods and, furthermore, justify the following methods for finding a local minimax:

$$\dot{x} = -F_x , \qquad \dot{y} = F_y + g_x^T x , \qquad (6.17)$$

$$\dot{x} = -F_x - g_x \dot{y} , \qquad \dot{y} = F_y , \qquad (6.18)$$

$$\dot{x} = -F_x - g_x F_y , \qquad \dot{y} = -F_{yy}^{-1} F_y . \qquad (6.19)$$

This is a discrete version of the method (6.17) exclusive:

$$x_{k+1}=x_k-\alpha F_x(x_k, y_k), \quad y_{k+1}=y_k+\alpha[F_y(x_k, y_k)+ \\ +F_{yy}^{-1}(x_k, y_k)F_{yx}(x_k, y_k)F_x(x_k, y_k)]. \tag{6.20}$$

According to the results of Section 1.5, the stationarity conditions (1.5.17) are necessary conditions for a local minimax in the problem (1.5.2). Applying the Newton method to solving the system (1.5.17), we obtain the following method for finding a local minimax:

$$\begin{aligned} x_{k+1}&=x_k-\Phi^{-1}(x_k, y_k)[F_x(x_k, y_k)- \\ &\quad -F_{xy}(x_k, y_k)F_{yy}^{-1}(x_k, y_k)F_y(x_k, y_k)], \\ y_{k+1}&=y_k-F_{yy}^{-1}(x_k, y_k)[F_y(x_k, y_k)- \\ &\quad -F_{yx}(x_k, y_k)(x_{k+1}-x_k)]. \end{aligned} \tag{6.21}$$

3. PROOF OF CONVERGENCE

Assuming that in all of the methods cited above g_x^T is calculated by the simplified formula (6.16), we state and prove the following theorem on convergence.

THEOREM 2.6.2. Let the sufficient conditions for a local minimax be satisfied at the point $z_* = [x_*, y_*]$ (see Theorem 1.5.7). Then

- 1. solutions of the systems (6.13), (6.14), (6.17) - (6.19) converge locally exponentially as $t \to \infty$ to the point z_*;
- 2. there exists $\bar{\alpha} > 0$ such that for any $0 < \alpha < \bar{\alpha}$ discrete versions of these systems of the form (6.20) converge locally linearly to the point z_*.

Proof. We use Theorem 2.1.1 on stability of the first-order approximation. Let

$$\delta x(t)=x(t)-x_*, \quad \delta y(t)=y(t)-y_*, \\ \delta z=[\delta x, \delta y]\in E^{n+m}.$$

Dropping second-order terms, we obtain a variational equation for the system (6.13):

$$\delta\dot{z} = B_1 \delta z \quad , \tag{6.22}$$

where B_1 denotes the square matrix of the dimension $(n+m)^2$, representable through four block-matrices:

$$B_1 = \left[\begin{array}{c|c} -\Phi(z_*) & 0_{nm} \\ \hline F_{yx}(z_*) + F_{yy}^{-1}(z_*) F_{yx}(z_*) \Phi(z_*) & F_{yy}(z_*) \end{array}\right].$$

Elements of all the matrices are calculated at the point z_*. The characteristic equation of the matrix B_1 has the form

$$|B_1 - \lambda I_{n+m}| = |-\Phi(z_*) - \lambda I_n| \cdot |F_{yy}(z_*) - \lambda I_m| = 0.$$

Therefore, the eigenvalues of the matrix B_1 consist of the eigenvalues of the matrix $F_{yy}(z_*)$ and $-\Phi(z_*)$. Both matrices are symmetric and negative definite. Therefore, all the eigenvalues of the matrix B_1 are real and strictly negative. Hence we conclude that the equilibrium point z_* for the system (6.13) is asymptotically stable.

The convergence of the other methods can be proved in a similar way. We denote by B_2, B_3, B_4, B_5 the matrices of the variational equations (6.22) for the methods (6.14), (6.17)-(6.19), respectively. It is easy to show that the characteristic equations are the following:

$$|B_2 - \lambda I_{n+m}| = \left|\begin{array}{c|c} -\Phi - \lambda I_n & 0_{nm} \\ \hline -F_{yy}^{-1} F_{yx} + F_{yy}^{-1} F_{yx} \Phi & -(1+\lambda) I_m \end{array}\right| =$$
$$= |-\Phi - \lambda I_n| \cdot |-(1+\lambda) I_m| = 0,$$

$$|B_3 - \lambda I_{n+m}| = \left|\begin{array}{c|c} -F_{xx} - \lambda I_n & -F_{xy} \\ \hline F_{yx} + F_{yy}^{-1} F_{yx} F_{xx} & F_{yy} + F_{yy}^{-1} F_{yx} F_{xy} - \lambda I_m \end{array}\right| = 0,$$

$$|B_4-\lambda I_{n+m}|=\left|\begin{array}{c|c}-\Phi-\lambda I_n & 0_{nm}\\ \hline F_{yx} & F_{yy}-\lambda I_m\end{array}\right|=$$
$$=|-\Phi-\lambda I_n|\cdot|F_{yy}-\lambda I_m|=0,$$
$$|B_5-\lambda I_{n+m}|=\left|\begin{array}{c|c}-\Phi-\lambda I_n & 0_{nm}\\ \hline -F_{yy}^{-1}F_{yx} & -(1+\lambda)I_m\end{array}\right|=$$
$$=|-\Phi-\lambda I_n|\cdot|-(1+\lambda)I_m|=0.$$

These relations imply that all roots of the characteristic equations for the matrices B_2, B_4, B_5 are real and negative. We multiply the upper row of the block matrix $B_3 - \lambda I_{n+m}$ on the left by $F_{yy}^{-1}F_{yx}$ and add the product to the lower row, the determinant of the matrix being unchanged in this case:

$$|B_3-\lambda I_{n+m}|=\left|\begin{array}{c|c}-F_{xx}-\lambda I_n & -F_{xy}\\ \hline \lambda g_x+F_{yx} & F_{yy}-\lambda I_m\end{array}\right|.$$

Next, we multiply the right column of the matrix obtained on the right by $F_{yy}^{-1}F_{yx}$ and subtract the product from the left column. Then

$$|B_3-\lambda I_{n+m}|=\left|\begin{array}{c|c}-\Phi-\lambda I_n & -F_{xy}\\ \hline 0_{mn} & F_{yy}-\lambda I_m\end{array}\right|=$$
$$=|-\Phi-\lambda I_n|\cdot|F_{yy}-\lambda I_m|=0.$$

From this we infer that the eigenvalues of the matrix B_3 are real, negative, and the method (6.17) converges exponentially to the stationary point z_*.

If α is sufficiently small, then by Theorem 2.3.7 the convergence of discrete version of the methods follows from the proofs given above of the convergence of continuous versions of the methods. ///

We note that one can weaken the conditions of the Theorem in the case of the methods (6.14) and (6.19) and require that

$F_{yy}(z_*)$ be nondegenerate instead of $F_{yy}(z_*)$ being negative definite. This property is, however, not essential; hence we shall not mention it in the sequel.

The convergence of the Newton method (6.21) follows from the general Theorem 2.5.3; according to the Remark we reformulate Theorem 2.5.3 as follows.

THEOREM 2.6.3. Let the function $F(z)$ be twice differentiable in the neighborhood of the stationary point z_*, and let the matrices $\Phi(z_*)$ and $F_{yy}(z_*)$ be nondegenerate and the matrix $F_{zz}(z)$ satisfy a Lipschitz condition in the neighborhood of Z_*. Then the iterations (6.21) converge to z_* locally, at a quadratic rate.

The methods described above can be used to obtain diverse modifications. For example, let $y = g(x)$ on the right-hand side of the first equation in (6.13). Then we obtain

$$\dot{x} = -F_x(x, g(x)) , \qquad \dot{y} = F_y(x,y) + g_x^T(x)\dot{x} . \tag{6.23}$$

In the methods (6.13), (6.14), (6.17) - (6.19) the derivative of the function $g(x)$ can be found from (6.15). When the conditions of Theorem 1.5.7 are satisfied, the methods will still converge to z_*. This follows from the fact that for the variational equations the eigenvalues are the same as those in the case where the formula (6.16) is used. Numerical realization of such methods is, however, much more cumbersome since for calculating the right-hand sides of systems of differential equations an additional problem of maximization of $F(x,y)$ in y must be solved.

The employment of the function $g(x)$ in (6.13), (6.14) does not improve the quality of the methods applied. In the method

(6.23), on the other hand, a property of global convergence develops, as well as in the following method:

$$\dot{x} = -F_x(x, g(x)), \qquad \dot{y} = g(x) - y. \tag{6.24}$$

<u>THEOREM 2.6.4</u>. Let $F(x,y)$ and $g(x)$ be everywhere continuously differentiable in all arguments. Let for any $x \in E^n$ the function $F(x,y)$ be strictly concave in y and the function $F(x, g(x))$ be strictly convex in x. Then the methods (6.23), (6.24) converge as $t \to \infty$ globally to the stationary point z_* which is the unique global solution of the problem (1.5.2).

<u>Proof</u>. We show that the stationary point z_* being the equilibrium point for both systems, is globally asymptotically stable. We define the positive definite functions

$$v_1(x, y) = \frac{1}{2}\|x - x_*\|^2 + \frac{1}{2}\|y - g(x)\|^2 \geqslant 0,$$
$$v_2(x, y) = \frac{1}{2}\|x - x_*\|^2 + F(x_*, y_*) - F(x_*, y) \geqslant 0,$$

which are zero at the stationary point $z = z_*$ only. Obviously, the function v_1 is infinitely large. The function v_2 is strictly convex in x and y, and, by Theorem 1.1.2, is also infinitely large.

Differentiating v_1 and v_2, using the systems (6.23) and (6.24), respectively, and using the convexity and concavity conditions, we obtain

$$\frac{dv_1}{dt} = \langle F_x(x, g(x)), x_* - x\rangle + \langle F_y(x, y), y - g(x)\rangle \leqslant$$
$$\leqslant F(x_*, y_*) - F(x, g(x)) + F(x, y) - F(x, g(x)) \leqslant 0,$$

$$\frac{dv_2}{dt} \leqslant F(x_*, y_*) - F(x, g(x)) + F(x_*, y) - F(x_*, g(x)) \leqslant 0,$$

$\dot{v}_1 = \dot{v}_2 = 0$ only for $x = x_*$, $y = y_* = g(x_*)$. Hence, by the Barbashin-Krassovskij method, the equilibrium point $z = z_*$ for the systems (6.23) and (6.24) are globally asymptotically stable; and therefore solutions of these systems will only converge to z_* for any initial Cauchy data.

To apply the last two methods it is necessary to determine values of the function $g(x)$. Multiple computation of $g(x)$ can be replaced by integration of a system of differential equations. Indeed, for any $y = g(x)$ a necessary condition for the extremum $F_y(x, g(x)) = 0$ must be satisfied. Differentiating this relation over x, we obtain that $y = g(x)$ is a solution of the following Cauchy problem:

$$\frac{dg}{dx} = -F_{yy}^{-1}(x, g(x)) F_{yx}(x, g(x)) ,$$

$$g(x_0) = y_0 . \tag{6.25}$$

These methods have two advantages: first, they are relatively simple and, second, they enable one to find a solution with a high accuracy. Their disadvantage, however, is the need to impose rigorous constraints on the function $F(x,y)$. If these methods are applied to arbitrary functions, the results obtained via these methods ought to be regarded only as preliminary. In order to assert that the solutions found thereby are global, it is necessary to study the problem more fully, or to make sequential calculations via global methods.

All of the methods given in this section can be used to find saddle points. Indeed, if at the point z_* sufficient conditions for the strict saddle are satisfied: $F_{yy}(z_*) < 0$, $F_{xx}(z_*) > 0$, then necessarily $\Phi(z_*) > 0$ at this point, and the conditions of Theorem 1.5.7 are a priori satisfied.

Next we give two methods intended specifically for finding saddle points:

$$\dot{x} = -F_x(x, g(x)), \qquad \dot{y} = F_y(d(y), y), \tag{6.26}$$

$$\dot{x} = d(y) - x, \qquad \dot{y} = g(x) - y. \tag{6.27}$$

Here $d(y) = \operatorname{Arg}\min_{x \in E^n} F(x,y)$.

Assume that for any $x \in E^n$, $y \in E^m$, $x \neq x_*$, $y \neq y_*$ we have the strict global saddle-point condition (1.5.16). To prove this, we construct the positive definite function

$$v(x,y) = F(x, g(x)) - F(d(y), y).$$

If the function F is strictly convex-concave, the functions $d(y)$ and $g(x)$ are differentiable. Computing the derivative of the function v by the systems (6.26) and (6.27), we obtain respectively

$$\begin{aligned}
&\dot{v} = -\|F_x(x, g(x))\|^2 - \|F_y(d(y), y)\|^2 \leqslant 0,\\
&\dot{v} = \langle F_x(x, g(x)), d(y) - x\rangle +\\
&+\langle F_y(d(y), y), y - g(x)\rangle \leqslant F(d(y), g(x)) - F(x, g(x)) +\\
&\qquad + F(d(y), y) - F(d(y), g(x)) \leqslant -v(x, y) \leqslant 0.
\end{aligned}$$

Therefore, the methods converge, and in the second method

$$v(x(t), y(t)) \leq v(x_0, y_0)e^{-t}.$$

4. METHODS FOR FINDING LOCAL SOLUTIONS IN x AND GLOBAL SOLUTIONS IN y

If in the problem (1.5.2) it becomes somehow possible to solve an interior problem for each $x \in E^n$, the problem of finding the minimax reduces to the problem of seeking the unconstrained minimum of the following function of many variables:

$$x_* \in \operatorname*{Arg\,min}_{x \in E^n} \phi(x) \quad . \tag{6.28}$$

The function $g(x)$ is determined either analytically or from the solution of the maximization problem, or from the integration of the system (6.25). In the case where the dimension of the vector y is small, it is possible to construct a grid on the set of admissible values of the vector y and find the values of the function $\phi(x)$ from the array of the values $F(x,y_i)$ on this grid.

Abstracting from solving an interior problem, we arrive at the problem (6.28), for the solution of which one has to construct a sequence of points x_k converging to x_*. First we consider the case where the function $\phi(x)$ is differentiable and its derivatives can be determined by the formulas (6.7) and (6.8). We denote by m_1, m_2, m_3, m_4 the smallest eigenvalues of the following symmetric matrices, respectively: $F_{yy}(z_*)$, $F_{xx}(z_*)$, $F_{xy}(z_*)F_{yx}(z_*)$, $\phi_{xx}(x_*)$.

<u>THEOREM 2.5.6</u>. Let the conditions of Theorem 1.5.7 be satisfied at the point $z_* = [x_*, y_*]$. Then we have the relation

$$m_4 \geq m_2 - \frac{m_3}{m_1} \geq m_2 \quad .$$

Proof. If the matrix $F_{yy}(z_*)$ is negative definite, we have for any nonzero vectors $y \in E^m$

$$m_1 \| y \|^2 \leqslant y^T F_{yy}(z_*) y < 0,$$
$$y^T F_{yy}^{-1}(z_*) y \leqslant \| y \|^2 / m_1 < 0.$$

Letting $y = F_{yx}(z_*)x$, we obtain

$$x^T \varphi_{xx}(z_*) x \geqslant m_2 \| x \|^2 - y^T F_{yy}^{-1}(z_*) y \geqslant \left(m_2 - \frac{m_3}{m_1}\right) \| x \|^2$$

implying that $m_4 \geq m_2 - \frac{m_3}{m_1}$. Taking into account that the smallest eigenvalue of the matrix $F_{xy}(z_*)F_{yx}(z_*)$ is nonnegative and that $m_1 < 0$, we arrive at the required inequality. ///

For many numerical minimization methods the convexity of the objective function $\phi(x)$ in the neighborhood of the point x_* is of importance. It follows from the inequality obtained that for $m_4 > 0$ the function $\phi(x)$ is convex. This property can hold even in those cases where the initial function $F(x,y_*)$ is not convex in the neighborhood of the point x_*. We shall use this property in Chapter 4 in solving nonlinear programming problems.

To solve (6.28), we apply the simplest methods of unconstrained minimization:

● 1. The gradient method with a constant step

$$x_{k+1} = x_k - \alpha F_x(x_k, g(x_k)) .$$

If the function $\phi(x)$ is differentiable, $m_4 > 0$, the method converges locally for any $0 < \alpha < \bar{\alpha}$. Here $\bar{\alpha} = \frac{2}{M}$, M being the largest eigenvalue of the matrix $\phi_{xx}(x_*)$;

• 2. The method of steepest descent

$$x_{k+1} = x_k - \alpha_k F_k(x_k, g(x_k)) \quad ,$$

where $\alpha_k = \operatorname{Arg}\min_{\alpha} \alpha(x_k - F(x_k, g(x_k)))$. Assuming that $\phi_x(x_k)$ is small and dropping third order quantities in α, we obtain

$$\varphi(x_k - \alpha\varphi_x(x_k)) = \\ = \varphi(x_k) - \alpha\|\varphi_x(x_k)\|^2 + \frac{\alpha^2}{2}\varphi_x^T(x_k)\varphi_{xx}(x_k)\varphi_x(x_k) \ ;$$

minimizing the right-hand side in α, we obtain

$$\alpha_* = \frac{\|\varphi_x(x_k)\|^2}{a_k}, \quad a_k = \varphi_{x_i}^T(x_k)\varphi_{xx}(x_k)\varphi_x(x_k).$$

This yields the estimate

$$\varphi(x_{k+1}) \leqslant \varphi(x_k) - \frac{\|\varphi_x(x_k)\|^4}{2a_k},$$

ensuring relaxation of the minimization process for a minimax definite function;

• 3. The Newton method

$$x_{k+1} = x_k - [\phi_{xx}(x_k)]^{-1}\ \phi_x(x_k) \quad .$$

If the matrix $\phi_{xx}(x)$ satisfies a Lipschitz condition and the conditions of Theorem 1.5.7 are satisfied, the Newton method converges locally at a quadratic rate;

• 4. The generalized gradient method (see the formula (4.5))

$$x_{k+1} = x_k - \alpha_k\gamma_k H_k \quad , \qquad H_k \in \partial\phi(x_k) \ .$$

To minimize $\phi(x)$, many other numerical methods of unconstrained minimization can be used.

In those cases when the interior problem has no unique solution, the function $\phi(x)$ is no longer differentiable, the method of steepest descent and the Newton method are not applicable. Then one has to use either general methods of nonsmooth optimization, or develop special versions for the minimization of maximum functions. The latter possibility has been discussed in V.F. Dem'yanov and V.N. Vasil'ev [1], [2].

Chapter 3

THE PENALTY FUNCTION METHOD

The penalty function method is one of the best-known numerical methods of nonlinear programming. The idea of the method is simple and quite universal; which explains why this method is widely used in solving various extremal problems. Many variations on the method have been suggested, and new ones still continue to pop up from time to time. A detailed bibliography of the non-Soviet literature can be found in A. Fiacco and G. McCormick [1].

The many modifications currently available notwithstanding, they still do not exhaust all the possible applications of this method; and more studies in this direction would be potentially fruitful. At the same time, it should be noted that practical computations for specific problems have revealed an essential disadvantage of the method: it is unsuited for solving problems where high accuracy is needed. The use of large values of the penalty coefficient leads to the minimization of ill-conditioned functions, which complicates the computations considerably. Many authors have justly criticized the method for this reason. One needs, however, to balance this with the strong points of the method. First, the domain of convergence is frequently of an essentially larger size than in other methods; second, the computational schemes for instrumenting the method are characterized by their extreme simplicity. For these reasons the penalty function method is indispensable for finding initial, approximate solutions. However, if the computations must provide a higher degree of accuracy of the solution, it is more appropriate to employ other methods of rapid convergence (for instance, the methods described in Chapter 4), using as an initial approximation the results of the penalty function method.

We start our presentation of the method with the more or less traditional version, and then go over to its modifications. As will be evident from our further discussion, the penalty function method can include also the cost-function parametrization method and its modifications, described in Section 3 of this chapter.

1. THE EXTERIOR PENALTY FUNCTION METHOD

1. THE GENERAL IDEA OF THE METHOD

We consider the nonlinear programming problem (1.6.1). We say that the function $S(x)$ defined on E^n is a penalty function if the following three conditions are satisfied:

- 1. the function $S(x)$ is everywhere continuous on E^n;
- 2. $S(x) = 0$ for any $x \in X$;
- 3. $S(x) > 0$ for any $x \notin X$.

In the case where the "feasible set" X is defined by the condition (1.6.2), the penalty function is usually constructed of the form

$$S(x) = \sum_{i=1}^{e} \varphi(g^i(x)) + \sum_{j=1}^{c} \varphi(h_+^j(x)). \qquad (1.1)$$

Here the continuous function $\phi(q)$ is such that $\phi(0) = 0$ and $\phi(q) > 0$ for all $q \neq 0$. Typical choices for $\phi(\cdot)$ are the functions

$$\varphi(q) = q^2, \quad \varphi(q) = q^4, \quad \varphi(q) = |q|, \quad \varphi(q) = e^{q^2} - 1.$$

We say that an additive penalty function of the form (1.1) is separable. A simple example of a non-separable function is given by

$$S^1(x) = \max_{i \in [1:e]} \max_{j \in [1:c]} [|g^i(x)|, h_+^j(x)]. \qquad (1.2)$$

We introduce now an auxiliary function:

$$P(x,\tau) = f(x) + \tau S(x) , \qquad (1.3)$$

where τ is a positive parameter referred to as the penalty coefficient.

The penalty function method is in essence the following. One chooses some monotonically increasing sequence $\tau_1 < \tau_2 < \tau_3 \cdots$ and solves the unconstrained minimization problem for the function $P(x,\tau_k)$ in x for $k = 1,2,3,\ldots$. One then obtains a sequence of the points $\{x_k\}$ satisfying the condition

$$x_k \in \operatorname{Arg} \min_{x \in E^n} P(x, \tau_k). \qquad (1.4)$$

If $\tau_k \to \infty$, under certain conditions each convergent subsequence of the sequence $\{x_k\}$ converges to a point of the set of solutions X_* of the problem (1.6.1). If for some finite value of τ_k

one obtains that $x_k \in X$, the initial problem (1.6.1) has been solved since in this case the point x_k belongs necessarily to the set of solutions X_*. Cases of this kind with special penalty functions used will be considered in Section 3.2. In the general case the sequence $\{x_k\}$ is infinite. A strict justification of the convergence method will follow from the proof, given in Subsection 1.3, of the convergence of the first, simplified version of the penalty function method.

The penalty function $S(x)$ introduced is nonzero everywhere outside the feasible set. Hence this function is usually called an exterior penalty function (exterior penalty), and the reduction of the problem (1.6.1) to the sequence of problems of the unconstrained minimization of auxiliary functions of the form (1.4) is called the exterior penalty function method or the exterior point method. Another type of penalties, called the interior penalties, will be described in Section 3.4 below.

Here is a simple example that illustrates the method described above. Let

$$f(x) = x, \qquad g(x) = x^2, \qquad e = 1, \qquad c = 0, \qquad n = 1 \; . \tag{1.5}$$

The solution of this problem is the point $x = 0$. We construct the penalty function as

$$P(x,\tau) = x + \tau[g(x)]^2 = x + \tau x^4 \;;$$

from the necessary minimum condition

$$P_x(x,\tau) = 1 + 4\tau x^3 = 0$$

we find the dependence $x(\tau) = -(4\tau)^{-1/3}$. This shows that $x(\tau) \to 0$ as $\tau \to \infty$, the method really leads to a solution of the problem.

2. COMPUTATIONAL ASPECTS

In the numerical implementation, the computing time goes mostly to finding the points x_k satisfying the condition (1.4). To solve the initial problem more accurately, one should increase the value of the penalty coefficient τ. However, the increase of τ leads to the situation that $P(x,\tau)$ as a function of x has the shape of a ridge since in the neighborhood of the boundary of the feasible set the function $\tau S(x)$ changes abruptly from zero (on the feasible set) to large values outside the feasible set. Any numerical minimization procedure for these functions is extremely complicated; hence it is preferable to increase slowly the penalty coefficient τ: for some τ_k find the point x_k, then, minimizing $P(x,\tau_{k+1})$, take $x = x_k$ as an initial point. The knowledge of a good initial approximation makes it easier to find the unconstrained minimum.

In practice it is also important that all the functions specifying the constraints be sufficiently well "fitted." The nonlinear programming problem (1.6.1) does not change character if in the conditions $g(x) = 0$, $h(x) \le 0$ some of the constraints are multiplied by an arbitrary large positive number. In numerical computations this operation means that these particular constraints are accounted for, while the remaining constraints are strongly violated. Hence, while using this method one should have the possibility of "scaling" the constraints by multiplication by suitable "weight" coefficients. In many cases, however, the increased "weight" of the constraints fails to reach the goal. For example, consider the problem where

$$f(x) = x^5, \qquad g(x) = x^2, \qquad e = 1, \qquad c = 0, \qquad n = 1 .$$

The point $x = 0$ is the solution of this problem. We use the simplest penalty function

$$P(x,\tau) = x^5 + \tau x^4 .$$

It is easy to see that for any τ

$$\inf_{x \in E^1} P(x,\tau) = -\infty .$$

Thus, in the auxiliary problem the minimum is unattainable for any penalty coefficients. In our example the cost function $f(x)$ on the unfeasible set decreases as $x \to -\infty$ considerably faster than the penalty function does. To make the method work, a more explicit penalty is needed. For example, one needs to let $P(x,\tau) = x^5 + \tau(e^{x^4} - 1)$ or $P(x,\tau) = x^5 + \tau x^8$; then the auxiliary problems will have solutions for $\tau > 0$, and the method ensures that the problem is solved.

As is evident from this example, to make the penalty function method work efficiently it may be necessary in some cases to make a preliminary analysis of the problem, and choose non-standard penalty functions. In the cases where the user (a skilled mathematician) has a program already available, it may be necessary to modify it. However, it is simpler to change the technique of specifying the functions in the initial problem. Assume for instance that in our last example the feasible set is given by the condition $g(x) = e^{x^2} - 1 = 0$ rather than by the condition $g(x) = x^2 = 0$. The feasible set does not change; and there is no need to make any changes in the program in use. At the same time, a more explicit dependence of $g(x)$ than that given in the initial problem, allows one to be sure that the auxiliary problem (1.4) has a solution.

If the functions defining the problem (1.6.1) are sufficiently smooth, it is then desirable that the function P possess the smoothness of the same degree as the former since in this case for their minimization one can use methods which have a high rate of convergence. On can assume, for example, that

$$P(x, \tau) = f(x) + \tau \left[\sum_{i=1}^{e} [g^i(x)]^2 + \sum_{j=1}^{c} [h_+^j(x)]^4 \right].$$

If the functions f, g, h are twice differentiable over x, the function $P(x,\tau)$ is also twice differentiable over x. A drawback of this function is that for large values of $|g^i|$ and h_+^j the nature of dependence of $P(x,\tau)$ on constraints of the equality type is different from that of the inequality type; to remove this difference one can construct a penalty function in a more artful way, letting, for example,

$$S(x) = \sum_{j=1}^{e} (g^i(x))^2 + \sum_{j=1}^{c} \psi(h^j(x)). \tag{1.6}$$

Here

$$\psi(y) = \begin{cases} 0 & \text{if} \quad y \le 0 \quad , \\ k_1 y^3 & \text{if} \quad 0 < y \le r \quad , \\ y^2 + k_2 y + k_3 & \text{if} \quad r < y \quad , \end{cases}$$

r being sufficiently small (usually $r = 10^{-4}$), k_1, k_2, k_3 being defined from the continuity conditions for the first and second derivatives of the functions ψ:

$$k_1 = \frac{1}{3} r , \qquad k_2 = -r , \qquad k_3 = \frac{r^2}{3} .$$

For large values of $|g^i|$, h^j_+ the dependence of S on $|g^i|$, h^j_+ is identical -- namely, quadratic. This function is widely used in numerical computations.

The penalty function method makes it possible to replace the initial problem of constrained minimization by that of solving a sequence of unconstrained minimization problems. This technique extends the domain for seeking the minimum in x, "removing" the constraints, which is quite convenient for finding local solutions. For finding global solutions, the method makes it necessary to find global minima of (1.4) on E^n, which is in general much more difficult than solving the problem (1.6.1). Indeed, if the set X is bounded, the initial nonlinear programming problem offers an opportunity to construct coverings of the set X, using some auxiliary sets (see Chapter 7), while using the penalty function method one needs to seek global minima in the entire space E^n, which complicates the whole process of seeking the extremum. Thus, the method of exterior penalties is effective only in finding local solutions, but has no potential application in finding global solutions.

For the nonlinear programming problem (1.6.29) the feasible set is the intersection of the sets X and U. When the penalty function method is used to solve problems of this kind, the function $S(x)$ "penalizes" as before for the violation of the condition $x \in X$; the auxiliary problem is to find

$$x(\tau) \in \operatorname{Arg} \min_{x \in U} [f(x) + \tau S(x)] \quad .$$

As usual, τ tends to go to infinity. Thus, the condition $x \in U$ is accounted for through the substitution of a minimization problem on the set U for the auxiliary problem of unconstrained minimization. The same generalization can be made for other versions of the penalty function method listed in this chapter.

3. THE FIRST SIMPLIFIED VERSION OF THE PENALTY FUNCTION METHOD

Intuitively it is seen that in numerical computations using the penalty function, auxiliary problems of unconstrained minimization can be solved approximately. It is then advisable to increase the accuracy of the computations as τ grows, as the points x_k approach the set X_*. This makes the implementation of the method somewhat simpler.

We consider now three nonnegative, continuous functions $\nu(t)$, $\mu(t)$, $\tau(t)$ of the scalar argument $t \in I_+$. We write the auxiliary function as

$$P(x,t) = \mu(t)f(x) + \tau(t)S(x) \quad . \tag{1.7}$$

The functions $\nu(t)$, $\mu(t)$, $\tau(t)$ for any $t \geq 0$ and $\delta > 0$ satisfy the following conditions:

$$\nu(t) \geqslant 0, \quad \mu(t) > 0, \quad \tau(t) > 0, \quad \lim_{t\to\infty} \frac{\nu(t)}{\mu(t)} = 0,$$
$$\lim_{t\to\infty} \frac{\tau(t)}{\mu(t)} = \infty, \quad \frac{\tau(t+\delta)}{\mu(t+\delta)} > \frac{\tau(t)}{\mu(t)}, \quad \frac{\mu(t+\delta)}{\nu(t+\delta)} > \frac{\mu(t)}{\nu(t)}. \tag{1.8}$$

Thus, the ratios $\tau(t)/\mu(t)$, $\mu(t)/\nu(t)$ monotonically tend to infinity as $t \to \infty$. We define the set

$$\Omega(t) = \{x(t)\colon\ P(x(t), t) \leqslant \min_{x \in E^n} P(x, t) + \nu(t)\},$$

which consists of the points furnishing the minimum of the auxiliary function $P(x,t)$ in x with an error $\nu(t)$ (with respect to the value of the function).

The first simplified version of the penalty function method is the following. A sequence $\{t_k\}$ is to be constructed whose elements satisfy the conditions

$$0 \leq t_0 < t_1 < \cdots < t_k \, , \qquad \lim_{k\to\infty} t_k = \infty \; .$$

For this sequence, the sequence of arbitrary points $x_k \in \Omega(t_k)$ is to be defined from which the convergent subsequences have to be culled. Under certain conditions, all the limit points belong to the set X_*.

We introduce the auxiliary set

$$B(t) = \{x \in E^n\colon\ P(x, t) \leqslant \mu(t) f(x_*) + \nu(t), \quad x_* \in X_*\}.$$

THEOREM 3.1.1. Let $f(x)$ and the penalty function $S(x)$ be continuous everywhere on E^n; let the set of solutions X_* of the problem (1.6.1) be non-empty; and let the set $B(0)$ be non-empty and bounded. Also, let the conditions (1.8) be satisfied for all $t \in I_+$. Then the first simplified version of the penalty function method converges to X_* globally.

Proof. We show that the sequence of sets $\{B(t)\}$ as $t \to \infty$ is contracting, i.e., we have the inclusion $B(t+\delta) \subset B(t)$ for any $\delta \geq 0$ and $t \in I_+$. Let $\tilde{x} \in B(t+\delta)$; then, using the condition (1.8), for any $x_* \in X_*$ we obtain

$$\mu(t) f(x_*) + \nu(t) = \mu(t)\left[f(x_*) + \frac{\nu(t)}{\mu(t)}\right] \geqslant$$
$$\geqslant \frac{\mu(t)}{\mu(t+\delta)}[\mu(t+\delta) f(x_*) + \nu(t+\delta)] \geqslant \frac{\mu(t)}{\mu(t+\delta)} P(\tilde{x}, t+\delta) \geqslant$$
$$\geqslant \mu(t) f(\tilde{x}) + \tau(t) S(\tilde{x}) = P(\tilde{x}, t).$$

Therefore, $\tilde{x} \in B(t)$. In particular, $B(t) \subset B(0)$ for all $t \geq 0$, the sequence of the sets $B(t)$ is bounded uniformly in all $t \in I_+$, we have the inclusion $X_* \subset B(t) \subset B(0)$, hence $B(t)$ for all $t \geq 0$ is non-empty and bounded, X_* also is bounded. The function $P(x,t)$ continuous in x attains its minimum on the compact set $B(t)$, where the quantity $p(t) = \min_{x \in B(t)} P(x,t)$ is defined, with the conditions

$$p(t) = \min_{x \in E^n} P(x, t) \leqslant \mu(t) f(x_*)$$

satisfied. Thus, the set $\Omega(t)$ also is non-empty and bounded. Because t is arbitrary for all $t \in I_+$, we have the inclusion $\Omega(t) \subset B(t)$.

By $\rho(t)$ we denote the distance between the sets $B(t)$ and X_*:

$$\rho(t) = \max_{x \in B(t)} \min_{y \in X_*} \|x - y\|. \tag{1.9}$$

The fact that the sequence of the sets $B(t)$ is contracting implies that $\rho(t)$ is a monotonically decreasing function of t; hence by Lemma 2.4.1, the limit $\lim_{t\to\infty} \rho(t) = a \geq 0$ exists. We prove that $a = 0$. For each $t \geq 0$ it is possible to define at least one point $\tilde{x}(t) \in B(t)$ at which the maximin condition is satisfied in (1.9). The set of all $B(t)$ for all possible values of t is bounded. Hence, from the bounded set of the points $\tilde{x}(t_k)$ we can extract a subsequence $\{x(t_{\bar{k}})\}$ converging to some point $\bar{x}$, corresponding to the monotonically increasing subsequence $\{t_{\bar{k}}\} \subset \{t_k\}$. In this case the conditions

$$\min_{y \in X_*} \|\bar{x} - y\| = a, \quad \lim_{t_{\bar{k}} \to \infty} \tilde{x}(t_{\bar{k}}) = \bar{x}, \quad \tilde{x}(t_k) \in B(t_k)$$

need to be satisfied. For all $t \geq 0$, $\tilde{x}(t) \in B(t)$ we have the inequality

$$f(\tilde{x}(t)) + \frac{\tau(t)}{\mu(t)} S(\tilde{x}(t)) \leqslant f(x_*) + \frac{\nu(t)}{\mu(t)}.$$

The ratio $\nu(t)/\mu(t)$ tends to zero as $t \to \infty$. Let $t_{\bar{k}}$ go to infinity. Then

$$f(\bar{x}) + d \leq f(x_*) \quad . \tag{1.10}$$

Here we have used

$$d = \lim_{t_{\bar{k}} \to \infty} \frac{\tau(t_{\bar{k}})}{\mu(t_{\bar{k}})} S(\tilde{x}(t_{\bar{k}})).$$

Let $d > 0$; then the fact that

$$\lim_{t_{\bar{k}} \to \infty} \tau(t_{\bar{k}})/\mu(t_{\bar{k}}) = \infty$$

yields $S(\bar{x}) = 0$. But the properties of the penalty function imply in this case that $\bar{x} \in X$; and from the inequality (1.10) we obtain that at the feasible point $\bar{x}$ the value of the objective function f is strictly smaller than the value of $f(x_*)$, where $x_* \in X_*$, which contradicts the definition of the set X_*; hence the case $d > 0$ is unfeasible. The quantity d cannot be negative since the functions $\tau(t)$, $\mu(t)$, $S(x(t))$ are nonnegative for $t \geq 0$, hence $d = 0$, $f(\bar{x}) = f(x_*)$, $\bar{x} \in X_*$. Therefore

$$\lim_{t_{\bar{k}} \to \infty} \rho(t_{\bar{k}}) = \min_{y \in X_*} \|\bar{x} - y\| = a = 0.$$

But if for the monotonically increasing function $\rho(t)$ the subsequence $\rho(t_{\bar{k}})$ has been found for which the limit equal to zero exists, then the subsequence $\rho(t_k)$ has the same limit. This implies that $\lim_{t_k \to \infty} B(t_k) = X_*$. For each $t \geq 0$ we have the inclusion $\Omega(t) \subset B(t)$, hence the convergence of $B(t)$ to X_* implies the convergence of $\Omega(t)$ to X_*. ///

4. THE SECOND SIMPLIFIED VERSION OF THE PENALTY FUNCTION METHOD

The calculations made by the formulas of Subsection 3.3.2 are simpler than those made by the initial scheme, but they have a very essential defect: it is frequently difficult to guarantee that the unconstrained minimum is found with the specified degree of accuracy. Hence the accuracy of solution of the problem (1.4) will be verified in a different way. For each fixed t_k the process of minimizing the auxiliary function $P(x,t_k)$ terminates as soon as some point x_k satisfying the condition

$$\|P_x(x_k, t_k)\| = \|\mu(t_k) f_x(x_k) + \tau(t_k) S_x(x_k)\| \leqslant \nu(t_k) \tag{1.11}$$

has been found. Here we need to assume that the functions $f(x)$ and $S(x)$ are differentiable, however, the condition (1.11) is easily verifiable. This technique has been suggested and justified, independently, by several authors: M. A. Kostina [1], R. Mifflin [1], Yu. G. Evtushenko [9]. Following the lines of [9], we state and prove a theorem on convergence.

We assume that the penalty function $S(x)$ is representable in the form (1.1), where the function $\phi(q)$ is continuously differentiable over $q \in E^1$ and such that

$$\begin{cases} \phi(q) = \phi'(q) = 0, & \text{if } q = 0, \\ \phi(q) > 0, \quad \phi'(q) > 0, & \text{if } q \neq 0. \end{cases} \tag{1.12}$$

If the function ϕ is substituted in (1.1), the conditions for the penalty function $S(x)$, as formulated in Subsection 3.3.1, will be satisfied. Let

$$\left.\begin{aligned} u_k^i &= \frac{\tau(t_k)\varphi'(g^i(x_k))}{\mu(t_k)}, & i &\in [1:e], \\ v_k^j &= \frac{\tau(t_k)\varphi'(h_+^j(x_k))}{\mu(t_k)}, & j &\in [1:c]. \end{aligned}\right\} \tag{1.13}$$

<u>THEOREM 3.1.2</u>. Let the functions $f(x)$ and $S(x)$ be continuously differentiable on E^n, let the penalty function $S(x)$ be separable, and let the conditions (1.8) and (1.12) hold. We

assume that the constraints $g(x) = 0$, $h(x) \leq 0$ satisfy the strengthened constraint qualification and that the sequence $\{x_k\}$ obtainable from (1.11) is bounded. Then the set of limit points of the sequence $\{x_k, u_k, v_k\}$ as $t_k \to \infty$ is not empty, each limit point is a Kuhn-Tucker point for the problem (1.6.1).

Proof. Using (1.13), we write the gradient of the auxiliary function as

$$P_x(x_k, t_k) = \mu(t_k)[f_x(x_k) + g_x(x_k)u_k + h_x(x_k)v_k],$$

$v_k \geq 0$ for all k.

The sequence $\{x_k\}$ is bounded, hence we can extract a subsequence $\{x_{\bar{k}}\}$ converging to $\bar{x}$. We show that the point $\bar{x}$ is feasible. Assume the opposite: $\bar{x} \notin X$. We divide both sides of the inequality (1.11) by $\tau(t_k)$ and, letting $t_k \to \infty$, we obtain that

$$\sum_{i=1}^{e} \varphi'(g^i(\bar{x}))\, g_x^i(\bar{x}) + \sum_{j=1}^{c} \varphi'(h_+^j(\bar{x}))\, h_x^j(\bar{x}) = 0. \tag{1.14}$$

Now we take the dot-product by vector z from the definition 1.7.5 of the strengthened constraint qualification. Simple transformations give us

$$\sum_{i=1}^{e} \varphi'(g^i(\bar{x}))\, g^i(\bar{x}) + \sum_{j \in \Delta(\bar{x})} \varphi'(h_+^j(\bar{x}))\, h^j(\bar{x}) \leqslant 0 \ , \tag{1.15}$$

where we have introduced the index set

$$\Delta(x) = \{j \in [1:c]:\ h^j(x) \geqslant 0\}.$$

At the same time the conditions (1.12) imply that the function

$\phi(q)$ is convex and $\phi(0) = 0$, hence

$$\varphi'(g^i(\bar{x}))\,g^i(\bar{x}) \geqslant 0, \quad \varphi'(h_+^j(\bar{x}))\,h^j(\bar{x}) \geqslant 0$$

for all $i \in [1:e]$, $j \in \Delta(\bar{x})$. Comparing these inequalities with (1.15), we come to the conclusion that for the i and j indicated above the equalities

$$\varphi'(g^i(\bar{x}))\,g^i(\bar{x}) = \varphi'(h_+^j(\bar{x}))\,h^j(\bar{x}) = 0 \tag{1.16}$$

are satisfied. It follows from the condition $\bar{x} \notin X$ that we can find at least either one i such that $g^i(x) \neq 0$, or $j \in \Delta(\bar{x})$ such that $h^j(\bar{x}) > 0$. But then, by (1.12), either $\phi'(g^i(\bar{x})) \times g^i(\bar{x}) > 0$ or $\phi'(h_+^j(\bar{x}))h^j(\bar{x}) > 0$, which contradicts (1.16). Hence $\bar{x} \in X$.

Next we prove the boundedness of the sequence $\{u_k, v_k\}$. Assume the opposite and let

$$w_k = \sum_{i=1}^{e} |u_k^i| + \sum_{j=1}^{c} v_k^j, \quad \bar{w} = \overline{\lim_{t_k \to \infty}}\, w_k.$$

We divide both sides of the inequality (1.11) by w_k and pass to the limit as $t_k \to \infty$. Then we can find α_i, β_j ($i \in [1:e]$, $j \in \sigma(\bar{x})$), of which at least one is not zero, and such that

$$\sum_{i=1}^{e} \alpha_i g_x^i(\bar{x}) + \sum_{j \in \sigma(\bar{x})} \beta_j h_x^j(\bar{x}) = 0.$$

This contradicts the linear independence of the vectors $g_x^i(\bar{x})$, $h_x^j(\bar{x})$, $i \in [1:e]$, $j \in \sigma(\bar{x})$. Hence all the limit points of the sequence $\{x_k, u_k, v_k\}$ are finite. If for some $j \in [1:e]$

$$\lim_{t_k \to \infty} h^j(x_k) = h^j(\bar{x}) < 0,$$

then, by (1.13), we have $\lim_{t_k \to \infty} v_k^j = 0$. Hence at the limit points we have the complementarity condition, and each limit point of the sequence $\{x_k, u_k, v_k\}$ is a Kunn-Tucker point. ///

Let (1.6.1) be a convex programming problem. Then it follows from Theorem 1.1.7 that the auxiliary function $P(x,t)$ is convex in x and, by Theorem 1.2.6, we have the inequality

$$P(x, t) - P(x_*, t) \leqslant \langle P_x(x, t), x - x_* \rangle.$$

Using the Cauchy-Bunyakowski inequality and noting (1.11), we obtain

$$P(x_k, t_k) - \mu(t_k) f(x_*) \leqslant \nu(t_k) \| x_k - x_* \|.$$

Because $\{x_k\}$ is bounded, it then follows that as $t_k \to \infty$ the function

$$\frac{\tau(t_k)}{\mu(t_k)} \left[\sum_{i=1}^{e} \varphi(g^i(x_k)) + \sum_{j=1}^{c} \varphi(h_+^j(x_k)) \right]$$

is bounded as well. But the fact that $\frac{\tau(t)}{\mu(t)} \to \infty$ implies that the quantity in the square brackets tends to zero. Hence $\bar{x} \in X$. Thus, for convex programming problems, in order to justify the convergence of the method there is no need to invoke the strengthened constraint qualification -- it suffices to introduce Slater's or Karlin's constraint qualification and require the vectors $g_x^i(x)$, $h_x^j(x)$ be linearly independent, $i \in [1:e]$, $j \in \sigma(x)$, $x \in X$.

5. THE THIRD MODIFICATION OF THE PENALTY FUNCTION METHOD

The two modifications given above simplify to a certain degree the computations; however, for the problems (1.6.1) in which the vector x has large dimension, repeated solving of the auxiliary problem (1.4) presents considerable difficulties. It is easier to implement the method if one gives up solving the problem (1.4) and, instead, seeks the limit points of the solutions of the following Cauchy problem:

$$\frac{dx}{dt} = -P_x(x, t), \quad x(0) = x_0, \tag{1.17}$$

where $P(x,t)$ is defined by the formula (1.7). This version of the method is similar to the Cauchy method (2.2.3): the difference is that now $\mu(t)$, $\tau(t)$ depend on t, so that the system (1.17) is non-autonomous. To justify the convergence of the method we introduce the set

$$B(t) = \{x \in E^n:\ P(x, t) \leq \mu(t) f(x_*)\}.$$

We assume that the functions $\mu(t)$, $\tau(t)$ are continuous and satisfy for $t \in I_+$ the conditions

$$\mu(t) > 0, \quad \tau(t) > 0, \quad \int_0^\infty \mu(t)\, dt = \infty. \tag{1.18}$$

<u>THEOREM 3.1.3</u>. Let $f(x)$ and $S(x)$ be convex functions continuously differentiable everywhere on E^n, let the set X_* be nonempty, let the set $B(0)$ be nonempty but bounded. Also, let the continuous functions $\mu(t)$, $\tau(t)$ for all $t \in I_+$ satisfy the condition (1.18), and let the ratio $\frac{\tau(t)}{\mu(t)} \to \infty$ be monotonic

as $t \to \infty$. Then the method (1.17) converges to X_* globally.

Proof. In the same way as in proving Theorem 3.1.1, we can show here that for any $t \in I_+$, $\delta \geq 0$ we have the inclusions $X_* \subset B(t+\delta) \subset B(t)$. Hence X_* is compact. We introduce the function

$$w(x) = [\text{dis}(x, X_*)]^2 = \min_{y \in X_*} \|x - y\|^2.$$

Next, we differentiate the function $w(x(t))$, using the system (1.17), and also exploit the convexity of $P(x,t)$ in x; we thus obtain that for any $x(x_0,t) \notin B(0)$ the inequalities

$$\frac{1}{2}\frac{dw(x)}{dt} \leqslant \mu(t) f(x_*) - P(x(x_0, t), t) \leqslant 0$$

are satisfied. The function $w(x)$ is bounded everywhere on $B(0)$. The inequality obtained implies that $w(x(x_0,t))$ is a nonincreasing function of t everywhere outside $B(0)$. Arguing in the same way as in proving Theorem 1.1.2, we can show that $w(x)$ is an infinitely large function. Taking into account Theorem 2.2.1 and Remark thereto, we conclude that the system (1.17) is Lagrange stable, i.e., each solution of this system is extendable (to the right) as $t \to \infty$ and bounded.

Now we use Theorem 2.2.7. Let

$$G_\varepsilon = \{x \in E^n : \text{dis}(x, X_*) < \varepsilon\},$$

$$\Phi(t) = \inf_{x \in E^n \setminus G_\varepsilon} \{P(x, t)/\mu(t) - f(x_*)\}.$$

Then for an arbitrary $\varepsilon > 0$ we have the estimate

$$\sup_{x \in E^n \setminus G_\varepsilon} \dot{w}(x) \leq -2\mu(t)\Phi(t) .$$

It is easy to show that for any $\varepsilon > 0$ there exists $t(\varepsilon)$ such that $B(t) \subset G_\varepsilon$ for all $t > t(\varepsilon)$, whence $\Phi(t) > 0$ for the same values of t. Since $\Phi(t)$ is an increasing function of t, for all $t > t(\varepsilon)$ we have

$$-\mu(t)\Phi(t) \leq -\mu(t)\Phi(t(\varepsilon)) < 0 ,$$

$$-\lim_{t\to\infty} \int_{t(\varepsilon)}^{t} \mu(t)\Phi(t)dt \leq -\lim_{t\to\infty} \Phi(t(\varepsilon)) \int_{t(\varepsilon)}^{t} \mu(t)dt = -\infty .$$

All of the conditions of Theorem 2.2.7 are satisfied, implying thereby that the method (1.17) converges to the set X_*. ///

Numerical implementation of the method (1.17) causes difficulties due to the choice of the functions $\mu(t)$, $\tau(t)$. Hence it is not recommendable to use this method for simple problems involving small dimensionality, which are solvable through the penalty function method invoking no simplifying techniques.

2. ESTIMATION OF ACCURACY

1. EXACT PENALTIES

There is a whole class of penalty functions having the remarkable property: for each fixed, sufficiently large value of the coefficient τ, the set X_* of solutions of the problem (1.6.1) coincides with the set of solutions of the auxiliary problem (1.4). We say that these functions are exact. The utilization of exact penalty functions permits us to reduce the constrained minimization problem (1.6.1) to the unconstrained minimization problem (1.4).

I. I. Eremin was the first who noticed in 1966 this property for convex programming problems [1], [2]; then W. Zangwill [1], V.D. Skarin [1], T. Pietrzykowski [1], C. Charalambous [1], S. Han and O. Mangasarian [1], and many other authors studied this question.

Let us introduce the auxiliary vector functions:

$$F(x)=[|g^1(x)|, \ \dots, \ |g^e(x)|, \ h_+^1(x), \ \dots, \ h_+^c(x)],$$
$$\Phi(x)=[g^1(x), \ \dots, \ g^e(x), \ h_+^1(x), \ \dots, \ h_+^c(x)],$$

mapping R^n into R^m, and combine the dual vectors, letting $y = [u,v] \in R^m$. If the problem (1.6.1) has a solution $x = x_*$ and the corresponding dual vectors $u_* \in R^e$, $v_* \in R_+^c$, we write the combination as $y_* = [u_*,v_*] \in R^m$.

We shall show in the sequel that the Hölder norm of the vector $F(x)$ is an exact penalty function (see Appendix II). The auxiliary function then has the form

$$P(x,\tau) = f(x) + \tau\|F(x)\|_p \ , \tag{2.1}$$

where $1 \le p \le \infty$. We consider the set

$$Z(\tau)=\operatorname*{Arg\,min}_{x\in R^n} P(x, \tau)$$

depending on τ as on a parameter. Let

$$\tau_* = \|y_*\|_q \ , \tag{2.2}$$

here $\|y_*\|_q$ denotes the norm of the vector y_*, which is dual to the norm $\|F(x)\|_p$ from (2.1).

<u>THEOREM 3.2.1</u>. In the problem (1.6.1) let there exist a saddle

point $[x_*,u_*,v_*]$ of the Lagrangian $L(x,u,v)$. Then for any values $\tau > \tau_*$ the sets X_* and $Z(\tau)$ coincide.

Proof. By Theorem 1.6.1 the point x_* belongs to the set of solutions X_* of the problem (1.6.1); at the point $[x_*,v_*]$ the complementarity condition is satisfied. Hence for any $x \in R^n$

$$P(x_*, \tau) = f(x_*) = L(x_*, u_*, v_*) \leqslant L(x, u_*, v_*).$$

The condition $v_* \geq 0$ implies the inequality $\langle h(x),v_*\rangle \leq \langle h_+(x),v_*\rangle$, using which we can find

$$P(x_*, \tau) \leqslant f(x) + \langle g(x), u_*\rangle + \langle h_+(x), v_*\rangle = \\ = f(x) + \langle y_*, \Phi(x)\rangle.$$

Using the Hölder inequality (see Appendix II), we obtain

$$P(x_*, \tau) \leqslant f(x) + \|y_*\|_q \, \|\Phi(x)\|_p = f(x) + \|y_*\|_q \, \|F(x)\|_p.$$

Noting that $\tau_* < \tau$ and taking into account the definition (2.1), we extend these estimates:

$$P(x_*,\tau) \leq P(x,\tau) \quad . \tag{2.3}$$

This inequality holds for any $x \in R^n$, $\tau \geq \tau_*$; hence the set $Z(\tau)$ is not empty and $X_* \subset Z(\tau)$. We prove that the sets X_* and $Z(\tau)$ coincide. We assume the opposite, that is, for some $\tau_1 > \tau_*$ we can find a point $x_1 \in Z(\tau_1)$ such that $x_1 \notin X_*$. Then

$$f(x_*) = P(x_*, \tau_1) = P(x_1, \tau_1) = f(x_1) + \tau_1 \|F(x_1)\|_p.$$

If $x_1 \in X$, then $x_1 \in X_*$ since in this case $F(x_1) = 0$ and $f(x_*) = f(x_1)$. If $x_1 \notin X$, then $\|F(x_1)\|_p \neq 0$ and for any

$\tau \in (\tau_*, \tau_1)$ we have the strict inequality

$$P(x_*, \tau) = f(x_*) > f(x_1) + \tau \|F(x_1)\|_p = P(x_1, \tau),$$

which contradicts the inequality (2.3). This implies that $x_1 \in X_*$, hence $Z(\tau) \subset X_*$. Noting the inclusion found before, $X_* \subset Z(\tau)$, we conclude that the sets X_* and $Z(\tau)$ coincide for all $\tau > \tau_*$. ///

By Theorem 1.6.7, the condition in Theorem 3.2.1 for the global saddle point of the Lagrangian to exist can be replaced in convex programming problems by the conditions for the problem (1.6.1) to have a solution, as well as by Slater's constraint qualification. In the general problem of nonlinear programming, the condition for the saddle point to exist in the problem (1.6.1) can be replaced by McCormick's sufficient minimum conditions. We have the following theorem.

<u>THEOREM 3.2.2</u>. Let the conditions of McCormick's Theorem 1.7.2 be satisfied at the Kunn-Tucker point $[x_*, u_*, v_*]$ for the problem (1.6.1). Then x_* is the local minimum point of the function $P(x,\tau)$ for any fixed $\tau > \tau_*$, where τ_* is defined from (2.2).

<u>Proof</u>. The function L_1 defined from (1.7.15) can be expressed in terms of

$$L_1(x,y) = f(x) + \langle y, F(x) \rangle .$$

For any $\tau > \tau_*$ one can find a vector $y = [u,v]$ such that $v > v_*$, $u^i > |u_*^i|$, for all $i \in [1:e]$ and

$$\tau = \|y\|_q > \|y_*\|_q = \tau_* . \tag{2.4}$$

By Lemma 1.7.5 there exists a neighborhood $G(x_*)$ of the point x_*, such that for all $x \in G(x_*)$, $x \neq x_*$ the conditions

$$P(x_*,\tau) = f(x_*) = L_1(x_*,u,v) < L_1(x,u,v) \tag{2.5}$$

are satisfied.

Using the Hölder conditions (2.4), (2.5), we obtain

$$P(x_*,\tau) \leq f(x_*) + \langle y,F(x)\rangle \leq f(x) + \|y\|_q \|F(x)\|_p = P(x,\tau).$$

We have arrived again at the inequality (2.3). The further considerations will be the same as those in proving Theorem 3.2.1. ///

Even when the functions defining the problem are everywhere differentiable, the function (2.1) is not differentiable at boundary points of the feasible set X. Hence such auxiliary functions are referred to as non-differentiable; the use of them simplifies the numerical implementation of the penalty function method since there is no more need to let the penalty coefficient go to infinity -- instead, it suffices to solve the problem (1.4) only once for $\tau > \tau_*$. But the quantity τ_* is usually unknown, and hence the problem (1.4) has to be solved for several values of τ. At the same time, the use of exact penalties makes the auxiliary problem (1.4) more complex, since in that case one needs to apply for an unconstrained minimization only slowly converging numerical methods which involve no differentiability of the auxiliary function. Hence exact penalty functions are usually applied in solving the problem in which the functions defining the problem are non-differentiable.

These results are easily transferrable to the more general

case where the auxiliary function has the form

$$P(x,\tau) = f(x) + B(x,\tau) \quad , \tag{2.6}$$

where $B(x,\tau)$ is a continuous function of both arguments, satisfying the three conditions:

- 1. $B(x,\tau) \geq \tau||F(x)||_p$ for all $x \in R^n$, $\tau > \tau_*$;
- 2. $B(x,\tau) = 0$ for all $x \in X$;
- 3. $B(x,\tau)$ is strictly increasing in τ for all $x \notin X$ and $\tau > \tau_*$.

The following functions, for example, satisfy these conditions:

$$B(x, \tau) = e^{\tau \| F(x) \|_p} - 1,$$
$$B(x, \tau) = \tau \| F(x) \|_p (1 + a \| F(x) \|_p), \ a \geqslant 0.$$

The assertions of Theorems 3.2.1 and 3.2.2 remain valid if the function $P(x,\tau)$ has the form (2.6).

Next we consider some particular cases. We assume that $B(x,\tau) = \tau||F(x)||_p$. If $p = 1$, then $q = \infty$, and the auxiliary function and the quantity τ_* are defined by the formulas

$$P_1(x, \tau) = f(x) + \tau S(x), \quad S(x) = \sum_{i=1}^{e} |g^i(x)| + \sum_{j=1}^{c} h^j_+(x),$$
$$\tau_{*1} = \max_{i \in [1:e]} \max_{j \in [1:c]} [|u^i_*|, v^j_*] = \|y_*\|_\infty. \tag{2.7}$$

We considered minimization of $P_1(x,\tau)$ (see the formula (1.6.14)), starting from different considerations.

The initial and dual norms coincide in the case of Euclidean norms when $p = 1 = 2$:

$$P_2(x, \tau) = f(x) + \tau \left[\sum_{i=1}^{e} [g^i(x)]^2 + \sum_{j=1}^{c} [h^j_+(x)]^2 \right]^{1/2},$$

$$\tau_{*2} = \left[\sum_{i=1}^{e} (u^i_*)^2 + \sum_{j=1}^{c} (v^j_*)^2 \right]^{1/2} = \| y_* \|_2 .$$

If we put $p = \infty$, then $q = 1$,

$$P_3(x, \tau) = f(x) + \tau S^1(x),$$

$$\tau_{*3} = \sum_{i=1}^{e} | u^i_* | + \sum_{j=1}^{c} v^j_* = \| y_* \|_1 ,$$

where the non-separable penalty function (1.2) is used.

Using the relation between distinct norms, given in Appendix II, we obtain that $\tau_{*1} \le \tau_{*2} \le \tau_{*3}$.

Therefore, among the three auxiliary functions above the function $P_1(x,\tau)$ has the smallest minimal penalty coefficient, the function P_3, whose penalty function is a Chebyshev norm of the vector $F(x)$, has the largest value of τ_*.

2. ESTIMATION OF ACCURACY

From Theorem 3.2.1 we can obtain an important result enabling us to estimate the errors arising in using the penalty function method.

THEOREM 3.2.3. In the problem (1.6.1) let there be a saddle point $[x_*, u_*, v_*]$ of the Lagrangian $L(x,u,v)$. Then for any $\tau > 0$, $p > 1$, $1/p + 1/q = 1$ we have the estimate

$$f(x_*) \le f(x) + \tau \| F(x) \|_p^2 + \frac{\| y_* \|_q^2}{4\tau} .$$

Proof. This estimate is well known in the literature for the case $p = q = 2$. It is proved, however, only for the case of convex programming, using some cumbersome computations involving convexity and differentiability. The result formulated in this theorem is of a more general kind and its justification is extremely elementary. Indeed, in proving Theorem 3.2.1 we obtained the inequality

$$f(x_*) \leq f(x) + ||F(x)||_p ||y_*||_q \, .$$

Using the well-known inequality $2ab \leq a^2 + b^2$, with

$$a\sqrt{2\tau} = ||y_*||_q \, , \qquad b = \sqrt{2\tau}\, ||F(x)||_p \quad ,$$

we immediately arrive at the required inequality. ///

The technique of our proof can be developed further. We use Young-Minkowski inequalities:

$$\langle a,b\rangle \leq \phi(a) + \phi^*(b) \; , \qquad \langle a,b\rangle \leq \psi(a)\, \psi^0(b) \quad .$$

Here $a,b \in R^s$, $\phi^*(b)$ is the conjugate of $\phi(a)$, $\psi^0(b)$ is the polar of the function $\psi(a)$:

$$\phi^*(b) = \sup_{a \in A} [\langle a,b\rangle - \phi(a)] \quad ,$$

$$\psi^0(b) = \inf \{\mu : \langle a,b\rangle \leq \mu\psi(a) \quad \forall\, a \in A\} \quad ,$$

where the set $A \in R^s$. The functions $\phi^*(b)$ and $\psi^0(b)$ are defined to be the "best" functions satisfying the above inequalities; we cannot replace $\phi^*(b)$ and $\psi^0(b)$ by any other functions of b, to strengthen these inequalities. For $p > 1$ these inequalities yield the following estimates:

$$\langle a,b\rangle \le \frac{1}{p}\|a\|_p^p + \frac{1}{q}\|b\|_q^q \quad ,$$

$$\langle a,b\rangle \le \|a\|_p \|b\|_q \quad ,$$

$$\langle a,b\rangle \le \sum_{i=1}^{s} [e^{a^i} + b^i(\ln b^i - 1)] \quad .$$

The second relation is a Hölder inequality. For $\alpha < 1$, $\alpha \neq 0$, $b > 0$, $a < 0$ we have

$$\langle a,b\rangle \le -\frac{1}{\alpha}\|a\|_\alpha^\alpha - \frac{1}{\beta}\|b\|_\beta^\beta \quad ,$$

$$\langle a,b\rangle \le -\|a\|_\alpha \|b\|_\beta \quad , \qquad \frac{1}{\alpha} + \frac{1}{\beta} = 1 \quad .$$

Here we need to explain in more detail the use of the sign of the norm for $\alpha < 0$. The first condition for the norm, $\|0\|_\alpha = 0$ is not satisfied herein. Hence the $\|a\|$ for $a \in R^S$ should be treated in the formal way, starting from the following definition:

$$\|a\|_\alpha = \left| \sum_{i=1}^{s} |a^i|^\alpha \right|^{1/\alpha} \quad .$$

Using the inequalities given above, we obtain

$$f(x_*) - \frac{1}{q}\left\|\frac{y^*}{\tau}\right\|_q^q \le f(x) + \frac{1}{p}\|\tau F(x)\|_p^p \quad .$$

If the problem involves no equality-type constraints, then we have

$$f(x_*) + \frac{1}{\tau}\sum_{i=1}^{c} v_*^i[1 - \ln v_*^i] \le f(x) + \frac{1}{\tau}\sum_{i=1}^{c} e^{\tau h^i(x)} = P(x,\tau) \quad ,$$

where the function $P(x,\tau)$ does not satisfy the conditions imposed on auxiliary functions of the form (1.3). For this function, the equality $f(x) = P(x,\tau)$ is not satisfied on the feasible set, but only in the limit as $\tau \to \infty$. Nevertheless, this class of "nearly penalty functions" can be successfully employed in solving nonlinear programming problems.

We may add Lagrangian to the auxiliary function. For example, let

$$Z(y,\tau) = \operatorname{Arg}\min_{x \in R^n} P(x,y,\tau) , \qquad y = [u,v] ,$$

$$P(x,y,\tau) = f(x) + \langle u,g(x)\rangle + \langle v,h_+(x)\rangle + \tau \|F(x)\|_p .$$

Suppose in the problem (1.6.1) there exists a saddle point $[x_*,u_*,v_*]$ of the Lagrangian and also $\tau > \|y - y_*\|_q$. It is easy to show that then $x_* \in X_*$, $f(x_*) \le P(x,y_*,\tau)$ for any x and, therefore, $X_* = Z(y,\tau)$. Thus, the use of Lagrangian multipliers enables us to create a new class of exact functions. From the computational point of view this result is quite useful because it demonstrates that even an approximate value of the dual vectors u and v allows us to decrease the minimal penalty coefficient (2.2). This result implies as a particular case the assertion of Theorem 3.2.1. Indeed, letting $y = 0$, we arrive at the function (2.1). If in the formula for P we take $\langle v,h(x)\rangle$ instead of $\langle v,h_+(x)\rangle$, to satisfy the given properties we need to add the condition $v_* \ge v$.

Similar estimates can be obtained for distinct combined penalties when the scalar products $\langle u_*,g(x)\rangle$ and $\langle v_*,h(x)\rangle$, are estimated through different formulas. The concepts of the conjugate and the polar functions are quite useful in studying nonlinear programming problems. For example, crucial estimates can be obtained in a simple way, without assuming the convexity or differentiability of the functions defining the problem. Further elaboration of this approach will be found in Section 3.4. Let

$$P(x,\tau) = f(x) + \tau\|F(x)\|_2^2 \tag{2.8}$$

be the auxiliary function. Here and throughout the next section we use the Euclidean norm. Then

$$P(x, \tau) \geqslant f(x_*) - \frac{1}{4\tau}\left[\sum_{i=1}^{e}(u_*^i)^2 + \sum_{j=1}^{c}(v_*^j)^2\right]. \tag{2.9}$$

This implies that $Z(\tau) \to X_*$ as $\tau \to \infty$, in other words, the error in defining the value of the function being minimized when the penalty function method is used tends to zero in proportion to the quantity $1/\tau$.

3. DIFFERENTIABLE PENALTIES

We consider the problem (1.6.1) where the feasible set is defined only via constraints of the equality type. We assume that the penalty coefficient is sufficiently large and $\varepsilon = 1/\tau$ can be regarded as a small parameter. Let the second simplified version (1.11) be implemented in which the point $x(\tau)$ satisfies the condition

$$P(x, \tau) \geqslant f(x_*) + \langle f_x(x_*),\ x - x_*\rangle + \tau\|F(x)\|^2.$$

Introducing the auxiliary vector $u \in E^e$ and the vector

$a \in E^n$, $\|a\| \leq 1$, we rewrite this condition in the form of a system of $n+e$ nonlinear equations

$$\begin{aligned} f_x(x) + g_x(x)\,u &= \varepsilon a, \\ 2g(x) - \varepsilon u &= 0. \end{aligned} \tag{2.10}$$

Now we investigate the dependence of x, u on the small parameter $\varepsilon > 0$, generated by this system. Assume that in the initial problem (1.6.1) there exists a Kuhn-Tucker point $[x_*, u_*]$ at which the sufficient minimum conditions given by Theorem 1.7.2 are satisfied, and also the constraint qualification holds at the point x_*.

We shall seek the solution to (2.10) for $\varepsilon > 0$, using the formal power series expansion:

$$\begin{aligned} x(\varepsilon) &= x_* + \varepsilon x_1 + \varepsilon^2 x_2 + \dots \\ u(\varepsilon) &= u_* + \varepsilon u_1 + \varepsilon^2 u_2 + \dots \end{aligned} \tag{2.11}$$

We substitute these series into (2.10), assuming that the functions f and g are differentiable as many times as required. We expand f and g in a series of powers of ε. Equating the expressions for equal powers of ε, we obtain a system of linear equations for defining the coefficients in (2.11). In the zero approximation we have

$$L_x(x_*, u_*) = f_x(x_*) + g_x(x_*)\,u_* = 0, \quad g(x_*) = 0, \tag{2.12}$$

that is, we obtain the Kuhn-Tucker conditions. For the first power terms we obtain

$$\begin{aligned} L_{xx}(x_*, u_*)\,x_1 + g_x(x_*)\,u_1 &= a, \\ 2g_x^T(x_*)\,x_1 &= u_* \end{aligned} \tag{2.13}$$

from which we find

$$
\begin{aligned}
x_* + \varepsilon x_1 &= x_* + \varepsilon L_{xx}^{-1}(x_*, u_*)[a - g_x(x_*)u_1], \\
u_* + \varepsilon u_1 &= u_* + \varepsilon [2g_x^T(x_*)L_{xx}^{-1}(x_*, u_*)g_x(x_*)]^{-1} \times \\
&\qquad \times [2g_x^T(x_*)L_{xx}^{-1}(x_*, u_*)a - u_*].
\end{aligned}
\tag{2.14}
$$

Let us investigate this approximation. We compute the values of f and g:

$$
\begin{aligned}
f(x_* + \varepsilon x_1) &= f(x_*) + \varepsilon f_x^T(x_*)x_1 + O(\varepsilon^2), \\
g(x_* + \varepsilon x_1) &= g(x_*) + \varepsilon g_x^T(x_*)x_1 + O(\varepsilon^2).
\end{aligned}
$$

Using (2.12), (2.13), we obtain

$$
\begin{aligned}
f(x_* + \varepsilon x_1) &= f(x_*) - \frac{\varepsilon}{2}\|u_*\|^2 + O(\varepsilon^2), \\
g(x_* + \varepsilon x_1) &= \frac{\varepsilon}{2}u_* + O(\varepsilon^2).
\end{aligned}
\tag{2.15}
$$

For the auxiliary function we have

$$
P\left(x_* + \varepsilon x_1, \frac{1}{\varepsilon}\right) = f(x_*) - \frac{\varepsilon}{4}\|u_*\|^2 + O(\varepsilon^2). \tag{2.16}
$$

We have arrived at the estimate close to (2.9). From the formulas (2.15), (2.16) it follows that the points $x(\varepsilon)$ are such that the computed values of the auxiliary and of objective functions are smaller than the values of $f(x_*)$ by quantities of the order ε. The signs of the components of the vector $g(x(\varepsilon))$ coincide with the signs of the corresponding components of the vector u_*. The choice of the concrete point x satisfying the condition (1.11) has no influence (up to second-order terms) on the values of $f(x)$ and $g(x)$.

We show that the formulas (2.14) give an asymptotic approximation of the solutions $x(\varepsilon)$ and $u(\varepsilon)$ of the system (2.10) as $\varepsilon \to 0$ up to second-order terms, i.e.,

$$\lim_{\varepsilon\to 0} \frac{1}{\varepsilon^2} \| x(\varepsilon) - x_* - \varepsilon x_1 \| < \infty, \quad \lim_{\varepsilon\to 0} \frac{1}{\varepsilon^2} \| u(\varepsilon) - u_* - \varepsilon u_1 \| < \infty. \tag{2.17}$$

<u>THEOREM 3.2.4</u>. Let $[x_*, u_*]$ denote the Kuhn-Tucker point for the problem (1.6.1), let at the point $x = x_*$ the constraints $g(x) = 0$ satisfy the constraint qualification; let the functions defining the problem be twice differentiable in the neighborhood of the point x_*, where their matrices of the second derivatives satisfy a Liptschitz condition, and let the matrix $L_{xx}(x_*, u_*)$ be non-singular. Then the formulas (2.14) yield an asymptotic approximation of solutions of the system (2.10) as $\varepsilon \to 0$ up to ε^2.

<u>Proof</u>. Let

$$x(\varepsilon) = x_* + \varepsilon x_1 + \delta x, \quad u(\varepsilon) = u_* + \varepsilon u_1 + \delta u,$$
$$\delta z = [\delta x, \ \delta u] \in E^{n+e}, \quad z = [\varepsilon x_1 + \delta x, \ \varepsilon u_1 + \delta u] \in E^{n+e}.$$

Using the differentiability property, we can write the system (2.10) as

$$\begin{aligned} f_x(x) + g_x(x)\, u &= L_{xx}(x_*, \ u_*)(\varepsilon x_1 + \delta x) + \\ &\quad + g_x(x_*)(u_* + \varepsilon u_1 + \delta u) + \gamma_1(z) = \varepsilon a, \\ 2g(x) - \varepsilon u &= 2g_x^T(x_*)(\varepsilon x_1 + \delta x) - \varepsilon(u_* + \varepsilon u_1 + \delta u) + \\ &\quad + \gamma_2(\varepsilon x_1 + \delta x). \end{aligned} \tag{2.18}$$

Here the functions $\gamma_1(z)$, $\gamma_2(x)$ are such that for any $z \in E^{n+e}$, $x \in E^n$ we have the inequalities $\gamma_1(z) \le c_1 \|z\|^2$, $\gamma_2(x) \le c_2 \|x\|^2$, where c_1, c_2 are constants. We can justify this representation in the same way as we prove the formula (5) in Appendix I. Let

$$N = \left[\begin{array}{c:c} L_{xx}(x_*, u_*) & g_x(x_*) \\ \hdashline 2g_x^T(x_*) & 0_{ee} \end{array}\right],$$

$$\Gamma(z) = \begin{bmatrix} -\gamma_1(z) \\ -\gamma_2(\varepsilon x_1 + \delta x) \end{bmatrix}, \quad z_1 = \begin{bmatrix} x_1 \\ u_1 \end{bmatrix}.$$

Then noting (2.13), we can write the relations (2.18) as

$$\delta z = N^{-1}\Gamma(\varepsilon z_1 + \delta z). \tag{2.19}$$

The matrix N^{-1} is bounded; hence there exists c such that $||N^{-1}\Gamma(\varepsilon z_1 + \delta z)|| < c||\varepsilon z_1 + \delta z||^2$.

By Lemma 2.3.2 the equation (2.19) has a unique solution if

$$\varepsilon \| z_1 \| \leqslant \frac{\sqrt{2}-1}{2c}.$$

It is always possible to obtain this inequality if ε is sufficiently small. Then the solution δz obtained from (2.19) satisfied the condition

$$\| \delta z \| \leqslant 2c\varepsilon^2 [\| x_1 \| + \| u_1 \|]^2.$$

Or, returning to the primary functions, we obtain

$$\frac{1}{2c\varepsilon^2}[\| x(\varepsilon) - x_* - \varepsilon x_1 \| + \| u(\varepsilon) - u_* - \varepsilon u_1 \|] \leqslant [\| x_1 \| + \| u_1 \|]^2,$$

where the right-hand side of the inequality does not depend on ε. Passing to the limit as $\varepsilon \to 0$, we obtain the conditions (2.17), proving the Theorem. ///

4. EXTRAPOLATION IN THE PENALTY FUNCTION METHOD

The results obtained in the previous subsection are quite useful for a qualitative study of the method. At the same time the approach can be exploited in the numerical implementation. If the quantities $\varepsilon_k = \frac{1}{\tau_k}$ change only slightly from one iteration to the other, an extrapolation of the vectors x_k is useful. Let the second simplified version be implemented for a known value of x_k satisfying the condition (1.11). Let

$$P_x\left(x_k, \frac{1}{\varepsilon_k}\right) = f_x(x_k) + \frac{2}{\varepsilon_k} g_x(x_k)\, g(x_k) = \varepsilon_k a_k. \tag{2.20}$$

Here $\|a_k\| \le 1$. For the approximated dual vector we take

$$u_k = \frac{2g(x_k)}{\varepsilon_k}. \tag{2.21}$$

We introduce a small parameter μ and consider the problem of solving the system

$$\begin{aligned} f_x(x) + g_x(x)\, u &= (\varepsilon_k - \mu) a_k, \\ 2g(x) &= (\varepsilon_k - \mu)\, u. \end{aligned} \tag{2.22}$$

The formal power series of μ is:

$$\begin{aligned} x(\mu) &= x_k + \mu \bar{x}_1 + \mu^2 \bar{x}_2 + \ldots, \\ u(\mu) &= u_k + \mu \bar{u}_1 + \mu^2 \bar{u}_2 + \ldots \end{aligned} \tag{2.23}$$

Substituting (2.23) into (2.22), we make a power series expansion in μ. Equating the expressions for equal powers of μ, we obtain a system of linear equations for defining the coefficients in (2.23). For $\mu = 0$ we obtain (2.20), (2.21) and next:

$$L_{xx}(x_k, u_k)\bar{x}_1 + g_x(x_k)\bar{u}_1 = -a_k,$$
$$2g_x^T(x_k)\bar{x}_1 - \varepsilon_k\bar{u}_1 = -u_k.$$

Having solved this system, we substitute into (2.23) the expressions for $\bar{x}_1$, $\bar{u}_1$ and obtain

$$x(\mu) = x_k - \mu L_{xx}^{-1}(x_k, u_k)[a_k + g_x(x_k)\bar{u}_1] + O(\mu^2),$$
$$u(\mu) = u_k + \mu\bar{u}_1 + O(\mu^2),$$
$$\bar{u}_1 = [\varepsilon_k + 2g_x^T(x_k)L_{xx}^{-1}(x_k, u_k)g_x(x_k)]^{-1}\times$$
$$\times(u_k - 2g_x^T(x_k)L_{xx}^{-1}(x_k, u_k)a_k).$$

Assuming $\mu = \varepsilon_k - \varepsilon_{k+1}$, we find the approximated vectors x_{k+1}, u_{k+1}; the former can be taken as an initial vector in finding the minimum $P(x, 1/\mu)$ in $x \in E^n$. If the computations via the penalty function method are terminated upon finding the vector x_k, then, for an approximate solution it is appropriate to take instead of the vectors x_k, u_k the vectors x_{k+1} and u_{k+1} obtained from the formulas found via an extrapolation to the point $\mu = \varepsilon_k$:

$$x_{k+1} = x_k - L_{xx}^{-1}(x_k, u_k)[L_x(x_k, u_k) + g_x(x_k)\tilde{u}], \tag{2.24}$$

$$u_{k+1} = u_k + \tilde{u},$$
$$\tilde{u} = [g_x^T(x_k)L_{xx}^{-1}(x_k, u_k)g_x(x_k)]^{-1}\times \tag{2.25}$$
$$\times[g(x_k) - g_x^T(x_k)L_{xx}^{-1}(x_k, u_k)L_x(x_k, u_k)],$$

where the following representation of the auxiliary function gradient has been used:

$$P_x\left(x_k, \frac{1}{\mu}\right) = L_x(x_k, u_k) = \mu a_k,$$

the validity of which follows from (2.20) and (2.21); furthermore, we have noted that ε_k is small, which made the expression for $\bar{u}_1$ simpler.

It is possible to abstract from the extrapolation problem and regard the formulas (2.24) and (2.25) as some numerical method for solving the problem (1.6.1); in this method neither the penalty function nor the penalty coefficient is explicit. It is easily seen that the formulas (2.24) and (2.25) define nothing but Newton's method applied to solving the system of $n+e$ equations $L_x(x,u) = 0$, $g(x) = 0$.

Methods of the form (2.24), (2.25) will be obtained in the next chapter on the basis of quite different arguments. The approach presented above points out the relationship between these methods which seem to belong to quite different classes.

From a computational point of view, the use of extrapolation is justified if to solve the auxiliary problem (1.11) the methods defining the matrix $L_{xx}(x_k,u_k)$ have been used; then the conversion of x and u is not difficult. Otherwise, the extrapolation can be simplified. According to the results obtained, we can assume that the function $x(\mu)$ depends linearly on μ, and hence for x_{k+1} we take the vector

$$x_{k+1} = x_k + (x_k - x_{k-1}) \frac{\varepsilon_{k+1} - \varepsilon_k}{\varepsilon_k - \varepsilon_{k-1}} . \tag{2.26}$$

5. CONTINUATION METHODS

We assume that the system (2.22) for each $\mu \in [0,\varepsilon_k]$ has the solution $[x(\mu), u(\mu)]$ depending continuously on μ. As the parameter μ goes from 0 to ε_k, the solution to (2.22), $[x(\mu), u(\mu)]$ describes some spatial curve in E^{n+e}, one endpoint of which is $[x(0), u(0)]$ being the solution to the system

(2.10), the other endpoint being the Kuhn-Tucker point $[x(\varepsilon_k), u(\varepsilon_k)] = [x_*, u_*]$.

If the conditions of Lemma 4.1.1 are satisfied, the matrix

$$N(x_*, u_*) = \begin{bmatrix} L_{xx}(x_*, u_*) & g_x(x_*) \\ 2g_x^T(x_*) & 0_{ee} \end{bmatrix}$$

is non-singular; and by the theorem on implicit functions, for sufficiently small ε_k the system (2.22) has a unique solution passing through the point $[x_*, u_*]$; moreover, the functions $x(\mu)$, $u(\mu)$ are differentiable and satisfy the following system of $n + e$ ordinary differential equations:

$$\begin{aligned} L_{xx}(x, u)\frac{dx}{d\mu} + g_x(x)\frac{du}{d\mu} &= -a_k, \\ 2g_x^T(x)\frac{dx}{d\mu} - \varepsilon(1-\mu)\frac{du}{d\mu} &= -u(\mu). \end{aligned} \tag{2.27}$$

Integrating this system over μ from $\mu = 0$ to $\mu = \varepsilon_k$ with the Cauchy initial data $x(0) = x_0$, $u(0) = u_0$, where x_0, u_0 denote the solutions of the system (2.10), we obtain the Kuhn-Tucker point sought.

The procedure described can be regarded as a numerical method of solving the problem (1.6.1). The method consists of two stages: an approximate minimization of the function $P(x, 1/\varepsilon)$ in x and a numerical solution of the Cauchy problem for (2.27). In contrast to the method (2.2.3), this method involves integration of the system (2.27) with high accuracy; hence if the Euler method is used, one needs to take a sufficiently small step of integration, or to use more exact methods (the Euler method with conversion, the Runge-Kutta method, and others). The integration errors for (2.27) entail the situation when the solutions $x(\mu)$, $u(\mu)$ do not satisfy the system (2.22) any longer and an additional correction is required.

This approach has been borrowed from the methods for solving nonlinear equations, known as "continuation" or "homotopy" method. A more detailed description of this method can be found in G. Ortega and W. Rheinboldt [1], and in Davidenko [1], [2]. This approach has not, however, been widely used for solving optimization problems.

3. THE COST FUNCTION PARAMETRIZATION METHOD

In this section we shall describe a class of methods for solving the problem (1.6.1), close to the penalty function method, based on the parametrization of the cost function; the main difference is that in the former method the auxiliary parameter changes automatically according to the rule prescribed, while in the penalty function method the variation of the penalty coefficients has to be preassigned. Various versions of the method are given in B. N. Pshenichnyj [1], V. V. Velichenko [1], [3], [4], Yu. P. Ivanilov [1], B. S. Razumikhin [1], D. Morrison [1], F. Lootsma [1], V. V. Ivanov and V. A. Lyudvichenko [1], and many other works. We shall dwell only on a few versions of the method.

1. PRELIMINARY RESULTS

Suppose we are solving the problem (1.6.1). Using the penalty function used in Section 3.1, we make up the auxiliary function

$$M(x,\eta) = (f(x) - \eta)^2 + S(x) . \quad (3.1)$$

We introduce the set

$$X(\eta) = \operatorname*{Arg\,min}_{x \in E^n} M(x,\eta) . \quad (3.2)$$

LEMMA 3.3.1. If $x_* \in X_*$, $f(x_*) > \eta$, $x_1 \in X(\eta)$, then $f(x_*) \geq f(x_1)$.

Proof. From the condition $x_1 \in X(\eta)$ we obtain

$$(f(x_*) - \eta)^2 = M(x_*,\eta) \geq M(x_1,\eta) \geq (f(x_1) - \eta)^2 . \quad (3.3)$$

If $f(x_1) \geq \eta$, it follows from (3.3) that

$$|f(x_*) - \eta| = f(x_*) - \eta \geq |f(x_1) - \eta| = f(x_1) - \eta .$$

We then have that $f(x_*) \geq f(x_1)$. If $f(x_1) < \eta$, then $f(x_*) \geq \eta > f(x_1)$; thus, we have arrived again at the required inequality. ///

LEMMA 3.3.2. Let $\eta = f(x_*)$, $x_* \in X_*$, then $X_* = X(\eta)$, $M(x_*,\eta) = 0$.

Proof. For any $x \in E^n$ we have the relation $M(x_*,\eta) = 0 \leq M(x,\eta)$ implying $x_* \in X(\eta)$.

Conversely, let $x_1 \in X(\eta)$. In this case for any $x \in E^n$

$$0 \leq M(x_1,\eta) \leq M(x,\eta) .$$

In particular,

$$0 \leq M(x_1,\eta) \leq M(x_*,\eta) = 0 .$$

This is possible only if $S(x_1) = 0$, that is the point x_1 is feasible and, in addition, such that $f(x_1) = \eta = f(x_*)$, hence

$x_1 \in X_*$. We then conclude that $X_* = X(\eta)$. ///

<u>LEMMA 3.3.3</u>. Let $\eta_1 < \eta_2$, $x_1 \in X(\eta_1)$, $x_2 \in X(\eta_2)$. Then $f(x_1) \le f(x_2)$.

<u>Proof</u>. For any x we have the inequalities

$$0 \leqslant (f(x_1)-\eta_1)^2 + S(x_1) \leqslant (f(x)-\eta_1)^2 + S(x),$$
$$0 \leqslant (f(x_2)-\eta_2)^2 + S(x_2) \leqslant (f(x)-\eta_2)^2 + S(x).$$

In (3.4) we take as x the point x_2, and in (3.5) the point x_1. We obtain

$$(f(x_1)-\eta_1)^2 + S(x_1) \leqslant (f(x_2)-\eta_1)^2 + S(x_2),$$
$$(f(x_2)-\eta_2)^2 + S(x_2) \leqslant (f(x_1)-\eta_2)^2 + S(x_1).$$

Adding them and making some transformations, we obtain

$$f(x_1)(\eta_2 - \eta_1) \le f(x_2)(\eta_2 - \eta_1)$$

yielding the required inequality: $f(x_1) \le f(x_2)$. ///

It follows from the Lemma that the value of the cost function $f(x)$ for $x \in X(\eta)$ is a monotonically increasing function of the parameter η.

2. THE FIRST VERSION DUE TO D. MORRISON

We assume that some lower estimate η_0 of the optimal cost function $f(x)$ is known, i.e., $\eta_0 \le f(x_*)$, $x_* \in X$. We can obtain this estimate, for example, by solving the auxiliary problem of the unconstrained minimization (1.4); by (2.9) it is then possible to put $\eta = f(x_k)$. The cost-function parametrization method consists in sequential determination of the points x_k and increasing the parameter η_k so that $\eta_k \to f(x_*)$. By the lemmas proved in Section 3.1, $X(\eta_k) \to X_*$, ensuring the convergence of the method to the set X_*. The key point here is the rule of changing the parameter from one iteration to another. D. Morrison [1] suggests the following version.

On the kth iteration, let some value of the parameter $\eta_k \le f(x_*)$ be known. From a solution of the auxiliary problem (3.2) we find the point $x_k \in X(\eta_k)$. If $x_k \in X$ or $M(x_k, \eta_k) = 0$, the calculations are over since in this case $S(x_k) = 0$, $x_k \in X_*$. Otherwise we put

$$\eta_{k+1} = \eta_k + \sqrt{M(x_k, \eta_k)} \quad , \tag{3.6}$$

and the iterative process continues. We prove the convergence of the version described.

LEMMA 3.3.4. If $\eta_k \le f(x_*)$, then $\eta_{k+1} \le f(x_*)$.

Proof. For any $x \in E^n$ we have

$$M(x_k, \eta_k) \le M(x, \eta_k) \quad .$$

In particular, for $x = x_*$ we obtain

$$M(x_k, \eta_k) \le (f(x_*) - \eta_k)^2 \quad .$$

But by hypothesis $\eta_k \le f(x_*)$, hence

$$\sqrt{M(x_k, \eta_k)} \leqslant f(x_*) - \eta_k,$$
$$\eta_{k+1} = \eta_k + \sqrt{M(x_k, \eta_k)} \leqslant f(x_*). \qquad ///$$

THEOREM 3.3.1. In the problem (1.6.1) let the set X_* be non-empty, let the function $M(x, \eta)$ be everywhere continuous, and let the value $\eta_0 < f(x_*)$ be known. Then the version suggested by

D. Morrison converges to the set X_*.

Proof. By Lemma 3.3.4 the monotonically increasing sequence $\{\eta_k\}$ is bounded from above by the quantity $f(x_*)$. Hence the limit exists:

$$\lim_{k\to\infty} \eta_k = \bar{\eta} \leq f(x_*) \quad . \tag{3.7}$$

We consider now the second auxiliary problem of finding

$$\min_{x\in X(\lambda)} f(x), X(\lambda) = \{x \in E^n : S(x) \leq \lambda\} \quad . \tag{3.8}$$

We denote by $X_*(\lambda)$ the set of solutions of this problem. Obviously, $X_* = X_*(0)$, $X = X(0)$. Let $F(\lambda) = f(\bar{x})$, where $\bar{x} \in X_*(\lambda)$. We prove our Lemma, assuming that the function $F(\lambda)$ is right continuous at the point $\lambda = 0$, i.e., for any $\varepsilon > 0$ there exists $\delta > 0$ such that if $0 \leq \lambda < \delta_F$, then $|F(\lambda) - F(0)| < \varepsilon$. Here $F(0) = f(x_*)$, $x_* \in X_*$.

To prove the assertion of the Theorem, we show that for any positive ε_1, ε_2 we can find N such that for all $k > N$ we have the inequalities

$$f(x_*) - f(x_k) < \varepsilon_1 , \qquad S(x_k) < \varepsilon_2 . \tag{3.9}$$

For $\varepsilon_1 > 0$ we take $\delta_1 > 0$ such that if $0 < \lambda \leq \delta_1$, then $|F(\lambda) - F(0)| < \varepsilon_1$. It follows from (3.7) that there exists N such that for all $k > N$ the condition

$$0 \leqslant \eta_{k+1} - \eta_k < \min[\sqrt{\varepsilon_2}, \sqrt{\delta_1}]$$

is satisfied, yielding

$$\sqrt{M(x_k,\ \eta_k)} < \min[\sqrt{\varepsilon_2},\ \sqrt{\delta_1}],$$
$$(f(x_k)-\eta_k)^2+S(x_k) \leqslant \min[\varepsilon_2,\ \delta_1],$$
$$S(x_k) \leqslant \min[\varepsilon_2,\ \delta_1].$$

Therefore, $S(x_k) < \delta_1$ for all $k > N$. Hence $|f(x_k) - f(x_*)| < \varepsilon_1$ for all $k > N$. Noting the assertion of Lemma 3.3.4, we arrive at the inequalities (3.9), thus proving the Theorem. ///

3. THE SECOND VERSION DUE TO D. MORRISON

In his article [1] published in 1968, D. Morrison outlined one more version of changing the parameter η:

$$\eta_{k+1} = \eta_k + \frac{M(x_k,\ \eta_k)}{f(x_k)-\eta_k}. \qquad (3.10)$$

Later, many authors studied this version. One of the first among those works was J. Kowalik, M. Osborne, and D. Ryan [1].

4. A VERSION DUE TO B.S. RAZUMIKHIN

In the book of B.S. Razumikhin [1] the following rule for defining the parameter is suggested:

$$\eta_{k+1} = f(x_k). \qquad (3.11)$$

5. COMPUTATIONAL ASPECTS

Let us compare the three versions of the method. We denote by Δ_1, Δ_2, Δ_3 the variations of the parameter η on the kth iteration according to the rules (3.6), (3.10) and (3.11), respectively:

$$\Delta_1 = \sqrt{M(x_k,\ \eta_k)},\quad \Delta_2 = \frac{M(x_k,\ \eta_k)}{f(x_k)-\eta_k},\quad \Delta_3 = f(x_k)-\eta_k.$$

It is obvious that $|\Delta_1| \geq |\Delta_3|$. Furthermore, from the representation

$$\Delta_2 = \Delta_1 + \frac{\sqrt{M(x_k, \eta_k)}}{f(x_k) - \eta_k} \left[\sqrt{(f(x_k) - \eta_k)^2 + S(x_k)} - [f(x_k) - \eta_k]\right]$$

it follows that for $f(x_k) > \eta_k$ we have the inequalities

$$\Delta_2 \geq \Delta_1 \geq \Delta_3 \quad .$$

Thus, the auxiliary parameter η increases maximally at each iteration for the version (3.10); minimally for the version (3.11). Hence one might expect that the version (3.10) has the maximum rate of convergence. The numerical computations confirm this suggestion; at the same time, however, the numerical tests make one notice the following. The properties of the cost-function parametrization method given above were obtained under the assumption that the auxiliary problem of minimization of the function $M(x, \eta_k)$ has an exact solution; because of numerical errors we can obtain $\eta_k > f(x_*)$. In that case the further calculations according to either of the formulas (3.6), (3.10), (3.11) will make no sense since they are justified only under the assumption that $\eta_k \leq f(x_*)$. Hence the accuracy of solution of the auxiliary problem must be maximal when the version (3.10) is realized, and minimal when the version (3.11) is realized. This circumstance "averages" somewhat the computational time when distinct versions are used.

The properties described in Section 3.1 imply that a solution of the problem (1.6.1) can be found by defining the minimum η_* for which the condition

$$\min_{x \in E^n} M(x, \eta_*) = 0 \qquad (3.12)$$

is satisfied. This reduction can be used for solving completely the problem (1.6.1), or partially, for example, in those cases where for any reasons the value $\eta_k > f(x_*)$ has been obtained, and the values of η_k need to be decreased until the conditions (3.12) are satisfied. Practical realization of this approach causes considerable difficulties due to two circumstances:

first, for $\eta > \eta_* = f(x_*)$ the problem (3.12) has as a rule, a continuum of solutions, and even for a slight variation of the parameter η the solution of (3.12) frequently changes considerably, which makes the computations more complex;

second, which is more important, in numerical computations it is often hard to verify that the condition (3.12) is satisfied.

The verification of equality to zero requires a high accuracy of minimization, which also complicates the computations.

6. THE COMPATIBILITY WITH THE PENALTY FUNCTION METHOD

Let the penalty function be additive and have the form

$$S(x) = \sum_{i=1}^{e} [g^i(x)]^2 + \sum_{j=1}^{c} [h^j_+(x)]^2.$$

Next we compare the two auxiliary problem of unconstrained minimization of the functions $P(x,\tau)$ and $M(x,\eta)$, specified by the formulas (1.3) and (3.1). We assume that the functions defining the problem (1.6.1) are everywhere differentiable. Then the necessary conditions for the minimum in x are as follows:

$$P_x(x,\ \tau)=f_x(x)+2\tau\left[\sum_{i=1}^{e} g^i(x)\,g_x^i(x)+\right.$$
$$\left.+\sum_{j=1}^{c} h_+^j(x)\,h_x^j(x)\right]=0, \qquad (3.13)$$
$$M_x(x,\ \eta)=2(f(x)-\eta)f_x(x)+$$
$$+2\left[\sum_{i=1}^{e} g^i(x)\,g_x^i(x)+\sum_{j=1}^{c} h_+^j(x)\,h_x^j(x)\right]=0. \qquad (3.14)$$

If for the penalty coefficient we take

$$\tau = \frac{1}{2(f(x)-\eta)}, \qquad (3.15)$$

the conditions (3.13) and (3.14) coincide. Implementing a version of the cost-function parametrization method, we obtain the sequence $\{\eta_k\}$, and we define the sequence of penalty coefficients τ_k, using (3.15). In this case the sequences of sets of solutions of the auxiliary problems coincide for both methods. The parametrization method, hence, differs essentially from the method of exterior penalty functions in the only fact that in the former method the policy of changing the auxiliary parameter η has been automatically defined, while in the method of the exterior penalty functions the user must define specifically the rule for variation of τ. Hence it is hard to compare these methods: each implementation of the cost-function parametrization method can be repeated via the method of exterior penalty functions, using the formula (3.15); and in a more special study of the problem in question one can choose a more advantageous policy of changing the coefficient τ, which will lead to better results if the penalty function method is used. On the other hand, for an inappropriate choice of the sequence $\{\tau_k\}$ the computational results obtained through the penalty function method will be inferior to those obtained through the cost-function parametrization method. These circumstances demonstrate how carefully one has to treat the so-called "numerical tests" of investigating the comparative effectiveness of various algorithms. If desired, it is possible to construct convincing examples of the problems solved, showing either the advantage of the cost-function parametrization method compared with the method of exterior penalty functions, or the examples "proving" the converse.

In conclusion, we make several general remarks. The cost-function parametrization method is more convenient than the method of exterior penalty functions because in the former, first, the rule of defining the subsequence $\{\eta_k\}$ is more concrete and, second, the auxiliary function M is bounded from below by zero; hence this removes one of the drawbacks of the penalty function method associated with the possibility that the auxiliary function is sometimes unbounded from below on an unfeasible set. At the same time, the parametrization method has two disadvantages: to implement this method, one needs to know the lower estimate of the value $f(x_*)$; the auxiliary problem of unconstrained minimization has to be usually solved more accurately than in the exterior penalty function method. In general, these methods are very similar, they are best suitable for preliminaty, coarse calculations; these methods hardly yield a high accuracy of solution.

7. GEOMETRIC INTERPRETATION

For convenience of representation, we express the function $M(x,\eta)$ as follows:

$$M(x,\eta) = (f(x)-\eta)^2 + R^2(x), \qquad R(x) = \sqrt{S(x)} .$$

In Figure 1 we diagram the system of co-ordinates in which R is measured along the abscissa and f along the ordinate.

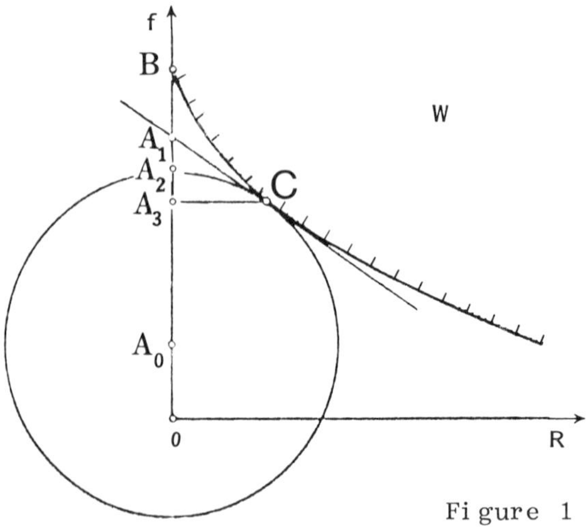

Figure 1

If for all possible values of x we consider the set of points with the coordinates $[R(x), f(x)]$, we see that they form some set W (in Figure 1, the shaded area). The point B having the co-ordinates $[0, f(x_*)]$ corresponds to a solution of the problem (1.6.1). On the plane $[R,f]$ the equation

$$(f - \eta_k)^2 + R^2 = \text{const} \tag{3.16}$$

is a circle centered at the point A_0, with the co-ordinates $[0,\eta_k]$. A solution of the auxiliary problem of minimizing $M(x,\eta_k)$ in $x \in E^n$ can be interpreted on the plane $[R,f]$ as finding the circle (3.16) of the minimum radius, having the common point C with the set W. The point C thus obtained has the co-ordinates $[R, (x_k), f(x_k)]$. In Figure 1 we also show: the point A_1 with the co-ordinates

$$[0, \eta_k + \sqrt{M(x_k, \eta_k)}] ,$$

the point A_2 with the co-ordinates

$$\left[0, \eta_k + \frac{M(x_k, \eta_k)}{f(x_k) - \eta_k}\right] ,$$

the point A_3 with the co-ordinates $[0, f(x_k)]$.

The point A_2 obtains from the condition for intersection of a tangent to the circle with the ordinate: at the point C. In the first version of the method, the ordinate of the point A_1 is taken as η_{k+1}; in the second version this is the ordinate of the point A_2; and in the third version, the ordinate of the point A_3. In each version, upon finding η_{k+1} it is projected once more onto the set W and so on, and so on, until the value of η_k becomes sufficiently close to $f(x_*)$.

4. THE INTERIOR PENALTY FUNCTION METHOD

1. THE GENERAL IDEA OF THE METHOD

We consider the particular case of the problem (1.6.1) where the feasible set is defined only through constraints of the inequality type. Suppose we are seeking

$$\min_{x \in X} f(x), \quad X = \{x \in E^n\colon\ h(x) \leqslant 0\}. \tag{4.1}$$

As before, we denote by X_* the set of solutions of this

problem. The set of interior points is $X_0 = \{x \in E^n : h(x) < 0\}$, $\Gamma = X \setminus X_0$ is the boundary of the feasible set.

We introduce next the auxiliary function

$$P(x,t) = \mu(t)f(x) + \tau(t)b(x) \quad .$$

The continuous functions $\mu(t)$, $\tau(t)$ are defined for any $t \geq 0$ and such that for $\delta > 0$ the conditions

$$\mu(t) > 0, \ \tau(t) > 0, \frac{\tau(t)}{\mu(t)} > \frac{\tau(t+\delta)}{\mu(t+\delta)}, \quad \lim_{t\to\infty} \frac{\tau(t)}{\mu(t)} = 0 \tag{4.2}$$

are satisfied. Everywhere on X_0 the function $b(x)$ is continuous and $b(x) > 0$, for any infinite sequence of the points $\{x_i\}$ belonging to X_0 and converging to a point of Γ the limit

$$\lim_{i\to\infty} b(x_i) = +\infty \quad . \tag{4.3}$$

As very simple examples of the functions $b(x)$ satisfying the above condition we can give

$$\sum_{j=1}^{c} \frac{1}{-h^j(x)}, \quad \sum_{j=1}^{c} \frac{1}{(h^j(x))^2} .$$

The numerical method for solving the problem (4.1) is the following. For an arbitrary increasing sequence $0 \leq t_1 < t_2 < \cdots$ which tends to infinity, the auxiliary problem of minimizing $P(x,t_i)$ is being solved. As a result, one obtains the sequence of points $\{x_k\}$ satisfying the condition

$$x_k \in \operatorname{Arg}\min_{x\in E^n} P(x,t_k) \quad . \tag{4.4}$$

Under certain conditions, each limit point of a converging

subsequence of the sequence $\{x_k\}$ belongs to X_*, i.e., we have the convergence of the method described.

The condition (4.3) yields that as the point x is approaching the boundary Γ, the values $P(x,t)$ tend to infinity. Hence the minimization of $P(x,t_k)$ in x yields the points x_k belonging to the set X. The auxiliary problems (4.4) can be solved using numerical methods of local minimization of functions in many variables, starting the search from an arbitrary point $x_0 \in X_0$. The function $b(x)$ is thus a peculiar "barrier," which justifies the name the 'method of barriers' or 'interior penalty function method.'

2. PROOF OF CONVERGENCE

For an arbitrary point $x_0 \in X_0$ we define two sets:

$$\Omega(t) = \{x \in E^n\colon\ P(x, t) \leqslant P(x_0, t),\ x \in X_0\},$$
$$B(t) = \{x \in E^n\colon\ \mu(t) f(x) \leqslant P(x_0, t),\ x \in X\}.$$

<u>THEOREM 3.4.1</u>. Let the function $f(x)$ be continuous on X, let the set of solutions X_* of the problem (4.1) be non-empty, and let for any $\varepsilon > 0$ there exist a ε-neighborhood $G_\varepsilon(X_*)$ of the set X_*, such that $N = G(X_*) \cap X_0 \neq \emptyset$; let the conditions (4.2) and (4.3) be satisfied, and let the set $B(0)$ be compact. Then the method of interior penalty functions converges to X_* on X_0.

<u>Proof</u>. It is seen that the inclusions $\Omega(t) \subset B(t) \subset B(0)$, $X_* \subset B(t+\delta) \subset B(t)$ hold for any $\delta \geq 0$. Therefore, the set X_* and for all $t \geq 0$ the sets $\Omega(t)$, $B(t)$ are bounded, and the auxiliary problems of minimizing (4.4) are solvable. For any $x_* \in X_*$, $x \in X$ we have $f(x_*) \leq f(x)$, all the more

$$f(x_*) \leqslant f(x) + \frac{\tau(t)}{\mu(t)} b(x). \tag{4.5}$$

All the elements of the sequence $\{x_k\}$ belong to the set $B(0)$. Hence the sequence $\{x_k\}$ is bounded and has a subsequence $\{x_{\bar{k}}\}$ converging to the point $\bar{x} \in X$. It follows from the inequality (4.5) that

$$f(x_*) \leqslant f(\bar{x}) + d, \quad d = \lim_{\bar{k} \to \infty} \frac{\tau(t_{\bar{k}})}{\mu(t_{\bar{k}})} b(x_{\bar{k}}).$$

Let $f(x_*) = f(\bar{x}) + d$. Then $d = 0$, $\bar{x} \in X_*$ since for $d > 0$ we would have $f(x_*) > f(\bar{x})$, $\bar{x} \in X$, which is impossible. We show that

$$f(x_*) < f(\bar{x}) + d \tag{4.6}$$

does not hold. Assume the opposite: (4.6) holds. By hypothesis of the Theorem there exists a neighborhood $G_\varepsilon(X_*)$ such that $N \neq \emptyset$. Let ε be so small that $\bar{x} \notin N$. We can always find the point $y \in N$ for which

$$f(x_*) \leq f(y) < f(\bar{x}) . \tag{4.7}$$

From the condition for defining the points x_k it follows that

$$f(x_k) + \frac{\tau(t_k)}{\mu(t_k)} b(x_k) \leqslant f(y) + \frac{\tau(t_k)}{\mu(t_k)} b(y).$$

Taking the values of t_k corresponding to the subsequence $\{x_{\bar{k}}\}$ and letting $k \to \infty$, we obtain $f(\bar{x}) + d \leq f(y)$, which contradicts (4.6), (4.7). Hence (4.6) does not hold. Each limit point of the sequence $\{x_k\}$ belongs to X_*. ///

3. COMPUTATIONAL ASPECTS

In this method the sequence of interior points of $\{x_k\}$ is constructed from a feasible set. This can be quite useful in those problems where it is not desirable to consider, for any reason, unfeasible points. For example, some of the functions defining the problem (4.1) do not need to be given on an unfeasible set. At the same time, this method is not applicable to the problem (1.6.1) when the functions $g(x)$ are nonlinear and the interior of the feasible set is empty. In computer implementation the auxiliary functions are usually combined so that equality-type constraints are accounted for through exterior penalty functions, and inequality-type constraints through interior penalty functions. For the problem (1.6.1) it is possible, for example, to use the function

$$P(x, t) = f(x) + \frac{1}{t} \sum_{i=1}^{e} [g^i(x)]^2 + t \sum_{j=1}^{c} \frac{1}{-h^j(x)} ,$$

where $t \to 0$ during the computational procedure. A detailed study of such combined penalty functions can be found in A.V. Fiacco and G.P. McCormick [1], and in E. Polak [1].

The Czechoslovakian mathematician M. Hamala shows in [1] that in implementing the interior penalty function method one can get rid of the condition for the function $b(x)$ to tend to infinity as $x \to \Gamma$, and require, instead, that the norm of the gradient of the function $b(x)$ tend to infinity as $x \to \Gamma$. In that case, as an auxiliary function one can, for example, use the function:

$$P(x, t) = f(x) + t \sum_{j=1}^{c} \sqrt{-h^j(x)}.$$

In Fiaccko and McCormick [1], and in Yu. G. Evtushenko [9], various "continuous" methods similar to (1.17) involving barrier funcions have been suggested.

In the numerical implementation of the method, approximate solutions of the auxiliary problem (4.4) are found, with simplified versions analogous to those of the exterior penalty function method.

4. ESTIMATION OF ACCURACY

We consider the auxiliary function

$$P(x, \varepsilon) = f(x) - \varepsilon \sum_{i=1}^{c} \ln(-h^i(x)).$$

We use the method of asymptotic expansions described in Section 3.2. For a small parameter we take ε, as an approximate solution to the problem (4.4) we take any point x satisfying the condition

$$\|P_x(x,\varepsilon)\| \leq \varepsilon .$$

For a particular point $x = x(\varepsilon)$ we have

$$\begin{aligned} f_x(x(\varepsilon)) + h_x(x(\varepsilon))\, v(\varepsilon) &= \varepsilon a, \\ D(v(\varepsilon))\, h(x(\varepsilon)) &= -\varepsilon I. \end{aligned}$$

Here $a \in E^n$, $\|a\| \leq 1$, $v \in E^c$, $I \in E^c$ denoting the vector all components of which are equal to unity.

We seek a solution to the system (4.8) as a series similar to (2.11). We substitute them into (4.8) and expand the functions as a power series of ε; equating the expressions for ε^0 and those for ε^1, we obtain

$$\begin{gathered} f_x(x_*)+h_x(x_*)v_*=0, \quad v_*^j h^j(x_*)=0, \\ L_{xx}(x_*, v_*)x_1+h_x(x_*)v_1=a, \\ D(v_*)h_x^T(x_*)x_1+D(h(x_*))v_1=-I. \end{gathered} \tag{4.9}$$

Assuming that in the problem (4.1) the conditions of Lemma 4.1.2 are satisfied, we solve the system (4.9) and obtain the following asymptotic estimates:

$$\begin{aligned} &x(\varepsilon)=x_*+\varepsilon L_{xx}^{-1}(x_*, v_*)(a-h_x(x_*)v_1)+O(\varepsilon^2), \\ &v(\varepsilon)=v_*+\varepsilon v_1+O(\varepsilon^2), \\ v_1=-&[D(h(x_*))-D(v_*)h_x^T(x_*)L_{xx}^{-1}(x_*, v_*)h_x(x_*)]^{-1}\times \\ &\qquad\qquad\times[I_c+D(v_*)h_x^T(x_*)L_{xx}^{-1}(x_*, v_*)a]. \end{aligned}$$

Dropping values of the order ε^2, we obtain that if $j \in \sigma(x_*)$, then $h^j(x_*) = 0$,

$$v_*^j \langle h_x^j(x_*), x_1\rangle=-1, \quad h^j(x_*+\varepsilon x_1)=-\frac{\varepsilon}{v_*^j}<0.$$

If $j \notin \sigma(x_*)$, we have

$$h^j(x_*)<0, \quad v_*^j=0, \quad v_1^j=-\frac{1}{h^j(x_*)}>0.$$

These formulas yield

$$f(x_*+\varepsilon x_1)=f(x_*)+\varepsilon N,$$

$$P(x_*, \varepsilon) = f(x_*)+\varepsilon\left[N+\sum_{j\in\sigma(x_*)}\ln\frac{v_*^j}{\varepsilon}-\sum_{j\notin\sigma(x_*)}\ln(-h^j(x_*))\right].$$

Here N denotes the number of active constraints at the point x_* (i.e., the number of those $h^j(x_*)$ for which $j \in \sigma(x_*)$).

The value of the objective function $f(x(\varepsilon))$ is therefore greater than that of $f(x_*)$ and tends to this value as $\varepsilon \to 0$. It is worth noting that here, as well as for exterior penalties, the values f, h, P do not depend on the particular choice of the vector a (up to second-order terms).

Many interesting estimates follow easily from the Young-Minkowski inequalities given in Subsection 3.2.2. Letting

$$\tau a = -h(x), \qquad b = \tau v_*, \qquad \alpha = -1, \qquad \beta = \tfrac{1}{2} \quad ,$$

we obtain

$$f(x_*) + 2\sqrt{\tau} \sum_{i=1}^{c} \sqrt{v_*^i} \le f(x) + \tau \sum_{i=1}^{c} [-h^i(x)]^{-1} \quad .$$

If

$$a = -h(x)\tau, \qquad \tau b = v_*, \qquad \alpha = \tfrac{1}{2}, \qquad \beta = -1 \quad ,$$

then

$$f(x_*) - \tau \sum_{i=1}^{c} [v_*^i]^{-1} \le f(x) - 2\sqrt{\tau} \sum_{i=1}^{c} \sqrt{-h^i(x)} \quad .$$

If

$$\tau a^i = -h^i(x) Z^i, \qquad Z^i b^i = \tau v_*^i, \qquad \alpha = -1, \qquad \beta = \tfrac{1}{2} \quad .$$

then

$$f(x_*) + 2\sqrt{\tau} \sum_{i=1}^{c} \sqrt{\frac{v_*^i}{Z^i}} \le f(x) + \tau \sum_{i=1}^{c} [-h^i(x) Z^i]^{-1} \quad .$$

It is possible to introduce the scaling vector Z in all other formulas as well, but we shall not do it because this vector makes the formulas very complex and carries no additional information. We can always assume that h is a scaled vector. Therefore instead of the vector $Z \in R^c$ we use the scalar τ everywhere.

For exterior penalties, the use of the Hölder inequality led us to exact penalties. We give a similar result for interior penalties. Let

$$a = h(x), \qquad b = v_*, \qquad \alpha = -1, \qquad \beta = \tfrac{1}{2} \quad .$$

Then for any $0 \le \tau \le \|v_*\|_{\frac{1}{2}}$, $x \in X$, we have

$$f(x_*) \le f(x) - \tau\left[\sum_{i=1}^{c} [-h^i(x)]^{-1}\right]^{-1} \quad .$$

For $\alpha = \frac{1}{2}$, $\beta = -1$ and any $0 \le \tau \le \|v_*\|_{-1}$, $x \in X$, we have

$$f(x_*) \le f(x) - \tau\left[\sum_{i=1}^{c} \sqrt{-h^i(x)}\right]^2 \quad .$$

Taking this approach, we can obtain many other estimates.

5. EXTRAPOLATION

The purpose of extrapolation in the implementation of the penalty function method was elucidated in Subsection 3.2.4. Thus we discuss it here only briefly. For a coefficient ε_k suppose that the vectors x_k, v_k have been found, which satisfy the conditions

$$L_x(x_k, v_k) = f_x(x_k) + h_x(x_k)v_k = \varepsilon_k a_k,$$
$$D(v_k)h(x_k) + \varepsilon_k I = 0,$$

where $\|a_k\| \le 1$.

Next, we introduce a small parameter μ and consider the system

$$f_x(x) + h_x(x)v = (\varepsilon_k - \mu)a_k,$$
$$D(v)h(x) = -(\varepsilon_k - \mu)I. \tag{4.10}$$

We seek a solution to (4.10) as a series analogous to (2.23). Making standard computations, we find the following asymptotic estimates of the solutions of the system (4.10);

$$\begin{aligned} x(\mu) &= x_k - \mu L_{xx}^{-1}(x_k, v_k)(a_k + h_x(x_k)\bar{v}_1) + O(\mu^2), \\ v(\mu) &= v_k + \mu\bar{v}_1 + O(\mu^2), \\ \bar{v}_1 &= [D(h(x_k)) - D(v_k)h_x^T(x_k)L_{xx}^{-1}(x_k, v_k)h_x(x_k)]^{-1} \times \\ &\quad \times (I_c + D(v_k)h_x^T(x_k)L_{xx}^{-1}(x_k, v_k)a_k). \end{aligned}$$

If the computational procedure following the interior penalty function method terminates upon finding x_k, then, by the formulas obtained, as the improved values of x and v we need to take

$$x_{k+1} = x_k - L_{xx}^{-1}(x_k, v_k)[L_x(x_k, v_k) + h_x(x_k)\tilde{v}], \tag{4.11}$$

$$\begin{aligned} v_{k+1} &= v_k + \tilde{v}, \\ \tilde{v} &= [D(h(x_k)) - D(v_k)h_x^T(x_k)L_{xx}^{-1}(x_k, v_k)h_x(x_k)]^{-1} \times \\ &\quad \times D(v_k)[h_x^T(x_k)L_{xx}^{-1}(x_k, v_k)L_x(x_k, v_k) - D(h(x_k))]. \end{aligned} \tag{4.12}$$

If we consider these formulas as some iterative method, we easily show that this method coincides with Newton's method applied to the solution of the system $L_x(x,v) = 0$, $D(v)h(x) = 0$.

We can now use the auxiliary function

$$P(x, \varepsilon) = f(x) + \varepsilon^2 \sum_{j=1}^{c} \frac{1}{-h^j(x)}. \tag{4.13}$$

Then the following system plays the role of (4.10):

$$\begin{aligned} f_x(x) + \sum_{j=1}^{c} (w^j)^2 h_x^j(x) &= (\varepsilon_k - \mu)a_k, \\ D(w)h(x) &= -(\varepsilon_k - \mu)I, \quad w \in E^c. \end{aligned}$$

It is not hard at all to obtain all the necessary computational formulas; an analog of the method (4.11) (4.12) in this case is given by the method (4.1.11), which will be obtained and studied in our next chapter, using, however, different considerations.

In the same way as in Section 3.2, we can simplify the extrapolation, using the formula (2.26). For the auxiliary function (4.13), the employment of the formula (2.26) will correspond to the linear dependence of x, v on the square root of the true penalty coefficient equal to ε^2 in (4.13).

5. THE LINEARIZATION METHOD

Currently, there are many versions of the linearization method; most of them can be interpreted as the implementations of the exterior penalty function method using non-differentiable penalties. This is the reason why we have described the linearization method in the chapter dealing with the exterior penalty functions. We shall give later several versions of the method, omitting, however, their justification. A detailed description of all the methods, including the proof of their convergence, can be found in the References. We shall consider this method for the problem (1.6.1), using the auxiliary function (2.1); although in very many works the problem (4.1) is considered, and either $P_1(x,\tau)$ or $P_3(x,\tau)$ are auxiliary functions.

1. THE GENERAL IDEA OF THE METHOD

For many mathematicians working in mechanical engineering, the most appropriate approach to solving the problem (1.6.1) is to use the linearization method. This method implies the following.

The functions defining the problem (1.6.1) are linearized at the point $x = x_k$; the linear programming problem is considered:

$$\min_{x \in X_k} \langle f_x(x_k), x - x_k \rangle,$$
$$X_k = \{x \in E^n : g(x_k) + g_x^T(x_k)(x - x_k) = 0, \quad h(x_k) + h_x^T(x_k)(x - x_k) \leq 0\}. \tag{5.1}$$

This problem can have unbounded solutions, hence usually additional constraints are introduced:

$$|x^i - x_k^i| \leq C, \qquad i \in [1:n].$$

Having solved the auxiliary problem, one finds the vector $x = \bar{x}_k$ and passes to the point

$$x_{k+1} = x_k + \alpha_k(\bar{x}_k - x_k), \qquad \alpha_k \geq 0. \tag{5.2}$$

This process is repeated at a new point x_{k+1}, and so on. Unfortunately, this method converges very slowly. It may seem that in order to obtain a high rate of convergence one needs the square approximation of all the functions defining the problem. It has been found that this is not quite true: a square rate of convergence can be obtained using linear approximations of the constraints plus taking a quadratic function as a cost function: instead of the problem (5.1) the problem

$$\min_{x \in X_k} \left[\langle f_x(x_k), x - x_k \rangle + \frac{1}{2}(x - x_k)^T N_k (x - x_k) \right] \tag{5.3}$$

needs to be solved.

Numerous versions of the linearization method differ in the

rules for determining the matrices N_k and the variables α_k. We list here several versions. We denote by $\bar{x}_k$ the vector x found by solving the problem (5.3), and denote by u_k and v_k the respective Lagrange multipliers. Also, we put $y_k = [u_k, v_k]$, $z_k = [x_k, y_k]$, $L(x_k, u_k, v_k) = L(z_k)$. We assume throughout that in the problem (1.6.1) there exists a saddle point $z_* = [x_*, u_*, v_*]$ of the Lagrangian $L(x,u,v)$.

2. A VERSION DUE TO P. WILSON

In (5.2) and (5.3) let

$$\alpha_k \equiv 1, \quad N_k = \frac{\partial^2 L(x_{k-1}, u_{k-1}, v_{k-1})}{\partial x^2}, \quad N_0 = I_n.$$

Under standard assumptions, Wilson [1] shows that this process has a local quadratic rate of convergence. To avoid the computation of matrices of the second derivatives, various simplified versions of the method have been devised.

3. VERSIONS DUE TO U. GARCIA PALOMARES AND O. MANGASARIAN [1]

These authors suggest the following methods for choosing α_k and N_k:

- 1) $\alpha_k \equiv 1$, the matrix N_k is such that $\|N_k - L_{xx}(z_k)\| \le \nu$;
- 2) $\alpha_k \equiv 1$, the matrix N_k is such that

$$\|N_k - L_{xx}(z_k)\| \leqslant \nu, \quad \lim_{k\to\infty} \frac{(N_k - L_{xx}(z_k))(x_{k+1} - x_k)}{\|z_{k+1} - z_k\|} = 0.$$

When the problem (5.3) has a non-unique solution, instead of z_{k+1} a point of the set of solutions is taken at which the norm $\|z_{k+1} - z_k\|$ is minimum. The variable ν is expressed in terms of derivatives of the functions defining the problem, at the point x_*. It is proved that if conditions similar to those of Lemma 4.1.1 are satisfied and the problem (1.6.1) has only inequality constraints, and furthermore, if the first rule is used, then the method has a linear rate of convergence; if the second rule is used, a superlinear rate of convergence.

4. THE FIRST VERSION DUE TO S. HAN

We assume that $\alpha \equiv 1$. We define the matrix N_k more precisely during the iteration procedure by trying to approximate the matrix $L_{xx}(z_k)$. To this end, we can exploit the various quasi-Newton formulas obtained in Section 2.5. We make use, for example, of Powell's symmetric version of Broyden's method (2.5.38). Let

$$\Delta x_k = x_{k+1} - x_k,$$
$$\Delta V_k = L_x(x_{k+1}, u_{k+1}, v_{k+1}) - L_x(x_k, u_{k+1}, v_{k+1}).$$

Then

$$N_{k+1} = N_k + \frac{(\Delta V_k - N_k \Delta x_k)\Delta x_k^T + \Delta x_k (\Delta V_k - N_k \Delta x_k)^T}{\langle \Delta x_k, \Delta x_k \rangle} - \frac{\langle \Delta V_k - N_k \Delta x_k, \Delta x_k \rangle}{[\langle \Delta x_k, \Delta x_k \rangle]^2} \Delta x_k \Delta x_k^T.$$

In a similar way it is possible to apply many other formulas for conversion of the matrix N_k. In [1] Han proves the superlinear rate of convergence of this version, making standard assumptions for the problem (4.1).

5. THE SECOND VERSION DUE TO S. HAN

All the versions described so far are only locally convergent. To expand the domain of convergence, one needs to introduce a special choice of the step α_k. To do this, the non-differentiable auxiliary function (2.1) can be used. We define α_k from the condition for approximate one-dimensional minimization of the auxiliary function in α_k:

$$P(x_{k+1}, \tau) \leqslant \min_{0 \leqslant \alpha_k \leqslant \delta} P(x_k + \alpha_k(\bar{x}_k - x_k), \tau) + \nu_k. \tag{5.4}$$

Here ν_k is a sequence of nonnegative numbers such that $\sum_{k=1}^{\infty} \nu_k < \infty$, τ, δ are positive constants; $\bar{x}_k$ denotes, as before, the solution to the problem (5.3), the matrices N_k are updated by any quasi-Newton formula. The implementation of this version of the method is not much more complicated than the preceding version: it includes only the line search (5.4). At the same time, as Han has shown in [2], his version of the method ensures the global convergence if (4.1) is a convex programming problem, Slater's condition is satisfied, τ is sufficiently large, and the function P_3 is substituted for P.

The computations described above can be regarded as the implementation of the exterior penalty function method, but instead of minimization of the non-differentiable function (2.1) in x the quadratic programming problems (5.3) are solved first and as the next step a one-dimensional minimization of (5.4) is done. On the other hand, this version can be regarded as a special method of unconstrained minimization of the non-differentiable function (2.1) in x.

A dual problem to the problem (5.3) (see the problem (1.6.7)) consists in finding

$$\max_{u\in E^e}\max_{v\in E_+^c}\min_{p\in E^n}\Big[\langle f_x(x_k),\, p\rangle+\frac{1}{2}p^T N_k p+ \\ +\langle u,\, g(x_k)+g_x^T(x_k)\,p\rangle+\langle v,\, h(x_k)+h_x^T(x_k)\,p\rangle\Big] \;, \tag{5.5}$$

where $p = x - x_k$. If the matrix N_k is not singular, the interior problem can easily be solved:

$$p = -N_k^{-1} L_x(x_k,u,v) \;. \tag{5.6}$$

Substituting this expression into (5.5), we arrive at the exterior problem of finding

$$\max_{u\in E^e}\max_{v\in E_+^c}\Big[L(x_k,\, u,\, v)-f(x_k)-\frac{1}{2}\|N_k^{-1}L_x(x_k,\, u,\, v)\|_2^2\Big]. \tag{5.7}$$

Let $[u_k,v_k]$ be the solution of the problem (5.7). Then, by (5.6) we have

$$\bar{x}_k - x_k = -N_k^{-1} L_x(x_k,u_k,v_k) \;.$$

Therefore, the quadratic programming problem (5.3) can be replaced by that of solving the maximization problem (5.7) which has only simple constraints.

6. A VERSION DUE TO B.N. PSHENICHNYJ

By hypothesis, $N_k \equiv I_n$. The step α_k is defined as follows. One finds the first value $s = 0,1,\ldots$ for which the inequality

$$P\left(x_k+\frac{1}{2^s}(\bar{x}_k-x_k),\, \tau_k\right)\leqslant P(x_k,\, \tau_k)-\frac{\varepsilon}{2^s}\|\bar{x}_k-x_k\|^2,\; 0<\varepsilon<1$$

is satisfied. If this inequality is satisfied for $s = s_1$ for the first time, it is assumed in (5.2) that $\alpha_s = 2^{-s_1}$. To simplify the auxiliary problem (5.3), only those constraints in which the feasibility has been violated maximally are taken into account.

To do this, one finds the quantity $S^1(x_k)$ (see the formula (1.2)); and the feasible set X_k in the problem (5.3) is given by

$$\begin{aligned} g^i(x_k)+\langle g_x^i(x_k),\ x-x_k\rangle &= 0, \quad i\in W_1(x_k),\\ h^j(x_k)+\langle h_x^j(x_k),\ x-x_k\rangle &\leqslant 0, \quad j\in W_2(x_k), \end{aligned} \tag{5.8}$$

where the index sets are used:

$$\begin{aligned} W_1(x_k) &= \{i\in[1:e]:\quad |g^i(x_k)|\geqslant S^1(x_k)-\delta_k\},\\ W_2(x_k) &= \{j\in[1:c]:\quad h^j(x_k)\geqslant S^1(x_k)-\delta_k\}. \end{aligned}$$

The initial value $\delta_0 > 0$ is specified by the user; next, it is halved each time when the auxiliary problem has no solution. The rule for the coefficient τ_k is as follows: if the sum of the moduli of the components of the dual vector, corresponding to the constraints (5.8), is less than τ_k, the value τ_k is doubled; otherwise $\tau_{k+1} = \tau_k$.

The convergence of the method is proved under the assumption that $P_3(x,\tau)$ is an auxiliary function and the gradients of the functions defining the problem satisfy a Lipschitz condition. The proof can be found in B. N. Pshenichnyj and Yu. M. Danilin [1].

7. VERSIONS DUE TO A.I. GOLIKOV AND V.G. ZHADAN [2]

These authors introduce the function $M(x)$ which is continuous on E^n and satisfies the condition $M(x) \geq 1 + \sqrt{n}\,\|f_x(x)\|_2$. In the first version, the following linear programming problem is solved: find

$$\min_x \min_q \ [\langle f_x(x_k),\ x-x_k\rangle + M(x_k)q] \quad , \tag{5.9}$$

where $q \in E^1_+$, the conditions (5.8) hold and, furthermore,

$$|x^j - x_k^j| \leq 1 + q , \qquad j \in [1:n] . \tag{5.10}$$

We denote by $\bar{x}_k$, q_k the solution to this problem, $p_k = \bar{x}_k - x_k$. The new point is defined by the formula similar to (5.2):

$$x_{k+1} = x_k + \alpha_k p_k . \tag{5.11}$$

The step α_k is chosen by reducing the initial step until the condition

$$P_3(x_{k+1}, \tau) \leqslant P_3(x_k, \tau) + \beta\alpha_k [\langle f_x(x_k), p_k\rangle - \tau S^1(x_k)] \tag{5.12}$$

is satisfied. The parameter $0 < \beta < 1$ and the initial step α are fixed.

In the second version, the following linear programming problem is solved: find

$$\min_a \min_b \min_q [\langle f_x(x_k), a-b\rangle + M(x_k) q]. \tag{5.13}$$

Here $a \in E_+^n$, $b \in E_+^n$, $q \in E_+^1$, the feasible set is given by

$$\begin{aligned} & g^i(x_k) + \langle g_x^i(x_k), a-b\rangle = 0, \quad i \in W_1(x_k), \\ & h^j(x_k) + \langle h_x^j(x_k), a-b\rangle \leqslant 0, \quad j \in W_2(x_k), \\ & \sum_{j=1}^n (a^j + b^j) \leqslant 1 + q. \end{aligned} \tag{5.14}$$

This problem is somewhat simpler than (5.9) since only one constraint (5.14) is used instead of n conditions (5.10); the new condition for a, b to be nonnegative is easily accountable. The point x_{k+1} is defined from (5.11), where $p_k = a - b$, the step being found from (5.12).

It is easy to verify that the dual problem to (5.9) is equivalent (in the sense of equivalence of optimal values of the dual variables corresponding to linearized constraints) to the following problem:

$$\max_{u,\,v\in U_k}\Bigg[\sum_{i\in W_1(x_k)} u^i g^i(x_k)+\sum_{j\in W_2(x_k)} v^j h^j(x_k)- \\ -\Big\| f_x(x_k)+\sum_{i\in W_1(x_k)} u^i g^i_x(x_k)+\sum_{j\in W_2(x_k)} v^j h^j_x(x_k)\Big\|_1\Bigg], \tag{5.15}$$

where

$$U_k=\Big\{u^i,\, v^j\geqslant 0:\ i\in W_1(x_k),\ j\in W_2(x_k), \\ \Big\| f_x(x_k)+\sum_{i\in W_1(x_k)} u^i g^i_x(x_k)+\sum_{j\in W_2(x_k)} v^j h^j_x(x_k)\Big\|_1\leqslant M(x_k)\Big\}.$$

The dual problem of (5.13) is the same as the last problem, with the only difference that instead of the norm $\|\cdot\|_1$ the Chebyshev norm $\|\cdot\|_\infty$ is needed. The cost functions in (5.7) and (5.15) are close to each other and differ only in the norms of the gradient $L_x(x,u,v)$. We will be discussing the methods of this kind in Section 4.6, where we use modified Lagrangians. The convergence of both versions can be proved under assumptions close to those used in Section 3.4. It can be found in A.I. Golikov and V.G. Zhadan [2].

Chapter 4

NUMERICAL METHODS FOR SOLVING NONLINEAR PROGRAMMING PROBLEMS USING MODIFIED LAGRANGIANS

The impediments in solving accurately many engineering problems using the penalty function method have impelled the development of numerical methods with geometric and quadratic convergence rates. The use of modified Lagrangians for solving nonlinear programming problems was suggested for the first time in Arrow, Hurwicz, and Uzawa [1]. Later on, this idea expanded and developed into a method with a geometric convergence rate. The first published works in this direction are Hestenes [1], Powell [1], Haarhoff and Buys [1], followed by Rockafellar [2], [3], Tret'yakov [1], Polyak and Tret'yakov [1], Gol'shtein and Tret'yakov [1], Bertsekas [1] - [3], Kort and Bertsekas [1], and many other authors. This method is frequently regarded as an independent method for solving nonlinear programming problems; Polyak and Tret'yakov call it "penalty-estimate method"; Rockafellar and Bertsekas call it the "method of multipliers," and other authors, the "augmented Lagrangian method."

We shall discuss this method in Section 4.3 (method 3.2), taking a still different approach, that is the reduction of the initial nonlinear programming problem (1.6.1) to that of finding a local maximin, minimax, saddle points and solving systems of nonlinear equations, using modified Lagrangians. This problem is then solved via the well-known methods for problems of this class. Under this approach, the method (3.21) is no more than simple iteration applied to solving a system of nonlinear equations. This reduction allows us to obtain a variety of methods for solving nonlinear programming problems.

1. THE SIMPLEST MODIFICATION OF THE LAGRANGIAN

1. PRELIMINARY RESULTS

Throughout this chapter we consider the general problem of nonlinear programming (1.6.1). By Theorem 1.6.1, under certain conditions, solving the problem (1.6.1) can be replaced by finding saddle points of the Lagrangian. The numerical methods of finding the saddle points are inappropriate for this purpose, because they are intended for finding unconstrained solutions, whereas in (1.6.9) we have the constraint $v \geqslant 0$ to account for. To overcome this difficulty, we modify the Lagrangian by introducing

$$F(x, u, w) = f(x) + \langle g(x), u \rangle + \sum_{j=1}^{c} (w^j)^2 h^j(x), \qquad u \in E^e, \quad w \in E^c. \tag{1.1}$$

Let

$$y = [u, w] \in E^m, \quad m = e + c, \quad z = [x, y] \in E^{n+m},$$
$$F(x, u, w) = F(x, y) = F(z).$$

Let the functions defining the problem be differentiable, and in the problem (1.6.1) let the Kuhn-Tucker point $[x_*, u_*, v_*]$ exist. Then we define $z_* = [x_*, u_*, w_*]$, where $w_*^j = \sqrt{v_*^j}$ for $j \in [1:c]$. From the Kuhn-Tucker conditions we see that z_* is a stationary point of $F(z)$ since

$$F_z(z_*) = 0_{n+m} . \tag{1.2}$$

This property allows us to reduce the problem (1.6.1) to finding solutions of (1.2). Repeating the proof of Theorem 1.6.1 almost verbatim we can show that if $z_* = [x_*, u_*, w_*]$ is an unconstrained saddle point of F, then $x_* \in X_*$ and $w_*^j h^j(x_*) = 0$. If the functions defining the problem are differentiable, then the saddle point z_* of F satisfies (1.2). However, the use of numerical methods of finding saddle points is complicated by the fact that F is not convex/concave, and this condition is really essential for ensuring convergence of many methods for finding saddle points. Hence it is appropriate to consider the problem of finding the local maximin

$$\max_{u \in E^e} \max_{w \in E^c} \min_{x \in E^n} F(x, u, w). \tag{1.3}$$

One can solve it using the methods in Section 2.6. We show that under standard assumptions, the matrix $F_{zz}(z_*)$ is nonsingular at the Kuhn-Tucker points and F has a local maximin. For this we introduce three auxiliary square matrices of orders $n + m$, $n + m$ and m, respectively:

$$F_{zz}(z) = \left[\begin{array}{c:c:c} F_{xx}(z) & g_x(x) & 2h_x(x)D(w) \\ \hdashline g_x^T(x) & 0_{ee} & 0_{ec} \\ \hdashline 2D(w)h_x^T(x) & 0_{ce} & 2D(h(x)) \end{array}\right],$$

$$H(z)=\left[\begin{array}{c|c|c} -F_{xx}(z) & -g_x(x) & -2h_x(x)D(w) \\ \hline g_x^T(x) & 0_{ee} & 0_{ec} \\ \hline 2D(w)h_x^T(x) & 0_{ce} & 2D(h(x)) \end{array}\right],$$
$$N(z)=[F_{xu}(z)\,|\,F_{xw}(z)]^T F^{-1}(z)\,[F_{xu}(z)\,|\,F_{xw}(z)]- \\ -\left[\begin{array}{c|c} F_{uu}(z) & F_{uw}(z) \\ \hline F_{wu}(z) & F_{ww}(z) \end{array}\right]. \qquad (1.4)$$

<u>LEMMA 4.1.1</u>. Let the sufficient conditions for a minimum given by Theorem 1.7.4 be satisfied at the Kuhn-Tucker point $[x_*,u_*,v_*]$ of the problem (1.6.1) and let the constraint qualification hold at x_*. Then the matrix $F_{zz}(z_*)$, where $z_* = [x_*,u_*,w_*]$, is nonsingular.

To show it is nonsingular, we have only to show that its null space is zero:

$$F_{zz}(z_*)\bar{z}=0, \quad \bar{z}=[\bar{x},\ \bar{u},\ \bar{w}]. \qquad (1.5)$$

Written out in detail this system is

$$F_{xx}(z_*)\bar{x}+g_x(x_*)\bar{u}+2h_x(x_*)D(w_*)\bar{w}=0, \qquad (1.6)$$
$$g_x^T(x_*)\bar{x}=0, \quad D(w_*)h_x^T(x_*)\bar{x}+D(h(x_*))\bar{w}=0. \qquad (1.7)$$

From (1.7) it follows that for all $j\in\sigma(x_*)$: $h^j(x_*) = \bar{x}^T h^j(x_*) = 0$, $w_*^j \neq 0$. For $j \notin \sigma(x_*)$ we have $h^j(x_*) < 0$, $\bar{w}^j = w_*^j = 0$. In both cases

$$h^j(x_*)\bar{w}^j=0, \ D(w_*)h_x^T(x_*)\bar{x}=0.$$

Assume $||\bar{x}|| \neq 0$. Then multiplying (1.6) on the left by $\bar{x}^T$ and noting (1.7), we obtain

$$\bar{x}^T F_{xx}(z_*)\bar{x} = 0 , \qquad (1.8)$$

where $\bar{x}$ satisfies the conditions

$$g_x^T(x_*)\bar{x} = 0, \quad [h_x^j(x_*)]^T\bar{x} = 0, \quad j \in \sigma(x_*).$$

To each Kuhn-Tucker point $[x_*, u_*, v_*]$ there corresponds a point $z_* = [x_*, u_*, w_*]$. Hence $F_{xx}(z_*) = L_{xx}(x_*, u_*, v_*)$ and from (1.8) we have

$$\bar{x}^T L_{xx}(x_*, u_*, v_*)\bar{x} = 0.$$

But this equality contradicts the positive definiteness of the matrix $L_{xx}(x_*, u_*, v_*)$ on $K_3(x_*)$ (see the condition (1.7.16)). Hence $\bar{x} = 0$. From (1.6) and (1.7) it then follows that

$$g_x(x_*)\bar{u} + 2 \sum_{j \in \sigma(x_*)} h_x^j(x_*) w_*^j \bar{w}^j = 0, \quad h^j(x_*)\bar{w}^j = 0.$$

By the strict complementarity condition, all $v_*^j \neq 0$ for $j \in \sigma(x_*)$, for these same values $w_*^j \neq 0$. Noting the constraint qualification, we conclude that $\bar{u} = 0$ and $\bar{w}^j = 0$ if $j \in \sigma(x_*)$. It was shown above that $\bar{w}^j = 0$ for all $j \notin \sigma(x_*)$. Hence all the the vectors $\bar{x}$, $\bar{u}$, $\bar{w}$ satisfying (1.5) are null and $F_{zz}(z_*)$ is nonsingular. ///

LEMMA 4.1.2. Let the constraint qualification and strict complementarity condition be satisfied at the Kuhn-Tucker point $[x_*, u_*, v_*]$. Also, let the functions defining the problem be twice continuously differentiable in a neighborhood of x_*, and the matrix $L_{xx}(x_*, u_*, v_*)$ be positive definite. Then:

• 1. the matrix $N(z_*)$ is positive definite;

• 2. all roots λ of the equation

$$|H(z_*) - \lambda I_{n+m}| = 0 \tag{1.9}$$

have strictly negative real parts;

• 3. the point $[x_*, y_*]$, where $y_* = [u_*, w_*]$, is a point of the strict local maximin of the problem (1.3);

• 4. the point $[x_*, y_*]$ is a saddle point (local in x, global in y) of the problem (1.3);

• 5. x_* is a local solution of the problem (1.6.1).

Proof. We can assume without loss of generality that $h^j(x_*) = 0$ for $1 \le j \le s$ and $h^j(x_*) < 0$ for $1+s \le j \le c$. Introduce the vectors

$$g = [u^1, \ldots, u^e, w^1, \ldots, w^s] \in E^k, \qquad k = e+s$$

and

$$h^b(x_*) = [h^{s+1}(x_*), \ldots, h^c(x_*)] \quad .$$

We can write the matrix $N(z_*)$ defined by the formula (1.4) as follows:

$$N(z_*) = \left[\begin{array}{c|c} F_{qx}(z_*)\,F_{xx}^{-1}(z_*)\,F_{xq}(z_*) & 0_{k(c-s)} \\ \hline 0_{(c-s)k} & -2D(h^b(x_*)) \end{array}\right].$$

By the strict complementarity condition all $w_*^j \ne 0$ for $1 \le j \le s$. Hence from the constraint qualification it follows that the columns of $F_{xq}(z_*)$ are linearly independent, the rank of this matrix is maximal, equal to k, and the matrix $F_{qx}(z_*)F_{xx}^{-1}(z_*)F_{xq}(z_*)$ of dimension k^2 is positive definite. The diagonal matrix $-D(h^b(x_*))$ is also positive definite by assumption. This implies that $N(z_*)$ is positive definite.

Let $H(z_*)\bar{z} = \bar{z}$ and let the vector $\tilde{z} = [\tilde{x}, \tilde{u}, \tilde{v}]$ be the complex conjugate of $\bar{z}$. We assume that $|\bar{z}| \ne 0$. Then

$$\operatorname{Re}\tilde{z}^T H(z_*)\bar{z} = \operatorname{Re}\lambda|\tilde{z}|^2 =$$
$$= \operatorname{Re}[-\tilde{x}^T F_{xx}(z_*)\bar{x} + 2\tilde{w}^T D(h(x_*))\bar{w}] \leqslant 0.$$

The last inequality follows from the positive definiteness of $F_{xx}(z_*)$ and the fact that $h(x_*) \leq 0$ at a feasible point. Let $\operatorname{Re}\lambda = 0$. Then

$$\operatorname{Re}[-\tilde{x}^T F_{xx}(z_*)\bar{x} + 2\tilde{w}^T D(h(x_*))\bar{w}] = 0$$

only if $\bar{x} = 0$ and $\bar{w}^j = 0$ for j such that $h^j(x_*) < 0$. From the system $H(z_*)\bar{z} = \bar{z}$ we obtain

$$\sum_{i=1}^{e} g_x^i(x_*)\bar{u}^i + 2\sum_{j=1}^{c} w_*^j \bar{w}^j h_x^j(x_*) = 0,$$

which may be rewritten as

$$\sum_{i=1}^{e} g_x^i(x_*)\bar{u}^i + 2\sum_{j\in\sigma(x_*)} w_*^j \bar{w}^j h_x^j(x_*) = 0.$$

By the strict complementarity condition, if $j \in \sigma(x_*)$, then all $w^j \neq 0$. From the constraint qualification we obtain that the vectors $g_x^i(x_*)$ and $h_x^j(x_*)$, where $j \in \sigma(x_*)$, are linearly independent. To prove that the above equation holds, it is necessary that $\bar{u} = 0$ and $\bar{w}^j = 0$ for all $j \in \sigma(x_*)$. But we showed above that $\bar{w}^j = 0$ for all $j \notin \sigma(x_*)$. Hence $\bar{u} = \bar{w} = 0$ and $\bar{z} = 0$, which contradicts the initial assumption $|\bar{z}| \neq 0$. Thus the case $\operatorname{Re}\lambda = 0$ does not hold, and all the roots of (1.9) have negative real parts.

At a stationary point $z_* = [x_*, y_*]$ the matrices F_{xx} and N are positive definite, therefore by Theorem 1.5.9, z_* is a strict local maximin point of the problem (1.3).

It is easy to show that z_* is a saddle point (local in x, global in y), i.e., there is a neighborhood $G(x_*)$ of x_* such that for any $x \in G(x_*)$, $x \neq x_*$, and any $y \in E^m$ the inequalities

$$F(x_*,y) \leq F(x_*,y_*) < F(x,y_*) \tag{1.10}$$

are satisfied.

The function $F(x,y)$ is not convex/concave (convex in x for any fixed y and concave in y for any fixed x). Concavity in w occurs only for fixed x with $h(x) \leqslant 0$.

From the positive definiteness of $F_{xx}(z_*)$ and the stationary condition (1.2) we have the right-hand inequality in (1.10). Noting that x_* is a feasible stationary point, we obtain

$$F(x_*, y) - F(x_*, y_*) = \sum_{i=1}^{e} g^i(x_*) u^i + \sum_{j=1}^{c} (w^j)^2 h^j(x_*) \leqslant 0$$

for any $u \in E^e$, $w \in E^c$, yielding in turn the left-hand side of (1.10).

For any $x \in X \cap G(x_*)$, from the right-hand inequality of (1.10) we have

$$f(x_*) = F(z_*) \leqslant F(x, y_*) \leqslant f(x),$$

i.e., x_* is a local solution of the initial problem (1.6.1).

For convex programming problems the function $F(x,y_*)$ is convex in x and from the condition $F_x(x_*,y_*) = 0$ it follows that $F(x_*,y_*) \leq F(x,y_*)$ for any $x \in E^n$.

If the conditions of the Lemma are satisfied, then the sufficient conditions for a minimum given by McCormick's theorem 1.7.2 hold automatically. Hence we also see that x_* is a local solution of the problem (1.6.1). ///

2. NEWTON'S METHOD

For finding solutions of (1.2) we use Newton's method

$$z_{k+1} = z_k - F_{zz}^{-1}(z_k) F_z(z_k). \tag{1.11}$$

To implement the method, we either have to invert the symmetric matrix F_{zz} of order $n+m$ or to solve the system of $n+m$ linear equations

$$F_{zz}(z_k)(z_{k+1} - z_k) = -F_z(z_k). \tag{1.12}$$

For problems of large dimensions one can use a block representation of the matrix of second derivatives

$$F_{zz}(z) = \left[\begin{array}{c|c} F_{xx}(z) & F_{xy}(z) \\ \hline F_{yx}(z) & F_{yy}(z) \end{array}\right].$$

We rewrite the system (1.12) in equivalent form:

$$\begin{aligned} F_{xx}(z_k)(x_{k+1} - x_k) + F_{xy}(z_k)(y_{k+1} - y_k) &= -F_x(z_k), \\ F_{yx}(z_k)(x_{k+1} - x_k) + F_{yy}(z_k)(y_{k+1} - y_k) &= -F_y(z_k). \end{aligned}$$

Supposing that $F_{xx}(z_k)$ is nonsingular, we extract from the first equation the difference $x_{k+1} - x_k$ and substitute the result into the second equation to obtain

$$y_{k+1} = y_k + N^{-1}(z_k)[F_y(z_k) - F_{yx}(z_k) F_{xx}^{-1}(z_k) F_x(z_k)], \tag{1.13}$$

$$x_{k+1} = x_k - F_{xx}^{-1}(z_k)[F_x(z_k) + F_{xy}(z_k)(y_{k+1} - y_k)]. \tag{1.14}$$

This process can be continued to obtain particular formulas for computing the vectors u_k and w_k. To this end, we rewrite (1.13) in the form

$$N(z_k)(y_{k+1} - y_k) = F_y(z_k) - F_{yx}(z_k) F_{xx}^{-1}(z_k) F_x(z_k). \tag{1.15}$$

Noting that the matrices of second derivatives F_{uu} and F_{uw} are null, we put N in block form

$$N = \left[\begin{array}{c|c} A & B \\ \hline B^T & C \end{array}\right].$$

Here we have set:

$$\begin{aligned} A &= F_{ux}F_{xx}^{-1}F_{xu} = g_x^T F_{xx}^{-1} g_x, \\ B &= F_{ux}F_{xx}^{-1}F_{xw} = 2g_x^T F_{xx}^{-1} h_x D(w), \\ C &= F_{wx}F_{xx}^{-1}F_{xw} - F_{ww} = 4D(w)\, h_x^T F_{xx}^{-1} h_x D(w) - 2D(h). \end{aligned}$$

Supposing $\Lambda(z_k)$ is nonsingular, from the system (1.5) we have

$$w_{k+1} = w_k + [C - B^T A^{-1} B]^{-1} [F_w - F_{wx}F_{xx}^{-1}F_x - B^T A^{-1} \times \times [F_u - F_{ux}F_{xx}^{-1}F_x]], \quad (1.16)$$

$$u_{k+1} = u_k + A^{-1} [F_u - F_{ux}F_{xx}^{-1}F_x - B[w_{k+1} - w_k]]. \quad (1.17)$$

Here all the matrices are computed at the point $z_k = [x_k, u_k, w_k]$. One can also obtain these formulas directly from (1.12) by using Frobenius' formula for inverting block matrices (see Appendix II).

Various modifications of the method are possible. One can determine x_k from solving the unconstrained minimization problem by taking

$$x_k \in \operatorname*{Arg\,min}_{x \in E^n} F(x, y_k). \quad (1.18)$$

We assume that in some neighborhood of y_* this procedure determines a unique, differentiable function $x(y)$. In this case we need to find a vector y being a solution of the system

$$\frac{d}{dy} F(x(y), y) = 0.$$

Using Newton's method for finding solutions to this equation, we obtain the formula

$$y_{k+1} = y_k + N^{-1}(x_k, y_k) F_y(x_k, y_k) \tag{1.19}$$

which follows also from (1.13) if we set there $F_x(x_k, y_k) = 0$. We can use the formula (1.14) to extrapolate x_{k+1} in solving the problem (1.18).

To show the convergence of Newton's method (1.11) to a solution of (1.12), we use Theorem 2.5.3. Using Lemma 4.1.1, we express the nonsingularity condition of $F_{zz}(z_*)$ in terms used in nonlinear programming theory and arrive at the following statement.

THEOREM 4.1.1. Let the conditions of Lemma 4.1.1 hold at the Kuhn-Tucker point $[x_*, u_*, v_*]$ and let $F_{zz}(z)$ satisfy a Lipschitz condition in a neighborhood of $z_* = [x_*, u_*, w_*]$. Then the method (1.11) converges locally to the point z_* with quadratic rate.

The method (1.11) thus ensures a quadratic rate of convergence in solving a general problem of nonlinear programming, which is unusually high for problems of this class. Later we will frequently use Newton's method for these problems. However, the method (1.11) will be special, since it provides the simplest rapidly converging computing formulas. The method is widely used for solving various problems of small dimension. As a rule, this method is used to finish the computations; it is applied after other, coarser methods have been used to get the initial approximation for the vector x and the dual vectors.

Just as was done in Section 2.5 in describing Newton's method, we can list here several disadvantages of the method (1.11) which may complicate solving the problem (1.6.1):

- 1. in some cases, the method diverges;
- 2. the method is not practical for solving problems where the dimension $n+m$ of the vector z is large because we then have to calculate and invert a high-order matrix;
- 3. for an insufficient, coarse initial approximation the method converges to dispensable stationary points (for example, to unfeasible x);
- 4. If any components of the vector w are zero in the initial iterations, they remain equal ot zero throughout all the iterations ("sticking" of the components of w occurs).

Drawback 1 can be avoided in some cases by introducing a variable step; the rules for adjusting the step are described in Section 2.5.

Drawback 2 can be removed by using the formulas (1.13) and (1.14), or (1.14), (1.16) and (1.17), or (1.18) and (1.19), instead of (1.11). One can also apply various quasi-Newton methods.

The last two drawbacks are characteristic of all methods based on the simplest modification of a Lagrangian of the form (1.1). One can try to add to F penalizing terms, e.g., by setting

$$P(z) = F(z) + \tau \left[\sum_{i=1}^{e} [g^i(x)]^2 + \sum_{j=1}^{c} [h_+^j(x)]^3 \right].$$

If z_* is a stationary feasible point of $F(z)$, then it will be a stationary point of $P(z)$. We can therefore find stationary points of $P(z)$. This technique somewhat expands the region of convergence; however it does not remove the possibility that for a fixed value of τ the process may converge to an unfeasible point.

3. ARTIFICIAL VECTORS

Another way of reducing the initial problem 1.6.1 to solving systems of equations is to introduce artificial variables making it possible to reduce (1.6.1) to a nonlinear programming problem in which the feasible set is given by equality-type constraints only. We used this technique in Subsection 1.7.4 and reduced the problem (1.6.1) to an equivalent problem (1.7.17) for which the Lagrangian L^1 has the form (1.7.18) and the necessary conditions of the minimum are given by the relations (1.7.20), or in more detailed form, (1.7.21). Using the notation of Section 1.7 we express Newton's method for finding stationary points of the function $L^1(z,y)$ as follows:

$$\left.\begin{aligned} L^1_{zz}(z_k, y_k)(z_{k+1}-z_k)+L^1_{zy}(z_k, y_k)(y_{k+1}-y_k)= \\ =-L^1_z(z_k, y_k), \\ L^1_{yz}(z_k, y_k)(z_{k+1}-z_k)=-L^1_y(z_k, y_k). \end{aligned}\right\} . \tag{1.20}$$

The matrix L^1_{zy} is defined similarly by (1.7.22):

$$L^1_{zy}(z, y)=\left[\begin{array}{c|c} g_x(x) & h_x(x) \\ \hline 0_{ce} & \frac{1}{2}D(p) \end{array}\right]=[L^1_{yz}(z, y)]^T.$$

Let

$$A(z, y)=\left[\begin{array}{c|c} L^1_{zz}(z, y) & L^1_{zy}(z, y) \\ \hline L^1_{yz}(z, y) & 0_{mm} \end{array}\right].$$

It is not hard to show that if the conditions of Lemma 4.1.1 are satisfied, the matrix $A(z_*,y_*)$ is nonsingular. Therefore the method (1.20) is locally convergent under standard assumptions.

The method (1.20) is harder to implement than (1.11) since

here we have to invert a matrix of order $n+m+c$. However, when using the method (1.20) we can assert that if the process converges to some point $[x_*, p_*, y_*]$, then the point x_* is feasible in the problem (1.6.1). The method (1.11) does not have this property. At the same time, we are not guaranteed here that the components of the dual vector y^{e+j}, $1 \le j \le c$, are nonnegative. In later sections we will make more radical modifications of the Lagrangian enabling us to avoid this drawback.

4. ITERATIVE METHODS

If the conditions of Lemmas 4.1.1 and 4.1.2 are satisfied at the Kuhn-Tucker point $[x_*, u_*, v_*]$, the corresponding point $[x_*, u_*, w_*]$ is a strict local maximin point of the function $F(z)$, and all the methods described in Section 2.6 can be used to solve the problem (1.6.1). For example, the methods (2.6.1), (2.6.17), (2.6.18), (2.6.24), and (2.2.3) applied to solving (1.3) are written in the form:

$$\dot{y} = \varepsilon F_y, \qquad \dot{x} = -F_x, \tag{1.21}$$

$$\dot{y} = F_y, \qquad \dot{x} = -F_x - F_{xx}^{-1} F_{xy} F_y, \tag{1.22}$$

$$\dot{y} = F_y - F_{yx} F_{xx}^{-1} F_x, \qquad \dot{x} = -F_x, \tag{1.23}$$

$$\dot{y} = F_y(x(y), y), \qquad \dot{x} = x(y) - x, \tag{1.24}$$

$$\dot{y} = F_y(x(y), y). \tag{1.25}$$

Here the dependence of $x(y)$ is defined from (1.18); ε is a small parameter $(0 < \varepsilon << 1)$; when the arguments of F are not shown it is assumed that the derivatives of this function are computed at $[x(t), y(t)]$.

Theorem 4.1.2. Let the conditions of Lemma 4.1.2 be satisfied at the Kuhn-Tucker point $[x_*, u_*, v_*]$. Then, there exists $\bar{\varepsilon} > 0$, $\bar{\alpha} > 0$ such that for any fixed $0 < \varepsilon < \bar{\varepsilon}$, $0 < \alpha < \bar{\alpha}$ and for $\varepsilon = 1$, the methods (1.21) - (1.25) and their discrete variants of the form (2.6.20) converge locally to the point $z_* = [x_*, u_*, w_*]$, $z_* = [x_*, u_*, w_*]$.

Compared with the theorems of Section 2.6, we have one more proof of the convergence of the system (1.21) for $\varepsilon = 1$. Hence we need to prove this assertion only. Form a variational equation. Let

$$\delta x = x(t) - x_*, \quad \delta u = u(t) - u_*, \quad \delta w = w(t) - w_*,$$
$$\delta z = [\delta x, \ \delta u, \ \delta w],$$

where $x(t)$, $u(t)$, $w(t)$ are solutions of the system (1.21) satisfying the initial conditions $x(0) = x_0$, $u(0) = u_0$, $w(0) = w_0$. Removing second-order terms in δz, we obtain

$$\delta \dot{z} = H(z_*) \, \delta z \ .$$

From the second assertion of Lemma 4.1.2 we have that the characteristic values of the matrix $H(z_*)$ have strictly negative real parts. This by Theorem 2.1.1 implies the asymptotic, exponential stability of z_* and therefore the local convergence of the method (1.21) and of its discrete variant.

If there are no inequality-type constraints in the problem (1.6.1), then for $\varepsilon = 1$ the method (1.21) is called the Arrow-Hurwicz method. It has a low rate of convergence and therefore is used only when other, rapidly converging methods are inapplicable (for example, in solving large-dimension problems). Note that to take into account inequality-type constraints one can use, rather than (1.21), a method analogous to (2.4.5):

$$x_{k+1} = x_k - \alpha_k L_x(x_k, u_k, v_k),$$
$$u_{k+1} = u_k + \alpha_k L_u(x_k, u_k, v_k),$$
$$v^j_{k+1} = \max[0, v^j_k + \alpha_k L_{v^j}(x_k, u_k, v_k)], \quad v^j_0 \geqslant 0, \quad j \in [1:c].$$

Here the step α_k is taken either to be constant or to vary according to the rules (2.4.6) - (2.4.8).

2. MODIFIED LAGRANGIANS

1. PRELIMINARY RESULTS

The reduction to solving systems of nonlinear equations, already expressed in Theorem 1.6.4, furnishes a rich class of numerical methods for solving nonlinear programming problems. We will develop this approach, working with the following modified Lagrangian:

$$H(x, u, v) = f(x) + \sum_{i=1}^{e} \varphi(g^i(x), u^i) + \sum_{j=1}^{c} \psi(h^j(x), v^j). \tag{2.1}$$

The functions ϕ and ψ must be such as to guarantee that (1.6.18) holds, which is written in the form

$$\sum_{j=1}^{c} \psi(h^j(x), v^j_*) \leqslant \sum_{j=1}^{c} \psi(h^j(x_*), v^j_*), \tag{2.2}$$

where $x \in X$. We will give below further conditions for the choice of ϕ and ψ.

For a fixed vector $[u,v]$ we define the set

$$X(u, v) = \operatorname*{Arg\,min}_{x \in E^n} H(x, u, v). \tag{2.3}$$

A necessary condition of the minimum of $H(x,u,v)$ in x is that

$$H_x(x, u, v) = f_x(x) + \sum_{i=1}^{e} \varphi_g(g^i(x), u^i) g_x^i(x) + \sum_{j=1}^{c} \psi_h(h^j(x), v^j) h_x^j(x) = 0 \tag{2.4}$$

holds.

Let us compare this with the necessary conditions for a minimum of the Lagrangian in x:

$$L_x(x, u, v) = f_x(x) + \sum_{i=1}^{e} u^i g_x^i(x) + \sum_{j=1}^{c} v^j h_x^j(x) = 0.$$

It makes sense to introduce the system of equations

$$\varphi_g(g^i, u^i) = u^i, \quad i \in [1:e], \tag{2.5}$$
$$\psi_h(h^j, v^j) = v^j, \quad j \in [1:c] \tag{2.6}$$

and to choose ϕ and ψ so that this system is solvable iff $g^i = 0$, $h^j \le 0$, $h^j v^j = 0$. If we can find x_*, u_*, v_* for which the system (2.5), (2.6) is solvable, $x_* \in X(u_*, v_*)$ and the conditions (2.2) hold, then by Theorem 1.6.4, the problem (1.6.1) will have a solution since $x_* \in X_*$.

Throughout below we will assume that the functions $\phi(g^i, x^i)$, $\psi(h^j, v^j)$ are continuous for any values of the arguments and at least continuously differentiable in g^i and h^j, respectively. Let $J(h^j)$ denote the set of real solutions to (2.6) depending on h^j as a parameter, and let the set of nonnegative elements in $J(h^j)$ be written $J_+(h^j)$. Let us impose additional conditions on ϕ and ψ:

- A_1. For any real u^j we have
  - 1. $\phi_g(0, u^i) = u^i$,
  - 2. if $g^i \neq 0$, then $\phi_g(g^i, u^i) \neq u^i$.

• A_2. For $h^j > 0$, $J(h^j) = \emptyset$; for $h^j < 0$ the set $J(h^j)$ consists of zero; $J(0)$ equals the set of all nonnegative numbers.

• A_3. For $h^j > 0$, $J_+(h^j) = \emptyset$; for $h^j < 0$ the set $J_+(h^j)$ consists of zero; any nonnegative number belongs to $J_+(0)$.

• A_4. If at $[x_*, v_*] \in E^{n+c}$ the complementarity condition is satisfied, $h(x_*) \le 0$, $0 \le v_*$, then for any $x \in X$ one has the inequality

$$\sum_{j=1}^{c} \psi(h^j(x), v_*^j) \leqslant \sum_{j=1}^{c} \psi(h^j(x_*), v_*^j).$$

• A_5. For any h^j and any $v^j \ge 0$ the derivative $\psi_h(h^j, v^j) \ge 0$.

Condition A_4 ensures the inequality (2.2). Condition A_5 will be useful for implementing the numerical methods. Let us give some functions satisfying A_1:

$$\varphi^1(g^i, u^i) = g^i u^i + \frac{1}{2}(g^i)^2, \quad \varphi^2(g^i, u^i) = g^i u^i + \operatorname{ch} g^i,$$

$$\varphi^3(g^i, u^i) = g^i(u^i - 1) + e^{g^i},$$

$$\varphi^4(g^i, u^i) = g^i u^i + \frac{1}{2\pi}[2g^i \operatorname{arctg} g^i - \ln[1+(g^i)^2]]\, e^{-(u^i)^2},$$

$$\varphi^5(g^i, u^i) = g^i u^i + \frac{(g^i)^2}{2}(1 + \varepsilon \cos u^i), \qquad -1 < \varepsilon < 1.$$

All these functions are of the class that can be represented in the form

$$\varphi^6(g^i, u^i) = g^i(u^i - \alpha'(0)\beta(u^i)) + \alpha(g^i)\beta(u^i).$$

Here $\alpha(g)$ is a differentiable function such that $\alpha'(0) \neq \alpha'(a)$ for any $a \neq 0$, the function $\beta(a)$ is differentiable and $\beta(a) \neq 0$ for any a. In even more general form,

$$\varphi^7(g^i, u^i) = g^i(u^i - \eta_g(0, u^i)) + \eta(g^i, u^i),$$

where for any u^i, the equality $\eta_g(0,u^i) = \eta_g(g^i,u^i)$ holds only for $g^i = 0$.

As simple examples for ψ we give the following:

$$\psi^1(h^j, v^j) = \frac{1}{2}[(h^j + v^j)_+]^2,$$
$$\psi^2(h^j, v^j) = \gamma(h^j_+) + v^j e^{h^j},$$
$$\psi^3(h^j, v^j) = \gamma(h^j_+) +$$
$$+ \frac{v^j}{r}\begin{cases} 1 + rh^j + \frac{r(r+1)}{2!}(h^j)^2 + \frac{r(r+1)(r+2)}{3!}(h^j)^3 & \text{if } h^j \geqslant 0, \\ (1-h^j)^{-r} & \text{if } h^j < 0, \end{cases}$$
$$\psi^4(h^j, v^j) = \gamma(h^j_+) + v^j[h^j - \gamma(-h^j_-)],$$
$$\psi^5(h^j, v^j) = r\left[\frac{1}{4}(h^j + v^j)^4_+ - h^j(v^j)^3\right] + v^j\begin{cases} h^j & \text{if } h^j \geqslant 0, \\ \operatorname{arctg} h^j & \text{if } h^j < 0. \end{cases}$$

Here $r > 0$, the function γ is such that $\gamma(0) = 0$, $\gamma(a) > 0$ and $\gamma'(a) > 0$ for $a > 0$, $\gamma'(a) = 0$ iff $a = 0$ (say, $\gamma(a) = a^2, a^3, a^4$).

The second and third functions can be written in the form

$$\psi^6(h^j, v^j) = \gamma(h^j_+) + v^j\alpha(h^j),$$

where $\alpha(a)$ is twice continuously differentiable, $0 \le \alpha'(a) < 1$ for $a < 0$, $1 < \alpha'(a)$ for $a > 0$.

The functions $\psi^1 - \psi^4$, ψ^6 satisfy A_3, A_4, A_5. Let us check A_3 for ψ^4. Let $h^j > 0$. Then (2.6) has the form $v^j + \gamma'(h^j_+) = v^j$, i.e., $\gamma'(h^j_+) = 0$, which holds only if $h^j \le 0$, but this contradicts the initial assumption. Therefore, for $h^j > 0$ (2.6) cannot be satisfied for v^j. Let $h^j < 0$. Then

$$\gamma(h^j_+) = \gamma'(h^j_+) = 0, \quad \gamma'(-h^j_-) > 0,$$
$$v^j + v^j\gamma'(-h^j_-) = v^j,$$

and therefore (2.6) is solvable only if $v^j = 0$. For $h^j = 0$ any value of v^j satisfies (2.6). Similarly we convince ourselves that A_3 holds for the remaining functions.

If the strict complementarity condition is satisfied at the point $[x_*, v_*]$, then condition A_4 follows from the inequalities

$$\psi^1(h^j(x), v_*^j) = \frac{1}{2}(h^j(x) + v_*^j)_+^2 \leqslant \frac{1}{2}(v_*^j)^2,$$
$$\psi^2(h^j(x), v_*^j) = v_*^j e^{h^j(x)} \leqslant \psi^2(h^j(x_*), v_*^j) = v_*^j,$$
$$\psi^3(h^j(x), v_*^j) = \frac{1}{r} v_*^j (1 - h^j(x))^{-r} \leqslant \psi^3(h^j(x_*), v_*^j) = \frac{1}{r} v_*^j,$$
$$\psi^4(h^j(x), v_*^j) = v_*^j h^j(x) - v_*^j \gamma(-h_-^j(x)) \leqslant \psi^4(h^j(x_*), v_*^j) = 0.$$

It is also obvious that A_5 holds for the functions $\psi^1 - \psi^4$, ψ^6.

Condition A_2 is satisfied for ψ^1 and ψ^5.

2. NUMERICAL METHODS USING MODIFIED LAGRANGIANS

For each fixed vector $[u,v]$ we determine $x(u,v) \in X(u,v)$ from (2.3), next we choose $[u,v]$ such that the following system of nonlinear equations is satisfied:

$$\varphi_g(g(x(u, v)), u) = u, \quad \psi_h(h(x(u, v)), v) = v. \tag{2.7}$$

Here

$$\varphi_g(g, u) = [\varphi_g(g^1, u^1), \ldots, \varphi_g(g^e, u^e)], \ \psi_h(h, v) = [\psi_h(h^1, v^1), \ldots, \psi(h^c, v^c)].$$

THEOREM 4.2.1. Let the functions ϕ and ψ be constructed according to conditions A_1 and A_2 and let there exist vectors u_*, v_*, $x_* = x(u_*, v_*) \in X(u_*, v_*)$ satisfying (2.7). Then in the problem (1.6.1):

• 1. if A_4 holds then $X_* \neq \emptyset$, $x_* \in X_*$;

• 2. if $H(x,u_*,v_*)$ is differentiable at x_*, then $[x_*,u_*,v_*]$ is a Kuhn-Tucker point.

Proof. From A_1 it follows that $g(x_*) = 0$, from A_2 that $h(x_*) \le 0$, $0 \le v_*$ and $v_*^j h^j(x_*) = 0$, $j \in [1:c]$. Hence the conditions (1.6.17) hold and when A_4 is satisfied it follows from Theorem 1.6.4 that $x_* \in X_*$. Noting that at $[x_*,u_*,v_*]$ the condition (2.4) holds, we conclude that $[x_*,u_*,v_*]$ is a Kuhn-Tucker point. ///

Theorem 4.2.1 opens many possibilities for using numerical methods of solving systems of equations in order to solve the initial nonlinear programming problem (1.6.1). Using, say, the method of simple iteration, we obtain the following method:

$$u_{k+1} = \varphi_g(g(x_k),\ u_k), \tag{2.8}$$
$$v_{k+1} = \psi_h(h(x_k),\ v_k), \tag{2.9}$$

where $x_k \in X(u_k,v_k)$. If this process converges to some point $[\bar{u},\bar{v}]$ and $\bar{x} \in X(\bar{u},\bar{v})$, then, when conditions A_1 and A_2 are satisfied the point $\bar{x}$ will be feasible, $[\bar{x},\bar{u},\bar{v}]$ will be a Kuhn-Tucker point, and when A_4 is satisfied $x \in X_*$. If we require A_3 to be satisfied instead of A_2, then there is no guarantee that $\bar{v} \ge 0$. To exclude this possibility, we introduce condition A_5 by which from $v_0 \ge 0$ it follows that in the process of iterations (2.8), (2.9) all the $v_k \ge 0$.

Introduce the abbreviated notation:

$$y = [u,\ v] \in E^m,\quad \Phi(x,\ y) = [\varphi_g(g(x),\ u),\ \psi_h(h(x),\ v)],$$
$$H(x,\ u,\ v) = H(x,\ y),\quad R_x(x) = [g_x(x),\ h_x(x)],$$
$$H_{xy}(x,\ y) = [g_x(x)\,\varphi_{gu}(g(x),\ u),\ h_x(x)\,\psi_{hv}(h(x),\ v)],$$

$$\Phi_y(x, y) = \left[\begin{array}{c|c} \varphi_{gu}(g(x), u) & 0 \\ \hline 0 & \psi_{hv}(h(x), v) \end{array}\right],$$
$$\Phi_x^T(x, y) = \left[\begin{array}{c} \varphi_{gg}(g(x), u) g_x^T(x) \\ \hline \psi_{hh}(h(x), v) h_x^T(x) \end{array}\right],$$
$$B(x, y) = \Phi_y(x, y) - \Phi_x^T(x, y) H_{xx}^{-1}(x, y) H_{xy}(x, y), \tag{2.10}$$

$$H_x(x, y) = f_x(x) + g_x(x)\varphi_g(g(x), u) + h_x(x)\psi_h(h(x), v),$$
$$\tilde{L}_{xx}(x, y) = f_{xx}(x) + \\ + \sum_{i=1}^{e} g_{xx}^i(x)\varphi_g(g^i(x), u^i) + \sum_{j=1}^{c} h_{xx}^j(x)\psi_h(h^j(x), v^j),$$
$$H_{xx}(x, y) = \tilde{L}_{xx}(x, y) + R_x(x)\Phi_x^T(x, y). \tag{2.11}$$

The matrices ϕ_{gg}, ψ_{hh} are diagonal with the diagonal elements

$$\frac{\partial^2 \varphi(g^i, u^i)}{\partial g^i \partial g^i}, \quad \frac{\partial^2 \psi(h^i, v^i)}{\partial h^i \partial h^i},$$

respectively. The matrices R_x, H_{xy}, Φ_y, Φ_x^T, B, $\tilde{L}_{xx}$, H_{xx} have dimension n×m, n×m, m×m, m×n, m×m, n×n, n×n. They are related as follows:

$$H_{xy}(x, y) = R_x(x)\Phi_y(x, y). \tag{2.12}$$

One can write the system (2.7) and the method (2.8), (2.9) in concise form:

$$y = \Phi(x(y), y), \tag{2.13}$$
$$y_{k+1} = \Phi(x(y_k), y_k). \tag{2.14}$$

Applying Newton's method to solving the system (2.13), we arrive at the process

$$y_{k+1} = y_k - [B(x(y_k), y_k) - I_m]^{-1}[\Phi(x(y_k), y_k) - y_k]. \tag{2.15}$$

We do not need to introduce the auxiliary problem (2.3); instead we apply Newton's method directly to solving the initial system of nonlinear equations (2.4) - (2.6). We then obtain the method

$$H_{xx}(x_k, y_k)(x_{k+1}-x_k)+H_{xy}(x_k, y_k)(y_{k+1}-y_k)= \\ =-H_x(x_k, y_k),$$
$$\Phi_x^T(x_k, y_k)(x_{k+1}-x_k)+[\Phi_y(x_k, y_k)-I_m](y_{k+1}-y_k)= \\ =-\Phi(x_k, y_k)+y_k.$$

Using Frobenius' formula, we find next that

$$y_{k+1}=y_k-(B-I_m)^{-1}[\Phi-y_k-\Phi_x^T H_{xx}^{-1}H_x], \tag{2.16}$$
$$x_{k+1}=x_k-H_{xx}^{-1}H_x-H_{xx}^{-1}H_{xy}(y_{k+1}-y_k). \tag{2.17}$$

Here all the functions and matrices are computed for $x = x_k$, $u = u_k$, $v = v_k$.

If the dependence of $x(y)$ is determined from (2.3), then note that (2.4), (2.15) follow from (2.16). If we set $\psi = \psi'$ in (2.15) and omit taking into account inequality-type constraints, then (2.15) turns into a method suggested and studied by a number of authors. We refer, say, to the work of Bertsekas [3] in which he studies the case when the auxiliary problem (2.3) is solved approximately: $\|H_x(x_k, y_k)\| \neq 0$ on each iteration and the formula (2.16) is used for computation.

3. PROOF OF CONVERGENCE FOR THE SIMPLE ITERATION METHOD

1. BASIC CONVERGENCE THEOREMS

Let

$$\Delta_1(x, u, v)=\min_{i\in[1:e]}\ \min_{j\in\sigma(x)}[\varphi_{gg}(g^i(x), u^i), \psi_{hh}(h^j(x), v^j)],$$
$$\Delta_2(x, v)=\min_{j\notin\sigma(x)}\psi_{hh}(h^j(x), v^j).$$

Let us impose another conditions on ϕ and ψ.

• A_6. At the Kuhn-Tucker point $[x_*, u_*, v_*]$, the system (2.5), (2.6) is satisfied and the quantity $\Delta_2(x_*, v_*) \geq 0$.

In the sequel we shall need the characteristic equation of the matrix B:

$$|B(x_*, u_*, v_*) - \lambda I_m| = 0 \quad . \tag{3.1}$$

Here and below, in proving the convergence, we assume that ϕ and ψ are twice differentiable in their arguments.

THEOREM 4.3.1. Let condition A_6 and the sufficient conditions for a minimum of Theorem 1.7.4 be satisfied at the Kuhn-Tucker point $z_* = [x_*, u_*, v_*]$. Furthermore, let the function $H(x,u,v)$ be twice continuously differentiable in a neighborhood of z_* and the modules of all roots of the equation (3.1) be strictly less than 1. Then, if $\Delta_1(x_*, u_*, v_*)$ is sufficiently large, the problem (2.3) has a local solution and the method (2.14) converges locally to z_*.

Proof. The system (2.5), (2.6) is satisfied at the Kuhn-Tucker point. Therefore, the formula (2.11) can be written in the following form:

$$H_{xx}(x_*, u_*, v_*) = L_{xx}(x_*, u_*, v_*) + \\ + g_x(x_*) \varphi_{gg}(0, u_*) g_x^T(x_*) + h_x(x_*) \psi_{hh}(h(x_*), v_*) h_x^T(x_*).$$

For the corresponding quadratic form we have the estimates

$$x^T H_{xx}(x_*, u_*, v_*) x \geqslant x^T \Big[L_{xx}(x_*, u_*, v_*) + \\ + \Delta_1(x_*, u_*, v_*) [g_x(x_*) g_x^T(x_*) + \sum_{j \in \sigma(x_*)} h_x^j(x_*) [h_x^j(x_*)]^T] \Big] x + \\ + \Delta_2(x_*, v_*) \sum_{j \notin \sigma(x_*)} [x^T h_x^j(x_*)]^2 \geqslant x^T L_{xx}(x_*, u_*, v_*) x + \\ + \Delta_1(x_*, u_*, v_*) x^T \Big[g_x(x_*) g_x^T(x_*) + \sum_{j \in \sigma(x_*)} h_x^j(x_*) [h_x^j(x_*)]^T \Big] x.$$

By Finsler's lemma 1.7.4 there is a $\bar{\tau}$ such that when $\Delta_1(x_*,u_*,v_*) > \bar{\tau}$ the matrix $H_{xx}(x_*,u_*,v_*)$ is positive definite. But then there exist neighborhoods $G(x_*)$ and $G(y_*)$ of x_* and $y_* = [u_*,v_*]$, respectively, such that for $x \in G(x_*)$, $y \in g(y_*)$, the matrix $H_{xx}(x_*,u_*,v_*)$ is positive definite, for any $y \in G(y_*)$ there exists an isolated local solution $x = x(y) \in G(x_*)$ of the problem (2.3) and the necessary condition of the minimum

$$H_x(x(y),\ y) = 0\ , \qquad x_* = x(y_*) \tag{3.2}$$

holds. On the other hand, according to the implicit function theorem we can find a neighborhood $G_1(x_*)$ of x_* such that on $G_1(x_*)$ there exists a continuously differentiable function $x(y)$ satisfying (3.2). Taking $G_1(x_*)$ so small that $G_1(x_*) \subset G(x_*)$, we obtain from (3.2) that

$$\frac{dx}{dy} = -H_{xx}^{-1}(x(y),\ y)\, H_{xy}(x(y),\ y).$$

Hence the composition function $\Phi(x(y),\ y)$ is differentiable at $y = y_*$ and

$$\frac{d\Phi(x(y_*),\ y_*)}{dy} = B(x_*,\ y_*),$$

where the matrix B is defined from (2.10). By Theorem 2.3.4, for the local convergence of the method (2.14) it suffices that the spectral radius of $B(x_*,y_*)$ be less than one, which is known to hold if the modules of the roots of the equation (3.1) are strictly less than 1. ///

Theorems 4.2.1 and 4.3.1 characterize different properties of the method (2.14). Theorem 4.3.1 is analogous to Theorem 4.1.2 and they both give sufficient conditions for the local convergence to a Kuhn-Tucker point. Theorem 4.2.1 shows the difference of (2.14) from the methods described in Section 4.1; by conditions $A_1 - A_5$ we are assured that in the case of convergence the limit points will be Kuhn-Tucker. The methods of Section 4.1 do not have these properties.

The analysis of the roots of (3.1) is difficult. Hence, introducing additional assumptions, we transform the equation (3.1) to another form. We write the vector h in the form

$$h(x) = [h^a(x),\ h^b(x)],\quad h^a(x) = [h^1(x),\ \ldots,\ h^s(x)],$$
$$h^b(x) = [h^{s+1}(x),\ \ldots,\ h^c(x)].$$

Analogously, we write the matrices

$$R_x^a(x) = [g_x(x),\ h_x^a(x)],\quad R_x(x) = [R_x^a(x),\ h_x^b(x)],$$
$$\Phi_y^a(x,\ y) = \left[\begin{array}{c:c} \varphi_{gu}(g(x),\ u) & 0 \\ \hdashline 0 & \psi_{hv}^a(h(x),\ v) \end{array}\right],$$
$$[\Phi_x^a(x,\ y)]^T = \left[\begin{array}{c} \varphi_{gg}(g(x),u)\, g_x^T(x) \\ \hdashline \psi_{hh}^a(h(x),v)\,[h_x^a(x)]^T \end{array}\right]$$

in block form. Here the matrices R_x^a, h_x^a, Φ_y^a, $[\Phi_x^a]^T$, ψ_{hh}^a have dimensions $n\times(e+s)$, $n\times(c-s)$, $(e+s)\times(e-s)$, $(e+s)\times n$, $s\times s$, respectively. Therefore, we have the representation

$$\Phi_y = \left[\begin{array}{c:c} \Phi_y^a & 0 \\ \hdashline 0 & \psi_{hv}^b(h(x),v) \end{array}\right],$$
$$[\Phi_x(x,\ y)]^T = \left[\begin{array}{c} [\Phi_x^a(x,\ y)]^T \\ \hdashline \psi_{hh}^b(h(x),\ v)\,[h_x^b(x)]^T \end{array}\right].$$

Let $[x_*, u_*, v_*]$ be a Kuhn-Tucker point. We can assume without loss of generality that the first s components of $h(x_*)$ are null. Let

$$|\Phi_y^a(x_*, y_*) - \lambda[\Phi_x^a(x_*, y_*)]^T L_{xx}^{-1}(x_*, y_*) R_x^a(x_*, y_*) - \lambda I_{e+s}| = 0. \quad (3.4)$$

<u>THEOREM 4.3.2</u>. Let condition A_6 and the conditions of Lemma 4.1.1 be satisfied at the Kuhn-Tucker point $z_* = [x_*, u_*, v_*]$, let the function $H(x,u,v)$ be twice continuously differentiable in a neighborhood of z_*, and let the modules of the roots of the equation (3.4) be strictly less than 1, and for all $j \in \sigma(x_*)$

$$\psi_{hh}^j(h(x_*), v_*^j) = 0 , \qquad -1 < \psi_{hv}^j(h(x_*), v_*^j) < 1 . \quad (3.5)$$

Then, if $\Delta_1(x_*, u_*, v_*)$ is sufficiently large, the problem (2.3) has a local solution and the method (2.14) converges locally to z_*.

<u>Proof</u>. Suppose that the matrices $H_{xx}(x,y)$ and $I_{xx}(x,y)$ are nonsingular. We multiply (2.11) on the left by $\Phi_x^T(x,y)\tilde{L}_{xx}^{-1}(x,y)$ and on the right by $H_{xx}^{-1}(x,y)H_{xy}(x,v)$. We obtain

$$\Phi_x^T(x, y)\tilde{L}_{xx}^{-1}(x, y) H_{xy}(x, y) = \\ = \Gamma(x, y)\Phi_x^T(x, y) H_{xx}^{-1}(x, y) H_{xy}(x, y).$$

Here we have set

$$\Gamma(x, y) = I_m + \Phi_x^T(x, y)\tilde{L}_{xx}^{-1}(x, y) R_x(x).$$

Using (2.12), (2.10), we transform this relation to the form

$$(\Gamma(x, y) - I_m)\Phi_y(x, y) = \Gamma(x, y)(\Phi_y(x, y) - B(x, y))$$

yielding

$$\Gamma(x,y)B(x,y) = \Phi_y(x,y) . \quad (3.6)$$

The solvability of (2.3) is proved just as in the preceding theorem. By Lemma 4.1.1 the matrix $\tilde{L}_{xx}(x_*, y_*) = L_{xx}(x_*, y_*)$ is nonsingular. Hence the matrix $\Gamma(x_*, y_*)$ is defined and by (3.6)

$$B(x_*, y_*) = \Gamma^{-1}(x_*, y_*)\Phi_y(x_*, y_*).$$

Hence the equation (3.1) is equivalent to

$$|\Phi_y(x_*, y_*) - \lambda\Gamma(x_*, y_*)| = 0,$$

which with (3.5) taken into account we can rewrite as

$$\left|\begin{array}{c|c} \Phi_y^a - \lambda[\Phi_x^a]^T L_{xx}^{-1} R_x^a - \lambda I_{e+s} & -\lambda[\Phi_x^a]^T L_{xx}^{-1} h_x^b \\ \hline 0 & \psi_{hv}^b - \lambda I_{c-s} \end{array}\right| = 0.$$

Thus the $c - s$ roots are found explicitly:

$$-1 < \lambda_j = \psi_{hv}(h^j(x_*), v_*^j) < 1, \quad j \in [1+s:c].$$

The remaining $m-(c-s)$ roots are found from (3.4). Thus, the modules of all the roots λ are strictly less than 1, which implies the convergence of the method. ///

2. SIMPLIFIED VERSIONS OF THE METHOD

The auxiliary problem (2.3) is usually solved only approximately. Just as was done in the second simplified version of the penalty function method, we will curtail the labor of minimizing $H(x, y_k)$ in x as soon as we find the point x_k satisfying

$$\|H_x(x_k, y_k)\| \leqslant e_k. \tag{3.7}$$

Here $\{e_k\} \to 0$ as $k \to \infty$. To prove the convergence of this version of the method, we assume that $\Phi(x, y)$ satisfies a Lipschitz condition in x:

$$\|\Phi(x_1, y) - \Phi(x_2, y)\| \leqslant l\|x_1 - x_2\|. \tag{3.8}$$

We bring in the lower limit of $H_{xx}(x,y)$:

$$d(x, y) = \min_{\|\bar{x}\|=1} \bar{x}^T H_{xx}(x, y)\bar{x}. \tag{3.9}$$

If all the conditions of Theorems 4.3.1 and 4.3.2 are satisfied, then $d(x,y) > 0$ and there exist neighborhoods $G(x_*)$ and $G(y_*)$ such that

$$d_* = \inf_{x \in G(x_*)} \inf_{y \in G(y_*)} d(x, y) > 0. \tag{3.10}$$

Let us determine the error on the kth iteration when using the method (2.14) with the simplified rule (3.7) for solving the auxiliary problem. We have

$$\begin{aligned} \|y_{k+1} - y_*\| = \|\Phi(x_k, y_k) - y_*\| = \|\Phi(x(y_k), y_k) - \\ - \Phi(x_*, y_*) + \Phi(x_k, y_k) - \Phi(x(y_k), y_k)\|. \end{aligned} \tag{3.11}$$

Using the Newton-Leibniz formula (see Appendix I), we obtain

$$\|\Phi(x(y_k), y_k) - \Phi(x_*, y_*)\| = \int_0^1 \left\langle \frac{d\Phi(x(y), y)}{dy}, y_k - y_* \right\rangle d\tau.$$

Here $y = y_* + \tau(y_k - y_*)$, the total derivative of Φ in y is computed as in (3.3):

$$\frac{d\Phi(x(y), y)}{dy} = B(x(y), y).$$

We use the Cauchy inequality and assume that the matrix norm is compatible with the Euclidean norm. Then

$$\|\Phi(x(y_k), y_k) - \Phi(x_*, y_*)\| \leq r\|y_k - y_*\|,$$

where

$$r = \sup_{y \in G(y_*)} \|B(x(y), y)\|.$$

From this and (3.8), (3.11) we obtain

$$\|y_{k+1}-y_*\| \leqslant r\|y_k-y_*\|+l\|x(y_k)-x_k\|. \tag{3.12}$$

Let us estimate the norm of $q = x(y_k) - x_k$. Using Taylor's formula, we have

$$H(x(y_k),\ y_k)= \\ =H(x_k,\ y_k)+\langle H_x(x_k,\ y_k),\ q\rangle+\frac{1}{2}q^T H_{xx}(\bar{x},\ y_k)q,$$

$$H(x_k,\ y_k)=H(x(y_k),\ y_k)+\frac{1}{2}q^T H_{xx}(\tilde{x},\ y_k)q,$$

where

$$\bar{x}=x_k+\theta_1 q,\quad \tilde{x}=x(y_k)-\theta_2 q,\quad 0<\theta_1<1,\quad 0<\theta_2<1$$

and the condition (2.3) is used. Let us combine the formulas. Then

$$\langle H_x(x_k,\ y_k),\ q\rangle=-\frac{1}{2}q^T[H_{xx}(\bar{x},\ y_k)+H_{xx}(\tilde{x},\ y_k)]q.$$

Using the Cauchy inequality and (3.9), (3.10), we obtain

$$d_*\|q\|\leqslant\|H_x(x_k,\ y_k)\|. \tag{3.13}$$

Noting (3.7), we arrive at the inequality

$$\|x(y_k)-x_k\|\leqslant\frac{e_k}{d_*}.$$

Substituting this into (3.12), we have the estimate

$$\|y_{k+1}-y_*\|\leqslant r\|y_k-y_*\|+\frac{l}{d_*}e_k. \tag{3.14}$$

This implies that to ensure rapid convergence it is necessary to decrease in a particular way the coefficient e_k, letting it go to zero.

Another way of adjusting the accuracy of solving the unconstrained minimization problem (2.3) is to find a point $x = x_k$

satisfying

$$\|H_x(x_k, y_k)\| \leqslant e_k \|\Phi(x_k, y_k) - y_k\|. \tag{3.15}$$

The following estimates hold:

$$\begin{aligned} C = \|\Phi(x_k, y_k) - y_k\| = \|\Phi(x(y_k), y_k) - \Phi(x_*, y_*) + \\ + \Phi(x_k, y_k) - \Phi(x(y_k), y_k) + y_* - y_k\| \leqslant \\ \leqslant (1+r)\|y_k - y_*\| + l\|x(y_k) - y_k\|. \end{aligned}$$

Using (3.13) and (3.15), we obtain

$$C \leqslant (1+r)\|y_k - y_*\| + \frac{l}{d_*} e_k C.$$

Whence

$$C \leqslant \frac{d_*(1+r)}{d_* - le_k}\|y_k - y_*\|.$$

Substituting this into (3.15), we have

$$\|H_x(x_k, y_k)\| \leqslant \frac{d_*(1+r)e_k}{d_* - le_k}\|y_k - y_*\|.$$

From (3.13) we obtain

$$\|x(y_k) - x_k\| \leqslant \frac{(1+r)e_k}{d_* - le_k}\|y_k - y_*\|.$$

From (3.12) we find the final formula

$$\|y_{k+1} - y_*\| \leqslant b\|y_k - y_*\|,\ b = \frac{rd_* + le_k}{d_* - le_k}. \tag{3.16}$$

For local convergence it suffices that $0 \le b < 1$, which will occur if e_k is sufficiently small:

$$e_k < \frac{d_*}{2l}(1-r), \quad 0 \leqslant r < 1.$$

When this condition is satisfied the method (2.14), with regulation of the accuracy for solving the auxiliary problem (3.15), converges with a geometric rate of convergence. In contrast to the rule (3.7), one does not need to have e_k tend to zero. If in (3.7), (3.15) $e_k \to 0$, then by (3.14), (3.16), Q is a multiplier of the process (2.14) equal to r at the point y_*, which corresponds to the formula (2.3.9).

The estimates obtained show that the method (2.14) is stable under errors in solving the auxiliary problem (2.3). This property and also the relatively high rate of convergence make the method (2.14) a very effective tool for solving nonlinear programming problems. The method is ordinarily used after a satisfactory estimate has been derived for the Lagrange multipliers.

In order to justify the method, the existence of a Kuhn-Tucker point in the problem (1.6.1) was postulated. Apparently, one can prove that the method converges under more relaxed assumptions without the requirement, in particular, of the existence of constrained Lagrangian multipliers. In this case, when (2.14) is used, $\|y_k\| \to \infty$ as $k \to \infty$, however, x_k will converge to X_*. As an illustration we consider the problem (3.1.5). If we use ϕ', then the method (2.14) leads to the process

$$u_{k+1} = u_k + x_k^2 ,$$

where x_k is determined from the condition

$$x_k = \operatorname{Arg}\min_x \left[x + u_k x^2 + \frac{1}{2} x^4 \right] ,$$

For small values $|x_k| << 1$ we have

$$x_k = x(u_k) \approx -\frac{1}{2u_k}, \quad u_{k+1} \approx u_k + \frac{1}{4u_k^2}.$$

As $u_k \to \infty$ we have $x_k \to 0 = X_*$.

3. THE SCALING

To guarantee the conditions of Theorem 4.3.1 and 4.3.2 for Δ_1 to be sufficiently large and preserve the elements of the matrix ψ_{hv} without changing the form of the functions ϕ and ψ, it is convenient to use instead of ϕ and ψ the "scaled" functions

$$\tilde{\varphi}(g(x), u) = \frac{\varphi(\tau g(x), u)}{\tau},$$
$$\tilde{\psi}(h(x), v) = \frac{\psi(\tau h(x), v)}{\tau}.$$

The derivatives of the new functions can simply be expressed in terms of those of the initial functions:

$$\tilde{\varphi}_g(g, u) = \varphi_{\bar{g}}(\bar{g}, u), \qquad \tilde{\psi}_{gu}(g, u) = \psi_{\bar{g}u}(\bar{g}, u),$$
$$\tilde{\varphi}_{gg}(g, u) = \tau\varphi_{\bar{g}\bar{g}}(\bar{g}, u), \quad \tilde{\varphi}_{gu}(g, u) = \varphi_{\bar{g}u}(\bar{g}, u).$$

Here $\bar{g}$ means that the initial function $\phi(g,u)$ is differentiated with respect to g, and as a next step, the value τg is taken for g in the resulting formula. The derivatives of ψ are computed similarly. Let us give now the basic computing formulas taking account of scaling:

$$H_{xx}(x_*, y_*) = L_{xx}(x_*, y_*) + \tau R_x(x_*)\Phi_x^T(x_*, y_*),$$
$$B(x_*, y_*) = \Phi_y(x_*, y_*) - \tau\Phi_x^T(x_*, y_*) H_{xx}^{-1}(x_*, y_*) H_{xy}(x_*, y_*),$$
$$\Phi_y(x_*, y_*) = \left[\begin{array}{c|c} \varphi_{gu}(0, u_*) & 0 \\ \hline 0 & \psi_{\bar{h}v}(\bar{h}(x_*), v_*) \end{array}\right],$$
$$\Phi_x^T(x_*, y_*) = \left[\begin{array}{c} \varphi_{gg}(0, u_*) g_x^T(x_*) \\ \hline \psi_{\bar{h}\bar{h}}(\bar{h}, v_*) h_x^T(x_*) \end{array}\right].$$

The equation (3.4) is replaced by

$$|\Phi_y^a(x_*, y_*) - \tilde{\lambda}\tau[\Phi_x^a(x_*, y_*)]^T L_{xx}^{-1}(x_*, y_*) R_x^a(x_*, y_*) - \tilde{\lambda} I_{e+s}| = 0. \tag{3.17}$$

The computing method (2.14) remains the same except that instead of g and h we take τg and τh, respectively.

Suppose there exists a scalar d such that for all $i \in [1:e]$, $j \in \sigma(x_*)$,

$$d = \varphi_{gu}(0, u_*^i) = \psi_{hv}(0, v_*^j).$$

Comparing the equations (3.4) and (3.17), we reach the conclusion that a simple relationship exists between their roots:

$$\tilde{\lambda} = \frac{\lambda d}{\lambda + \tau(d - \lambda)}, \tag{3.18}$$

and there is no root $\lambda = d$. Hence taking τ sufficiently large, we can guarantee that the condition $|\tilde{\lambda}| < 1$ be satisfied. This is known to hold if

$$\tau > \max \frac{1 + d}{\left|\frac{d}{\lambda_i} - 1\right|} ,$$

where λ_i are the roots of (3.4).

From (3.18) we see that as $\tau \to \infty$ the $|\tilde{\lambda}|$ tend to zero, i.e., the convergence is superlinear. However, one should bear in mind that for large values τ the value of H has many valleys and ridges, thus making the solution of the problem (2.3) more time-consuming, and the improvement is slow. Hence it is not practical to take large values of τ. At the same time, for small values of τ the conditions of Theorems 4.3.1 and 4.3.2 may be violated.

We give the results of numerical computations of the nonlinear programming problem (7.5.1). In Fig. 2 we show the

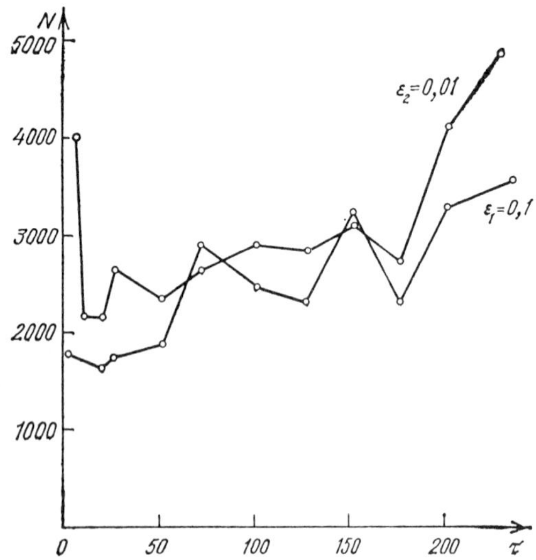

Figure 2

dependence of N, that is the number of function evaluations, on the parameter τ which is constant in each separate computation. The value N is proportional to the computation time. Using the method (2.14) with the functions ϕ^1 and ψ^5, the computations were made for 12 different values of τ and interpolation was carried out at intermediate points. The accuracy of solving the auxiliary problem (2.3) on each iteration was given by the condition $||H_x(x_k, u_k, v_k)|| \le \varepsilon_i$, where $\varepsilon_1 = 0.1$; $\varepsilon_2 = 0.01$. For each value τ the same initial point x_0 was taken, the same parameters of the methods were used, and the problem (2.3) was solved by the conjugate gradient method. The dependence was not monotonic.

The problem was solved most quickly when the values of τ were medium range. At the same time, in the problems in which one cannot achieve a high accuracy of $x \in X$, the coefficient τ ought to be increased; its role will be analogous to that of a penalty coefficient.

4. PARTICULAR CASES

For the functions ψ introduced above we have the following:

$$\psi^2_{hh}(h^j(x_*),\ v^j_*) = v^j_* e^{h^j(x_*)},$$
$$\psi^2_{hv}(h^j(x_*),\ v^j_*) = e^{h^j(x_*)},$$
$$\psi^3_{hh}(h^j(x_*),\ v^j_*) = v^j_*[1-h^j(x_*)]^{r-2}(1+r),$$
$$\psi^3_{hv}(h^j(x_*),\ v^j_*) = (1-h^j(x_*))^{-1-r},$$
$$\psi^5_{hh}(h^j(x_*), v^j_*) = 3r[h^j(x_*)+v^j_*]^2_+ - 2v^j_* h^j_-(x_*)[1+[h^j(x_*)]^2]^{-2},$$
$$\psi^5_{hv}(h^j(x_*), v^j_*) = 3r[[h^j(x_*)+v^j_*]^2_+ - (v^j_*)^2] + [1+[h^j(x_*)]^2]^{-1}.$$

Let the strict complementarity condition be satisfied at the point $[x_*, v_*]$. For these functions, if $j \in \sigma(x_*)$, then all $\psi_{hh}(h^j(x_*),\ v^j_*) > 0$ and if $j \notin \sigma(x_*)$ then all $0 < \psi_{hv}(h^j(x_*),\ v^j_*) < 1$, $\psi_{hh}(h^j(x_*),\ v^j_*) = 0$, as is required in the conditions of Theorem 4.3.2.

For ϕ^1, ϕ^2, ϕ^3 we have

$$\Delta_1(x_*,\ u_*,\ v_*) = \varphi_{gg}(g^i(x_*),\ u^i_*) = 1,$$
$$\Delta_2(x_*,\ v_*) = \varphi_{gu}(g^i(x_*),\ u^i_*) = 1.$$

For the remaining functions we have

$$\varphi^4_{gg}(g^i(x_*),\ u^i_*) = \frac{1}{\pi}e^{-(u^i_*)^2}, \qquad \varphi^4_{gu}(g^i(x_*),\ u^i_*) = 1,$$
$$\varphi^5_{gg}(g^i(x_*),\ u^i_*) = 1+\varepsilon\cos u^i_*, \qquad \varphi^5_{gu}(g^i(x_*),\ u^i_*) = 1,$$
$$\varphi^6_{gg}(g^i(x_*),\ u^i_*) = \alpha''(0)\beta(u^i_*), \qquad \varphi^6_{gu}(g^i(x_*),\ u^i_*) = 1,$$
$$\varphi^7_{gg}(g^i(x_*),\ u^i_*) = \gamma_{gg}(g^i(x_*),\ u^i_*), \qquad \varphi^7_{gu}(g^i(x_*),\ u^i_*) = 1.$$

This implies that for all these functions the conditions of Theorems 4.3.1 and 4.3.2, $\Delta_1(x_*,u_*,v_*) > 0$, $\Delta_2(x_*,v_*) \geqslant 0$, are satisfied, and that the quantity Δ_1 can be made arbitrarily large by using scaling.

If for ϕ, ψ we take ϕ^1 and ψ^5, respectively, then

$$H(x, u, v) = f(x) + \sum_{i=1}^{c} [u^i + \frac{\tau}{2} g^i(x)] g^i(x) + \\ + \frac{r}{4\tau} \sum_{j=1}^{c} [(v^j + \tau h^j(x))_+^4 - 4\tau h^j(x)(v^j)^3] + \\ + \sum_{j=1}^{c} \frac{v^j}{\tau} \begin{cases} \tau h^j(x) & \text{if} \quad h^j(x) \geqslant 0, \\ \operatorname{arctg} \tau h^j(x) & \text{if} \quad h^j(x) < 0. \end{cases}$$

The method (2.14) with scaling takes the form

$$\left.\begin{aligned} u_{k+1}^i &= u_k^i + \tau g^i(x_k), \\ v_{k+1}^j &= r[(v_k^j + \tau h^j(x_k))_+^3 - (v_k^j)^3] + \\ &\qquad + v_k^j [1 + [\tau h_-^j(x_k)]^2]^{-1}. \end{aligned}\right\} \tag{3.19}$$

If we use ϕ^1 and ψ^3, taking $r = 1$ and $\gamma(a) = a^4$, then we obtain

$$H(x, u, v) = \\ = f(x) + \sum_{i=1}^{e} \left[u^i + \frac{\tau}{2} g^i(x)\right] g^i(x) + \sum_{j=1}^{c} \left[\tau^3 (h_+^j(x))^4 + \\ + \frac{v^j}{\tau} \begin{cases} 1 + \tau h^j(x) + (\tau h^j(x))^2 + (\tau h^j(x))^3 & \text{if} \quad h^j(x) \geqslant 0, \\ (1 - \tau h^j(x))^{-1}, & \text{if} \quad h^j(x) < 0 \end{cases}\right],$$

$$\left.\begin{aligned} & u_{k+1}^i = u_k^i + \tau g^i(x_k), \quad v_{k+1}^j = 4[\tau h_+^j(x_k)]^3 + \\ & + v_k^j \begin{cases} 1 + 2\tau h^j(x_k) + 3(\tau h^j(x_k))^2 & \text{if} \quad h^j(x_k) \geqslant 0, \\ [1 - \tau h^j(x_k)]^{-2} & \text{if} \quad h^j(x_k) < 0. \end{cases} \end{aligned}\right\} \tag{3.20}$$

The simplest computations result when ϕ^1 and ψ^1 are used:

$$H(x, u, v)=f(x)+ \\ +\sum_{i=1}^{e}\left[u^i+\frac{\tau}{2}g^i(x)\right]g^i(x)+\frac{1}{2\tau}\sum_{j=1}^{c}(v^j+\tau h^j(x))_+^2, \tag{3.21}$$
$$u_{k+1}^i=u_k^i+\tau g^i(x_k), \qquad v_{k+1}^j=(v_k^j+\tau h^j(x_k))_+.$$

This case is extensively investigated in the literature on modified Lagrangians. In contrast to the preceding two cases, the function H has no second derivatives, which narrows the class of methods of unconstrained minimization applicable to solving the auxiliary problem (2.3); in particular, we have to give up Newton's method. The methods (3.19) - (3.21) have been used widely for solving nonlinear programming problems and are now part of the standard program library for DISO.

4. SOLUTION OF CONVEX PROGRAMMING PROBLEMS

1. THE USE OF THE SIMPLE ITERATION METHOD TO SOLVE CONVEX PROGRAMMING PROBLEMS

In most works on modified Lagrangians, the auxiliary functions H(x,u,v) are constructed such as to the Kuhn-Tucker point be a saddle point. This requirement is essential when, in order to solve the initial problem, one intends to use the saddle-point methods. We will use the simple iteration method to solve the problem (1.6.1); therefore the condition for a modified Lagrangian to have a saddle point is not necessary. In particular, if we use the function ψ^2 defined in Section 4.2, then the modified Lagrangian will have the form

$$H(x, v)=f(x)+\sum_{j=1}^{c}[\gamma(h_+^j(x))+v^j e^{h^j(x)}].$$

It is easily seen that for any vectors x

$$\sup_{v \in E^c} H(x, v) = \sup_{v \in E^c_+} H(x, v) = +\infty$$

and, therefore, the left-hand inequality in the saddle-point condition (1.6.9) never holds; nevertheless, this function is quite efficient in numerical calculations.

Let (1.6.1) be a convex programming problem. The set X_0 defined by (1.6.3) is not empty and the function $f(x)$ is uniformly convex in x, i.e., there exists a scalar ω such that for any $x, y \in E^n$ and $f_* \in \partial f(x)$ we have the inequality

$$f(y) \geqslant f(x) + \langle f_*, y - x \rangle + \omega \| y - x \|^2. \tag{4.1}$$

To simplify the presentation we assume that in (1.6.1) there are no equality-type constraints. Then the basic computing formulas (2.3) and (2.14) will become

$$v_{k+1} = \psi_h(h(x_k), v_k), \tag{4.2}$$

$$x_k \in \operatorname{Arg}\min_{x \in E^n} H(x, v), \tag{4.3}$$

$$H(x, v) = f(x) + \sum_{j=1}^{c} \psi(h^j(x), v^j).$$

In the first simplified version of the method we take as x_k an arbitrary vector satisfying the condition

$$\| \partial_x H(x_k, v_k) \|^2 \leqslant \delta \sum_{j=1}^{c} h^j(x_k) [\psi_h(h^j(x_k), v_k^j) - v_k^j] = \Gamma_k^2. \tag{4.4}$$

In the second version, x_k satisfies the inequality

$$\| \partial_x H(x_k, v_k) \|^2 \leqslant \delta \sum_{j=1}^{c} [\psi_h(h^j(x_k), v_k^j) h^j(x_k) - \psi(h^j(x_k), v_k^j) + \psi(0, v_k^j)] = \Gamma_k^1, \tag{4.5}$$

where δ is some nonnegative number. We will show in the sequel that

$$0 \leq \Gamma_k^1 \leq \Gamma_k^2 . \quad (4.6)$$

From these inequalities it follows that each point satisfying (4.5) will simultaneously satisfy the condition (4.4). This makes it easier to solve the problem of finding x_k for (4.4).

In all versions of the method, if we get $v_{k+1} = v_k$, the computing process is terminated and the corresponding vector x_k will be a solution of the problem (1.6.1).

2. PROOF OF CONVERGENCE

Consider the problem of finding

$$V_1 = \sup_{v \in E_+^c} \inf_{x \in E^n} L(x, v).$$

If in the problem (1.6.1) the set X_* is nonempty and Slater's CQ is satisfied, then the set

$$W = \left\{v:\ V_1 = \inf_{x \in E^n} L(x, v),\ v \geqslant 0\right\}$$

is also nonempty.

THEOREM 4.4.1. In the convex programming problem (1.6.1) let there exist a point $x_* \in X_*$, let conditions A_3, A_5 be satisfied, let (4.1) hold, for any $v \geq 0$ let the function $\psi(h,v)$ be differentiable and convex in h, let $0 \leq \delta < 4\omega$, and let any vector having nonnegative components be taken as v_0. Then the sequence x_k resulting from the simple iteration method (4.2) for any rule (4.3) - (4.5) for finding x_k, either terminates within a finite number of steps with x_* or has x_* a limit point as $k \to \infty$. If Slater's CQ is satisfied, then the sequence $\{v_k\}$ is bounded, all of its limit points belong to W and at least one such point exists.

Proof. From the condition $v_0 \geqslant 0$ and A_5, it follows that all $v_k \geqslant 0$. We shall extensively use this property in the sequel without mentioning it. The function $\psi(h,v)$ is convex in h and therefore, using (1.2.14), for any h^j and $\bar{h}^j$ we have

$$\psi_h(h^j, v^j)(\bar{h}^j - h^j) \leqslant \psi(\bar{h}^j, v^j) - \psi(h^j, v^j) \leqslant \psi_h(\bar{h}^j, v^j)(\bar{h}^j - h^j).$$

Setting $\bar{h}^j = 0$ in these inequalities and noting that by A_3, $\psi_h(0, v^j) = v^j$, we obtain

$$\psi_h(h^j, v^j)\, h^j \geqslant \psi(h^j, v^j) - \psi(0, v^j) \geqslant v^j h^j. \tag{4.7}$$

It follows that for all $j \in [1:c]$,

$$h^j[\psi_h(h^j, v^j) - v^j] \geqslant 0.$$

Substituting the estimate for $v^j h^j$ from (4.7), we obtain

$$h^j[\psi_h(h^j, v^j) - v^j] \geqslant \psi_h(h^j, v^j)\, h^j + \psi(0, v^j) - \psi(h^j, v^j) \geqslant 0.$$

Summing up these inequalities for $j \in [1:c]$, we arrive at the inequalities (4.6). From (4.7) and A_5 it follows that

$$0 \leqslant \psi_h(h^j, v^j) \leqslant v^j \quad \text{for} \quad h^j < 0,$$
$$0 \leqslant v^j \leqslant \psi_h(h^j, v^j) \quad \text{for} \quad h^j > 0.$$

By A_3, if $v^j \geq 0$ then

$$(\psi_h(h^j, v^j) - v^j)\, h^j = 0$$

holds iff

$$h^j \leq 0, \quad v^j h^j = 0.$$

If the last two conditions are satisfied, then $\psi_h(h^j, v^j) = v^j$.

By A_5, $\psi(h^j, v_k^j)$ is a nondecreasing function of h^j. Then $\psi(h(x), v_k)$ is convex in x. This follows from the obvious inequalities

$$\begin{gathered} h(\lambda x_1 + (1-\lambda)x_2) \leq \lambda h(x_1) + (1-\lambda) h(x_2), \\ \psi(h(\lambda x_1 + (1-\lambda) x_2), v_k) \leq \psi(\lambda h(x_1) + (1-\lambda) h(x_2), v_k) \leq \\ \leq \lambda \psi(h(x_1), v_k) + (1-\lambda) \psi(h(x_2), v_k). \end{gathered}$$

Hence $H(x, v_k)$ is also convex in x and by (4.1) is uniformly strictly convex. For any x, $\bar{x}$, $H_* \in \partial_x H(\bar{x}, v_k)$ the inequality

$$H(x, v_k) \geq H(\bar{x}, v_k) + \langle H_*, x - \bar{x} \rangle + \omega \| x - \bar{x} \|^2$$

holds. We conclude that on each iteration (4.2) the function $H(x, v_k)$ attains its lower bound in x at a unique point. Hence the rule (4.3) uniquely determines x_k. From (4.6) it follows that we can find x_k satisfying (4.4) and (4.5).

The following formulas are obvious:

$$\begin{gathered} L(x_k, v_{k+1}) = L(x_k, v_k) + \xi(x_k, v_k), \\ \xi(x_k, v_k) = \sum_{j=1}^{c} h^j(x_k) \left[\psi_h(h^j(x_k), v_k^j) - v_k^j \right], \\ \partial_x H(x_k, v_k) = \partial_x L(x_k, v_{k+1}). \end{gathered} \tag{4.8}$$

From the first equality and (4.7) it follows that

$$L(x_k, v_{k+1}) \geq L(x_k, v_k) \geq R(v_k) = \inf_{x \in E^n} L(x, v_k), \tag{4.9}$$

where $L(x_k, v_{k+1}) > L(x_k, v_k)$ iff strict inequality holds in at least one of the inequalities (4.7).

Consider the method (4.2) in case x_k is determined by the rule (4.3). The functions $H(x, v_k)$ and $L(x, v_{k+1})$ are convex in

x, hence we can write that

$$\langle \partial_x H(x_k, v_k), x - x_k \rangle \leqslant H(x, v_k) - H(x_k, v_k),$$
$$\langle \partial_x L(x_k, v_{k+1}), x - x_k \rangle \leqslant L(x, v_{k+1}) - L(x_k, v_{k+1}).$$

The rule (4.2) for changing the vector v implies the equality (4.8), according to which the set of subgradients of H at the point $[x_k, v_k]$ and the set of subgradients of the function L at the point $[x_k, v_{k+1}]$ coincide. The null vector belongs to the set $\partial_x H(x_k, v_k)$ since by (4.3), the minimum of the function $H(x, v_k)$ is attained at x_k. Hence the null vector lies in $\partial_x L(x_k, v_{k+1})$. This implies an important assertion:

$$\operatorname{Arg}\min_{x \in E^n} H(x, v_k) = \operatorname{Arg}\min_{x \in E^n} L(x, v_{k+1}). \tag{4.10}$$

For any v_{k+1} we have

$$f(x_*) \geqslant f(x_*) + \sum_{i=1}^{c} v_{k+1}^i h^i(x_*) = L(x_*, v_{k+1})$$

which together with (4.9) and (4.10) yields

$$\begin{aligned} f(x_*) \geqslant L(x_*, v_{k+1}) \geqslant R(v_{k+1}) = L(x_k, v_{k+1}) = \\ = L(x_k, v_k) + \xi(x_k, v_k) \geqslant L(x_k, v_k) \geqslant R(v_k). \end{aligned} \tag{4.11}$$

Thus the sequence $\{R(v_k)\}$ is monotonically increasing. We now consider the problem whether this sequence is strictly monotonically increasing.

Suppose $v_{k+1} = v_k$ on the kth step of the iterative process (4.2). Then

$$v_k = \psi_h(h(x_k), v_k) \tag{4.12}$$

and by A_3, A_5 the conditions

$$h(x_k) \leq 0, \qquad 0 \leq v_k, \qquad h^j(x_k)v_k^j = 0 \tag{4.13}$$

are satisfied. Furthermore, it follows from (4.11) that

$$R(v_{k+1}) = R(v_k) = \min_{x \in E^n} L(x, v_k) \tag{4.14}$$

and for any $x \in X$ the inequality

$$\sum_{j=1}^{c} h^j(x) v_k^j \leqslant \sum_{j=1}^{c} h^j(x_k) v_k^j = 0$$

is satisfied. Hence using Theorem 1.6.4 we have that $x_k \in X_*$, i.e., the method (4.2) solves (1.6.1) in k steps. Similarly, if $R(v_{k+1}) = R(v_k)$, then by (4.11) for all $j \in [1:c]$ we have

$$h^j(x_k)\left[\psi_h(h^j(x_k), v_k^j) - v_k^j\right] = 0.$$

But, as was shown above, these equalities imply (4.12) and $x_k \in X_*$. Hence, if (4.2) generates an infinite sequence, then the sequence $\{R(v_k)\}$ is strictly monotonically increasing and is bounded from above by $f(x_*)$. Hence it is necessary that the values of $R(v_k)$ tend to some limit d while remaining less than this limit for each k. All terms of the function $\xi(x_k, v_k)$ are nonnegative; therefore they must tend to zero. We thus obtain

$$\begin{gathered} \lim_{k \to \infty} R(v_k) = \lim_{k \to \infty} L(x_k, v_k) = d \leqslant f(x_*), \\ \lim_{k \to \infty} \xi(x_k, v_k) = 0, \quad \lim_{k \to \infty} h^j(x_k) v_k^j = 0. \end{gathered} \tag{4.15}$$

From (4.1) it follows that $L(x, v_{k+1})$ is strictly convex in x and for any $x \in E^n$ and $L_* \in \partial_x L(x_k, v_{k+1})$ we have the inequality

$$L(x_k, v_{k+1}) + \langle L_*, x - x_k \rangle + \omega \|x - x_k\|^2 \leqslant L(x, v_{k+1}). \tag{4.16}$$

For L_* we take the null vector, and for x the vector x_*. Then

$$\omega \| x_* - x_k \|^2 \leq L(x_*, v_{k+1}) - L(x_k, v_{k+1}) = f(x_*) - R(v_k).$$

Thus, the sequence $\{x_k\}$ is bounded.

Let us show that $d = f(x_*)$. Suppose that some subsequence $\{x_{\bar{k}}\}$ converges to a point $\bar{x} \neq x_*$. If $h^j(\bar{x}) \neq 0$ then $\lim_{\bar{k}\to\infty} v^j_{\bar{k}} = 0$, and the case $h^j(\bar{x}) = a > 0$ is impossible since then we should have $\psi_h(a,0) = 0$, which contradicts A_3. Hence $\bar{x} \in X$, automatically yielding $\bar{x} \in X_*$, $f(\bar{x}) = d$. The components v^j_k for which $h^j(x_*) = 0$ can tend to infinity. We shall show below that they are bounded if Slater's CQ is satisfied.

Now we consider the method (4.2) when x_k is found by (4.4). In (4.16), we fix the vector L_* and minimize the left and right sides of this inequality with respect to x. The minimum obtains on the left-hand side for

$$x = x_k - \frac{1}{2\omega} L_*.$$

Hence from (4.29) it follows that

$$L(x_*, v_{k+1}) \geq R(v_{k+1}) \geq L(x_k, v_{k+1}) - \frac{1}{4\omega} \| L_* \|^2.$$

For L_* we take the vector in $\partial_x H(x_k, v_k)$ for which (4.4) is satisfied. Noting (4.9), we obtain

$$R(v_{k+1}) \geq L(x_k, v_k) + \left(1 - \frac{\delta}{4\omega}\right) \xi(x_k, v_k) \geq L(x_k, v_k) \geq R(v_k).$$

Just as in the preceding case, if $v_{k+1} = v_k$ or $R(v_{k+1}) = R(v_k)$, the problem (1.6.1) has a solution. If the sequence $\{x_k, v_k\}$ is infinite, then $\{R(v_k)\}$ is strictly monotonically increasing and conditions (4.15) hold.

Setting $x = x_k$ in (4.16), we have

$$\omega \|x_* - x_k\|^2 \leqslant L(x_*, v_{k+1}) - L(x_k, v_{k+1}) + \langle L_*, x_k - x_* \rangle \leqslant$$
$$\leqslant R(v_*) - R(v_k) + \|L_*\| \|x_* - x_k\|.$$

Noting (4.8), (4,4), we find that

$$\omega \|x_* - x_k\|^2 \leqslant R(v_*) - R(v_k) + \|x_* - x_k\| \sqrt{\delta \xi(x_k, v_k)}.$$

Noting (4.15), we conclude that $\lim_{k\to\infty} x_k = x_*$.

Let $X_0 \neq \emptyset$ and let us show that $\{v_k\}$ is bounded. From the conditions

$$\lim_{k\to\infty} h(x_k) \leqslant 0, \quad \lim_{k\to\infty} h^j(x_k) v_k^j = 0$$

it follows that if for some subsequence $\{v_{\bar{k}}^j\}$ the condition $\lim_{\bar{k}\to\infty} v_{\bar{k}}^j = \infty$ is satisfied, then $\lim_{\bar{k}\to\infty} h^j(x_{\bar{k}}) = h^j(x_*) = 0$, i.e., $j \in \sigma(x_*)$ (see Definition 1.7.3 of the set $\sigma(x)$). By (4.10) for an arbitrary vector x we have

$$f(x_{k-1}) + \sum_{j=1}^{c} v_k^j h^j(x_{k-1}) \leqslant f(x) + \sum_{j=1}^{c} v_k^j h^j(x).$$

Let us divide both sides of this inequality by $\|v_k\|$. If $\lim_{\bar{k}\to\infty} \|v_{\bar{k}}\| = \infty$, then we can find a set of numbers a_j, $\sum_{j\in\sigma(x_*)} \alpha_j = 1$, such that

$$0 = \sum_{j\in\sigma(x_*)} \alpha_j h^j(x_*) \leqslant \sum_{j\in\sigma(x_*)} \alpha_j h^j(x)$$

for any x. But this inequality contradicts Karlin's constraint qualification (see Definition 1.6.7), which in this case is equivalent to Slater's constraint qualification (see Lemma 1.6.2). Hence the sequence $\{\|v_k\|\}$ is bounded. At each limit point of the sequence $\{x_k, v_k\}$, the Kuhn-Tucker necessary (which are also sufficient in this problem) conditions for a minimum are satisfied. The inequalities (4.6) imply the convergence of the method (4.2) for the x_k chosen from the condition (4.5). ///

5. REDUCTION TO A MAXIMIN PROBLEM

1. *PRELIMINARY RESULTS*

The methods described in Sections 4.2 - 4.4 are based on Theorem 1.6.4 and the function (1.6.15) is such that the vector $y \in E^m$ at the Kuhn-Tucker point coincides with the Lagrange multipliers and, furthermore, the numerical methods are based on the solution of the system (1.6.17). In this section, we reduce our problem to a maximin problem analogous to (1.3), but in contrast to the latter, we construct a modified Lagrangian instead of $F(x,u,w)$ so as to guarantee that for each point of the local maximin its corresponding point $[x,u,v]$ is a Kuhn-Tucker point for the initial problem (1.6.1). In this case the role of dual vectors is played by the vectors which do not coincide with the Lagrange multipliers but are somehow expressed through them. We introduce the vectors

$$\lambda \in E^e, \quad \varphi(g, \lambda) = [\varphi(g^1, \lambda^1), \ldots, \varphi(g^e, \lambda^e)] \in E^e,$$
$$\mu \in E^c, \quad \psi(h, \mu) = [\psi(h^1, \mu^1), \ldots, \psi(h^c, \mu^c)] \in E^c.$$

Just as was done before, ϕ_λ, ϕ_g, $\phi_{\lambda\lambda}$, $\phi_{\lambda g}$, ψ_μ, ψ_h, $\psi_{\mu h}$, $\psi_{\mu\mu}$ are the matrices of the first and second derivatives. Obviously, the matrices of the second derivatives are diagonal. A concrete element of the matrix is obtained if we give its scalar arguments, say, $\phi_g(g^i,\lambda^i)$.

Let us introduce a modified Lagrangian in the form (2.1):

$$M(x,\lambda,\mu)=f(x)+\sum_{i=1}^{e}\varphi(g^i(x),\lambda^i)+\sum_{j=1}^{c}\psi(h^j(x),\mu^j). \qquad (5.1)$$

In this section the index i takes on integer values from 1 to e, the index j takes on integer values from 1 to c. Throughout we assume that λ and μ have only real components. We impose the following conditions on ϕ and ψ.

• A_6'. The equation

$$\phi_\lambda(g^i,\lambda^i) = 0 \qquad (5.2)$$

can be satisfied iff $g^i = 0$; the equation (5.2) together with the condition

$$u^i = \phi_g(0,\lambda^i) \qquad (5.3)$$

uniquely determines λ^i for any given $u \in E^1$.

• A_7. The equation

$$\psi_\mu(h^j,\mu^j) = 0 \qquad (5.4)$$

can be satisfied iff

$$h^j \le 0, \qquad v^j = \psi_h(h^j,\mu^j) \ge 0, \qquad h^j v^j = 0, \qquad (5.5)$$

the equation (5.4) together with the condition

$$v^j = \psi_h(h^j,\mu^j) \qquad (5.6)$$

uniquely determine μ^j for any given values h^j and v^j satisfying (5.5).

• A_8. For any x_* and μ_* such that for $h^j(x_*)$ and μ_*^j, $j \in [1:c]$, the condition (5.5) is satisfied and for any $x \in X$ we introduce the functions

$$\sum_{j=1}^{c} \psi(h^j(x), \mu_*^j) \leqslant \sum_{j=1}^{c} \psi(h^j(x_*), \mu_*^j).$$

Condition A_6' is satisfied, for example, by the functions $\phi^1(g^i, \lambda^i)$, $\phi^2(g^i, \lambda^i)$, $\phi^3(g^i, \lambda^i)$ defined in Section 4.2. Also, we give the functions

$$\varphi^8(g^i, \lambda^i) = g^i \operatorname{sh} \lambda^i + (g^i)^2/2,$$
$$\varphi^9(g^i, \lambda^i) = g^i(\lambda^i + \varepsilon \sin \lambda^i) + e^{g^i}, \quad -1 < \varepsilon < 1.$$

From (5.3) it follows that for the functions ϕ^1, ϕ^2, ϕ^3 one has $\lambda = u$, i.e., the vector λ coincides with the Lagrange multipliers. For ϕ^8 and ϕ^9 the equation (5.3) has the form

$$u^i = \operatorname{sh} \lambda^i, \quad u^i = \lambda^i + \varepsilon \sin \lambda^i,$$

respectively, where the new vector λ does not coincide with u. A simple function satisfying A_7 and A_8 is

$$\psi^7(h^j, \mu^j) = \frac{1}{2}[(h^j + \mu^j)_+^2 - (\mu^j)^2]$$

for which the equation (5.4) has the form

$$(h^j + \mu^j)_+ = \mu^j \quad . \tag{5.7}$$

If $h^j > 0$, then for $h^i + \mu^2 \geq 0$ we have $h^j = 0$, which contradicts the assumption $h^j > 0$; for $h^j + \mu^j < 0$ we have $\mu^j = 0$, which contradicts the condition $h^j + \mu^j < 0$. Hence (5.7) cannot be satisfied for $h^j > 0$. If $h^j < 0$ then for $h^j + \mu^j \leq 0$ we have $\mu^j = 0$; if $h^j + \mu^j > 0$ then $h^j = 0$, which is impossible.

Thus, (5.7) can be satisfied only for $h^j \le 0$, and if $h^j < 0$ then

$$\mu^j = 0, \qquad v^j = 0, \qquad h^j v^j = 0,$$

if $h^j = 0$ then

$$\mu^j \ge 0, \qquad v^j = (\mu^j_+)^2 = (\mu^j)^2, \qquad h^j v^j = 0.$$

Thus if (5.7) has a solution, the conditions (5.5) are satisfied, the relation (5.6) determines the dependence $\mu^j = \sqrt{v^j}$ for $v^j \ge 0$. Hence, here the vector μ is also different from the Lagrange multipliers v.

Let

$$q = [x, \lambda, \mu], \qquad q_* = [x_*, \lambda_*, \mu_*], \qquad M(q) = M(x, \lambda, \mu) .$$

Then q_* is called a stationary point of $M(q)$ if

$$M_x(q_*) = 0_{n1}, \quad M_\lambda(q_*) = 0_{e1}, \quad M_\mu(q_*) = 0_{c1}.$$

LEMMA 4.5.1. Let the functions $f(x)$, $g(x)$, $h(x)$, ϕ and ψ be continuously differentiable in all their arguments and let conditions A_6' and A_7 be satisfied. If $z_* = [x_*, u_*, v_*]$ is a Kuhn-Tucker point in the problem (1.6.1), then the system

$$u_* = \varphi_g(g(x_*), \lambda_*), \quad v_* = \psi_h(h(x_*), \mu_*), \tag{5.8}$$

$$\varphi_\lambda(g(x_*), \lambda_*) = 0, \quad \psi_\mu(h(x_*), \mu_*) = 0 \tag{5.9}$$

uniquely determines λ_* and μ_* such that $q_* = [x_*, \lambda_*, \mu_*]$ is a stationary point of the function $M(q)$.

Proof. Let $z_* = [x_*, u_*, v_*]$ be a Kuhn-Tucker point. Then

$$g(x_*) = 0, \quad h(x_*) \leqslant 0, \quad L_x(z_*) = 0,$$
$$v_* \geqslant 0, \quad v_*^j h^j(x_*) = 0.$$

By A_6' and A_7 in this case the systems (5.8) and (5.9) uniquely determine λ_* and μ_* such that $M_\lambda(q_*) = M_\mu(q_*) = 0$. Differentiating $M(q)$ over x at the point $q = q_*$ and noting (5.8) and (1.7.1), we obtain

$$\begin{aligned} M_x(q_*) &= \\ &= f_x(x_*) + g_x(x_*)\varphi_g(g(x_*),\lambda_*) + h_x(x_*)\psi_h(h(x_*),\mu_*) = \\ &= f_x(x_*) + g_x(x_*)u_* + h_x(x_*)v_* = L_x(z_*) = 0. \end{aligned} \tag{5.10}$$

Thus for the Kuhn-Tucker point z_* the stationary point $q_* = [x_*,\lambda_*,\mu_*]$ of $M(q)$ is uniquely determined. ///

2. INVESTIGATION OF THE MAXIMIN PROBLEM

We replace the initial problem (1.6.1) by the constrained maximin problem for $M(q)$:

$$\max_{\lambda\in E^e}\max_{\mu\in E^c}\min_{x\in E^n} M(x,\lambda,\mu). \tag{5.11}$$

The next theorem is an analog of Theorem 4.2.1.

<u>THEOREM 4.5.1</u>. Let the function $M(x,\lambda,\mu)$ be continuously differentiable in its arguments, let the functions ϕ and ψ satisfy conditions A_6', A_7 and A_8, and let the vector $q_* = [x_*,\lambda_*,\mu_*]$ be a strict local maximin point of (5.11). Then for the problem (1.6.1)

- 1. the point x_* is a local solution;
- 2. the vectors u_* and v_* defined by (5.8) are such that $z_* = [x_*,u_*,v_*]$ is a Kuhn-Tucker point.

<u>Proof</u>. Reformulating the necessary minimax conditions (see Theorem 1.5.7) relative to the maximin problem (5.11), we obtain that the conditions (5.9) are satisfied for q_*. By conditions A_6' and

A_7 it then follows that $x_* \in X$. For the interior problem in (5.11) for any $x \in G(x_*)$ we have

$$M(q_*) = f(x_*) + \sum_{i=1}^{e} \varphi(g^i(x_*), \lambda_*^i) + \sum_{j=1}^{c} \psi(h^j(x_*), \mu_*^j) \leqslant$$
$$\leqslant M(x, \lambda_*, \mu_*) = f(x) + \sum_{i=1}^{e} \varphi(g^i(x), \lambda_*^i) + \sum_{j=1}^{c} \psi(h^j(x), \mu_*^j).$$

Using condition A_8 we obtain that $f(x_*) \leq f(x)$ for all $x \in G(x_*) \cap X$, whence we conclude that x_* is a local solution of (1.6.1).

The vector q_* is a stationary point of $M(q)$. Hence the conditions (5.9) and (5.10) hold implying that z_* is a Kuhn-Tucker point in (1.6.1). ///

The condition of the Theorem that q_* is a solution of the maximin problem can be weakened by requiring that $q_* = [x_*, \lambda_*, \mu_*]$ satisfy the relations

$$\varphi_\lambda(g(x_*), \lambda_*) = 0, \quad \psi_\mu(h(x_*), \mu_*) = 0,$$
$$x_* \in \operatorname*{Arg\,min}_{x \in E^n} M(x, \lambda_*, \mu_*).$$

To solve this system, one may use the approach described in Section 4.2. Here we limit ourselves to a reduction to the problem (5.11).

To use the methods of finding the local maximin given in Section 2.6, we need to show that the function M has properties similar to those formulated for the function $F(x,u,v)$ in Lemmas 4.1.1 and 4.1.2. To this end, we introduce the following additional assumptions concerning the choice of the functions ϕ, ψ and partially concerning the class of problems (1.6.1).

• A_9. For any values of the arguments, the function $\phi(g,\lambda)$ is twice continuously differentiable, satisfies A_6', and at the stationary point q_* of $M(q)$ we have the inequalities

$$\delta_1^i = \varphi_{gg}(0, \lambda_*^i) > 0, \quad \varphi_{\lambda g}(0, \lambda_*^i) \neq 0, \quad \varphi_{\lambda\lambda}(0, \lambda_*^i) \leqslant 0.$$

• A_{10}. For any values of the arguments, the function $\psi(h,\mu)$ is twice continuously differentiable, satisfies A_7, and at the stationary point q_* of $M(q)$ for $j \in \sigma(x_*)$ we have the inequalities

$$\delta_2^j = \psi_{hh}(0, \mu_*^j) > 0, \quad \psi_{\mu h}(0, \mu_*^j) \neq 0, \quad \psi_{\mu\mu}(0, \mu_*^j) \leqslant 0,$$

and for $j \notin \sigma(x_*)$ we have the relations

$$\psi_{hh}(h^j(x_*), \mu_*^j) \geqslant 0, \quad \psi_{\mu h}(h^j(x_*), \mu_*^j) = 0,$$
$$\psi_{\mu\mu}(h^j(x_*), \mu_*^j) < 0.$$

Conditions A_9 and A_{10} impose restrictions on ϕ and ψ as well as on the dual variables for the specific problem (1.6.1). Condition A_{10} makes it possible to avoid later the strict complementarity condition.

Let us construct three square matrices of order $n+m$, $n+m$ and m, respectively:

$$M_{qq}(q_*) = \begin{bmatrix} M_{xx} & g_x\varphi_{g\lambda} & h_x\psi_{h\mu} \\ \varphi_{\lambda g}g_x^T & \varphi_{\lambda\lambda} & 0_{ec} \\ \psi_{\mu h}h_x^T & 0_{ce} & \psi_{\mu\mu} \end{bmatrix},$$

$$H_1(q_*) = \begin{bmatrix} -M_{xx} & -g_x\varphi_{g\lambda} & -h_x\psi_{h\mu} \\ \varphi_{\lambda g}g_x^T & \varphi_{\lambda\lambda} & 0_{ec} \\ \psi_{\mu h}h_x^T & 0_{ce} & \psi_{\mu\mu} \end{bmatrix},$$

$$N_1(q_*) = [g_x\varphi_{g\lambda} \mid h_x\psi_{\mu h}]^T M_{xx}^{-1} [g_x\varphi_{g\lambda} \mid h_x\psi_{h\mu}] - \begin{bmatrix} \varphi_{\lambda\lambda} & 0_{ec} \\ 0_{ce} & \psi_{\mu\mu} \end{bmatrix}.$$

Here all the coefficients of the matrices are computed at the stationary point q_* of the function $M(q)$.

Just as in Section 4.3, we set

$$\Delta_1(x, \lambda, \mu) = \min_{i \in [1:e]} \min_{j \in [1:c]} [\varphi_{gg}(g^i(x), \lambda^i), \psi_{hh}(h^j(x), \mu^j)].$$

LEMMA 4.5.2. Let the sufficient conditions of Theorem 1.7.4 be satisfied at the Kuhn-Tucker point $z_* = [x_*, u_*, v_*]$ for the problem (1.6.1). Let the constraint qualification be satisfied at the point x_*. Let a stationary point $q_* = [x_*, \lambda_*, \mu_*]$ of $M(q)$ correspond to x_* and let conditions A_9 and A_{10} hold. If the quantity $\Delta_1(x_*, \lambda_*, \mu_*]$ is sufficiently large, then

• 1. the matrices $M_{xx}(q_*)$ and $N_1(q_*)$ are positive definite, the matrix $M_{qq}(q_*)$ is nonsingular;

• 2. all the roots λ of the equation

$$|H_1(q_*) - \lambda I_{n+m}| = 0$$

have strictly negative real parts;

• 3. the point q_* is a strict local maximin point for the problem (5.11).

Proof. Differentiating $M(q)$ in x and using the notation introduced in conditions A_9 and A_{10} leads to the formula

$$\begin{aligned} M_{xx}(q_*) = L_{xx}(z_*) &+ \sum_{i=1}^{e} \delta_1^i g_x^i(x_*)[g_x^i(x_*)]^T + \\ &+ \sum_{j \in \sigma(x_*)} \delta_2^j h_x^j(x_*)[h_x^j(x_*)]^T + \\ &+ \sum_{j \notin \sigma(x_*)} h_x^j(x_*)\psi_{hh}(h^j(x_*), \mu_*^j)[h_x^j(x_*)]^T. \end{aligned}$$

By A_{10} the last sum is nonnegative definite. From Finsler's

Lemma 1.7.4 we obtain that the first three terms determine a positive definite matrix if $\Delta_1(x_*, \lambda_*, \mu_*)$ is sufficiently large. This implies that $M_{xx}(q_*)$ is positive definite.

One can assume without loss of generality that $h^j(x_*) = 0$ for $1 \le j \le s$ and $h^j(x_*) < 0$ for $1+s \le j \le c$. Introduce the auxiliary vectors:

$$
\begin{aligned}
V = [&\varphi_{\lambda g}(0, \lambda_*^1), \ldots \\
&\ldots, \varphi_{\lambda g}(0, \lambda_*^e), \psi_{\mu h}(0, \mu_*^1), \ldots, \psi_{\mu h}(0, \mu_*^s)] \in E^k, \\
\Gamma_1 = [&\psi_{\mu\mu}(h^1(x_*), \mu_*^1), \ldots, \psi_{\mu\mu}(h^s(x_*), \mu_*^s)] \in E^s, \quad k = e + s, \\
\Gamma_2 = [&\psi_{\mu\mu}(h^{s+1}(x_*), \mu_*^{s+1}), \ldots, \psi_{\mu\mu}(h^c(x_*), \mu_*^c)] \in E^{c-s}.
\end{aligned}
$$

Introduce also the $n \times k$ matrix

$$R_x^a = [g_x^1(x_*), \ldots, g_x^e(x_*), h_x^1(x_*), \ldots, h_x^s(x_*)].$$

Then we can write $N_1(q_*)$ in the form

$$N_1(q_*) = \left[\begin{array}{c|c} D(V)(R_x^a)^T M_{xx}^{-1}(q_*) R_x^a D(V) + D(-\Gamma_1) & 0 \\ \hline 0 & -D(\Gamma_2) \end{array}\right].$$

By A_{10} the diagonal matrix $-D(\Gamma_2)$ is positive definite and the diagonal matrix $-D(\Gamma_1)$ is nonnegative definite. From A_9 and A_{10} it follows that all the coordinates of V are not equal to zero, and therefore the rank of $R_x^a D(V)$ is the same as that of R_x^a, that is maximal and equal to k. Therefore, the upper left block of the representation of $N_1(q_*)$ is the sum of two matrices, the first matrix being positive definite and the other being nonnegative definite. Their sum is a positive definite matrix, hence the whole matrix $N_1(q_*)$ is positive definite.

Using the partition of $M_{qq}(q_*)$ into blocks one can show that its determinant is expressed in terms of determinants of the matrices $M_{xx}(q_*)$ and $N_1(q_*)$ by the formula

$$|M_{qq}(q_*)| = |M_{xx}(q_*)| \cdot |-N_1(q_*)|$$

implying that $M_{qq}(q_*)$ is nonsingular.

The proof of the fact that the characteristic values of the matrix $H_1(q_*)$ have strictly negative real parts is an almost verbatim proof of the same property of $H(z)$ (see Lemma 4.1.2). By Theorem 1.5.9, q_* is a strict local maximin point of the problem (5.11). ///

We use the methods (1.21) - (1.25) to find the local maximin, taking as $F(x,y)$ the function $M(x,\lambda,\mu)$ and as y the vector $[\lambda,\mu] \in E^m$. Let us reformulate the convergence Theorem 4.1.2 as follows.

THEOREM 4.5.2. Let the conditions of Lemma 4.5.2 be satisfied at the Kuhn-Tucker point $z_* = [x_*,u_*,v_*]$. Then there exist $\bar{\varepsilon} > 0$, $\bar{\alpha} > 0$ such that for any fixed $0 < \varepsilon < \bar{\varepsilon}$, $0 < \alpha < \bar{\alpha}$ and for $\varepsilon = 1$, the methods (1.21) - (1.25) and their discrete variants of the form (2.6.20) converge locally to the stationary point q_* of the function $M(q)$.

One can analogously use Newton's method and give sufficient convergence conditions.

3. PARTICULAR CASES

The class of functions ϕ satisfying A_9 is extremely rich. For e.g., ϕ^1, ϕ^2, ϕ^3, ϕ^8, ϕ^9. For ϕ^1, ϕ^2, ϕ^3 the condition A_6' is satisfied and, in addition,

$$\varphi_{gg}(0,\lambda^i) = \varphi_{g\lambda}(0,\lambda^i) = 1, \quad \varphi_{\lambda\lambda}(0,\lambda^i) = 0,$$

i.e., conditions A_9 hold. For ϕ^8, ϕ^9 conditions A_9 are also valid since here

$$\varphi^8_{gg}(0, \lambda^i) = \varphi^9_{gg}(0, \lambda^i) = 1, \quad \varphi^8_{\lambda g}(0, \lambda^i) = \operatorname{ch} \lambda^i \neq 0,$$
$$\varphi^9_{\lambda g}(0, \lambda^i) = 1 + \varepsilon \cos \lambda^i \neq 0, \quad \varphi^8_{\lambda\lambda}(0, \lambda^i) = \varphi^9_{\lambda\lambda}(0, \lambda^i) = 0.$$

The twice differentiable functions ψ are more complicated. Let us give an example of such a function:

$$\psi^8(h^j, \mu^j) = \frac{1}{4}\left[(h^j + \mu^j)^4_+ - (\mu^j)^4 - (\mu^j)^2 \frac{(h^j_-)^4}{1 + (h^j_-)^4}\right].$$

It is easy to see that ψ^8 satisfies A_7, and $v^j = (\mu^j)^3_+$ for $h^j = 0$. If $h^j < 0$ then $\mu^j = v^j = 0$.

Calculating the second derivatives of ψ^8, for $j \in \sigma(x_*)$ we have

$$\psi^8_{hh}(0, \mu^j_*) = \psi^8_{\mu h}(0, \mu^j_*) = 3(\mu^j_*)^2_+ \geqslant 0,$$
$$\psi^8_{\mu\mu}(0, \mu^j_*) = 3[(\mu^j_*)^2_+ - (\mu^j_*)^2] \leqslant 0.$$

For $j \notin \sigma(x_*)$ we obtain

$$\psi^8_{hh}(h^j(x_*), \mu^j_*) = \psi^8_{\mu h}(h^j(x_*), \mu^j_*) = 0,$$
$$\psi^8_{hh}(h^j(x_*), \mu^j_*) = \frac{-(h^j_-(x_*))^4}{2[1 + [h^j_-(x_*)]^4]} < 0.$$

If at the Kuhn-Tucker point the strict complementarity condition is satisfied, then $\mu^j_* > 0$ for $j \in \sigma(x_*)$ and thus condition A_{10} is satisfied. To ensure the condition of Lemma 4.5.2 that the value of $\Delta_1(x_*, \lambda_*, \mu_*)$ is sufficiently large, one can use the scaling described in Subsection 4.2.3.

6. REDUCTION TO A MINIMAX PROBLEM

In Section 1.6 we examined the minimax problem (1.6.8). Using it for numerical solution of (1.6.1) is complicated by the fact that for unfeasible points the interior problem has infinitely large dual variables as solution. Hence it is appropriate to modify the Lagrangian so that in finding the minimax the interior problem has bounded solutions. We have been able to construct such functions, but we had, however, to introduce more complex modifications depending on the vector x and the dual vector, as well as on the gradients of the functions defining the problem (1.6.1). The numerical methods obtained through this approach possess some new characteristics and complement the methods described in the preceding sections of this chapter. We will go into more detail on these properties at the end of the section.

1. *PRELIMINARY RESULTS*

Introduce the following modified Lagrangian:

$$H(x, u, w) = F(x, u, w) - \tau \sum_{i=1}^{n} \varphi\left(\frac{\partial F(x, u, w)}{\partial x^i}\right), \tag{6.1}$$

where F is defined by (1.1), τ is a positive parameter, $\phi(t)$ satisfies A_{11}. The function $\phi(t)$ of a scalar argument is twice continuously differentiable on E^1, $\phi(t) \neq 0$ for any $t \neq 0$, $\phi'(0) = 0$, $\phi''(0) = 1$.

As the simplest examples of ϕ we use

$$\varphi_1(t) = \frac{1}{2} t^2, \qquad \varphi_2(t) = \operatorname{ch} t,$$

$$\varphi_3(t) = t \operatorname{arctg} t - \frac{1}{2} \ln(1 + t^2), \quad \varphi_4(t) = e^t - t.$$

Let us calculate the derivatives of H:

$$\begin{aligned} H_x(x,u,w) &= F_x(x,u,w) - \tau F_{xx}(x,u,w)\varphi'(F_x(x,u,w)), \\ H_u(x,u,w) &= F_u(x,u,w) - \tau g_x^T(x)\varphi'(F_x(x,u,w)), \\ H_w(x,u,w) &= F_w(x,u,w) - 2\tau D(w)h_x^T(x)\varphi'(F_x(x,u,w)). \end{aligned} \tag{6.2}$$

Here and below $\phi'(f_x)$ denotes the n-dimensional vector column with coordinates $\phi'(\partial F/\partial x^i)$, $i \in [1:n]$.

From the formulas (6.2) and condition A_{11} it follows that every stationary point of F is a stationary point of H. In general, the converse is false.

Let

$$y=[u,w], \quad z=[x,y]\in E^{n+m}, \quad F(x,u,w)=F(x,y)=F(z),$$
$$H(x,u,w)=H(x,y)=H(z).$$

We write out (6.2) in the form

$$\begin{aligned} H_x(x,y) &= F_x(x,y) - \tau F_{xx}(x,y)\varphi'(F_x(x,y)), \\ H_y(x,y) &= F_y(x,y) - \tau F_{yx}(x,y)\varphi'(F_x(x,y)). \end{aligned} \tag{6.3}$$

Later on we will need the following lemma.

LEMMA 4.6.1. Let $z_* = [x_*, y_*]$ be a stationary point of the function $F(x,y)$; let the functions defining the problem be twice continuously differentiable in a neighborhood of x_* and let A_{11} be satisfied. Then the function $B(x) = \sum_{i=1}^{n} \phi(\partial F(x,y_*)/\partial x^i)$ is twice differentiable with respect to x at x_* and we have

$$\frac{d^2}{dx^2}B(x_*,y_*) = F_{xx}^2(x_*,y_*). \tag{6.4}$$

Here and below $F_{xx}^2(x,y) = F_{xx}(x,y)F_{xx}(x,y)$.

Proof. The function $B(x)$ is continuously differentiable in a neighborhood of the point x_* being a stationary point of $B(x)$.

Let $B_x(x)$ denote the gradient of $B(x)$ at the point x_0. Also, let us estimate the norm

$$\Delta(x) = \| B_x(x) - B_x(x_*) - F^2_{xx}(z_*)(x - x_*) \| =$$
$$= \| F_{xx}(x, y_*) \varphi'(F_x(x, y_*)) - F^2_{xx}(z_*)(x - x_*) \| \leqslant$$
$$\leqslant \| F_{xx}(x, y_*)[\varphi'(F_x(x, y_*)) - \varphi'(F_x(z_*)) - F_{xx}(z_*)(x - x_*)] \| +$$
$$+ \| [F_{xx}(x, y_*) - F_{xx}(z_*)] F_{xx}(z_*)(x - x_*) \|.$$

Set $\beta = \|F_{xx}(z_*)\|$. Since the functions $f(x)$, $g(x)$ and $h(x)$ are twice continuously differentiable, for any $\varepsilon > 0$ one can find a neighborhood $G(x_*)$ of x_* such that

$$\| F_{xx}(x, y_*) - F_{xx}(x_*, y_*) \| \leqslant \varepsilon \quad \forall x \in G(x_*).$$

For these same x the inequality $\|F_{xx}(x, y_*)\| \leq \beta + \varepsilon$ will be satisfied.

Next, by the differentiability of the vector function $\phi'(F_x(x,y))$, noting that $\phi''(0) = 1$, we can take the neighborhood $G(x_*)$ so small that

$$\| \varphi'(F_x(x, y_*)) - \varphi'(F_x(z_*)) - F_{xx}(z_*)(x - x_*) \| \leqslant \varepsilon \| x - x_* \|.$$

Then for all $x \in G(x_*)$ we have: $\Delta(x) \leq (2\beta + \varepsilon)\varepsilon\|x - x_*\|$. And since ε is arbitrary, from the last inequality we obtain the formula (6.4). ///

2. INVESTIGATION OF THE MINIMAX PROBLEM

Consider the auxiliary minimax problem

$$\min_{x \in E^n} \max_{y \in E^m} H(x, y). \tag{6.5}$$

THEOREM 4.6.1. Let the conditions of Theorem 1.7.4 be satisfied at the Kuhn-Tucker point $[x_*, u_*, v_*]$ for the problem (1.6.1) and let the constraint qualification be satisfied at the point x_*.

Furthermore, let the functions defining the problem (1.6.1) be twice continuously differentiable in a neighborhood of x_* and let A_{11} be satisfied. Then there exists a $\bar{\tau} > 0$ such that for all $0 < \tau < \bar{\tau}$ the point $z_* = [x_*, u_*, w_*]$, where $w_*^j = \sqrt{v_*^j}$, $j \in [1:c]$, will be a strict local minimax point of the problem (6.5).

Proof. To the Kuhn-Tucker point $[x_*, u_*, v_*]$ there corresponds a point $z_* = [x_*, u_*, w_*]$ which is a stationary point of H. By Theorem 1.5.7, for z_* to be a strict local minimax point it is sufficient that the matrix $H_{yy}(x_*)$ be negative definite and the matrix

$$\Phi(z_*, \tau) = H_{xx}(z_*) - H_{xy}(z_*) H_{yy}^{-1}(z_*) H_{yx}(z_*) \tag{6.6}$$

be positive definite.

We obtain these matrices of second derivatives by differentiating (6.3), noting the stationary conditions and the assertion of Lemma 4.6.1:

$$\begin{aligned} H_{xx}(z_*) &= (I_n - \tau F_{xx}(z_*)) F_{xx}(z_*), \\ H_{yy}(z_*) &= F_{yy}(z_*) - \tau F_{yx}(z_*) F_{xy}(z_*), \\ H_{yx}(z_*) &= F_{yx}(z_*) (I_n - \tau F_{xx}(z_*)). \end{aligned}$$

One can assume with loss of generality that $h^j(x_*) = 0$ for $1 \le j \le s$ and $h^j(x_*) < 0$ for $1 + s \le j \le c$. Let

$$R_x^a(x) = [g_x^1(x), \ldots, g_x^e(x), h_x^1(x), \ldots, h_x^s(x)],$$
$$h^a(x) = [h^1(x), \ldots, h^s(x)], \quad h^b(x) = [h^{s+1}(x), \ldots, h^c(x)].$$

Noting the strict complementarity condition we write the matrix $H_{yy}(z_*)$ in block form

$$H_{yy}(z_*) = \left[\begin{array}{c|c} -\tau N_1 & 0_{(e+s)(c-s)} \\ \hline 0_{(c-s)n} & -N_2 \end{array}\right],$$

$$N_1 = D(\rho)[R_x^a(x_*)]^T R_x^a(x_*) D(\rho), \quad N_2 = -2D(h^b(x_*)).$$

Here ρ is an (e+s)-dimensional vector whose first e coordinates are ones, each subsequent $(e+i)^{th}$ coordinate is $2w_*^i$. All diagonal elements of the diagonal matrices $D(\rho)$ and N_2 are strictly positive. Therefore, noting the linear independence of the vectors $g_x(x_*)$, $h_x^a(x_*)$, we reach the conclusion that $H_{yy}(z_*)$ is negative definite and therefore has a negative definite inverse which can be written in the form

$$H_{yy}^{-1}(z_*) = \left[\begin{array}{c|c} -\frac{1}{\tau} N_1^{-1} & 0_{(e+s)(c-s)} \\ \hline 0_{(c-s)(e+s)} & -N_2^{-1} \end{array}\right]. \tag{6.7}$$

If we substitute this expression into (6.6) and note that N_2^{-1} will be multiplied on the left and right by matrices which by the strict complementarity condition are null, we obtain

$$\Phi(z_*) = (I_n - \tau\Gamma)\left[\Gamma + \frac{1}{\tau} W(I_n - \tau\Gamma)\right],$$

where

$$W = R_x^a(x_*) D(\rho) N_1^{-1} D(\rho)[R_x^a(x_*)]^T, \quad \Gamma = F_{xx}(z_*).$$

One can show that for sufficiently small τ the matrix $\Phi(z_*)$ is positive definite. Using Theorem 1.5.7, we then conclude that z_* is a strict local minimax point of the problem (6.5). ///

If in the conditions of the Theorem we additionally require $L_{xx}(x_*, u_*, v_*)$ to be positive definite, then for sufficiently small τ the matrix $H_{xx}(z_*)$ will be positive definite and, therefore, $H(z)$ has a strict local saddle at the point z_*. This is the case, for instance, for a convex programming problem with strictly convex $f(x)$.

3. NUMERICAL METHODS

If the conditions of Theorem 1.7.4 are satisfied, then to solve the problem (6.5) one can use methods for finding local minimax points. However, the function H depends on both the functions defining the problem (1.6.1) and their derivatives, which essentially complicates the numerical implementation of the methods given in Section 2.6. Let us construct methods analogous to (1.25) without this drawback.

For fixed x we solve the auxiliary problem of finding the value of the vector function $y(x) = [u(x), w(x)]$ from the condition

$$y(x) \in \operatorname{Arg} \max_{y \in E^m} H(x, y). \tag{6.8}$$

We iterate by any one of the following schemes:

$$x_{k+1} = x_k - \tau F_x(x_k, y_k) , \tag{6.9}$$

$$x_{k+1} = x_k - \tau \phi'(F_x(x_k, y_k)) , \tag{6.10}$$

where $y_k \in y(x_k)$. In this case, the analog of Theorem 4.2.1 is

<u>THEOREM 4.6.2</u>. Let the function ϕ satisfy condition A_{11} and the sequence $\{x_k, y_k\}$ obtained from (6.8) and from (6.9) or (6.10) converge to the point $[x_*, u_*, w_*]$. Then $[x_*, u_*, v_*]$, where $v_*^j = [w_*^j]^2$, $j \in [1:c]$, is a Kuhn-Tucker point of the problem (1.6.1).

Proof. By A_{11}, the convergence of $\{x_k, y_k\}$ implies that the condition $F_x(x_*, u_* w_*) = 0$ is satisfied at the limit point. Moreover, at this point the necessary condition for a maximum of $H(x_*, u, w)$ in u and w must hold. Hence $F_u(x_*, u_*, w_*) = 0$, $F_w(x_*, u_*, w_*) = 0$. Whence $g(x_*) = 0$ and $w_*^j h^j(x_*) = 0$ for all $j \in [1:c]$. By Theorem 1.3.1 is it necessary that the matrix $F_{yy}(x_*, u_*, w_*)$ be positive definite, so that $x_* \in X$. ///

THEOREM 4.6.3. Let the conditions of Theorem 4.6.1 be satisfied. Then we can find a $\tilde{\tau} > 0$ such that for all $0 < \tau < \tilde{\tau}$ the iterations given by (6.9), (6.10) converge locally to x_* at a linear rate.

We prove only (6.9); (6.10) is proved similarly. The necessary condition of the maximum in (6.8) is

$$H_y(x, y(x)) = 0 .$$

In proving Theorem 4.6.1 we showed that the matrix $H_{yy}(x_*, y_*)$ is negative definite. Hence there exists a neighborhood of x_*, where (6.8) has a local solution, the function $y(x)$ is single-valued, differentiable at the point $x = x_*$ and

$$\frac{dy(x_*)}{dx} = -H_{yy}^{-1}(x_*, y_*) H_{yx}(x_*, y_*). \tag{6.11}$$

Let us differentiate the composite function

$$T(x, \tau) = x - \tau F_x(x, y(x))$$

over x at the point $x = x_*$. Noting (6.11), we have

$$T_x(x_*, \tau) = \frac{dT(x_*, \tau)}{dx} =$$

$$= I_n - \tau F_{xx}(z_*) + \tau F_{xy}(z_*) H_{yy}^{-1}(z_*) H_{yx}(z_*).$$

Substituting here the formula (6.7), we obtain

$$T_x(x_*, \tau) = P(I_n - \tau F_{xx}(z_*)),$$

where $P = I_n - W$. Next we consider

$$\eta(\tau) = \|V(\tau)\| = \max_{q \neq 0} \frac{q^T V(\tau) q}{\langle q, q \rangle}, \quad V(\tau) = T_x^T(x_*, \tau) T_x(x_*, \tau).$$

We see by a direct check that the symmetric matrix P is such that $PP = P$. Hence

$$\begin{aligned} \eta(0) &= \|P\|^2 = \max_{q \neq 0} \frac{\langle Pq, Pq \rangle}{\langle q, q \rangle} = 1, \\ V'(+0) &= -[F_{xx}(z_*)P + PF_{xx}(z_*)]. \end{aligned} \tag{6.12}$$

The maximum in (6.12) obtains only for those $\bar{q}$ for which $P\bar{q} = \bar{q}$, i.e., when $\bar{q} \in K_3(x_*)$ (see the formula 1.7.16)). We use Theorem 1.5.3 on differentiating of the function of the maximum and obtain

$$\eta'(+0) = 2 \max_{\|\bar{q}\|=1} \bar{q}^T V'(+0) \bar{q} = -2 \min_{\|\bar{q}\|=1} \bar{q}^T F_{xx}(z_*) \bar{q} < 0 ,$$

which together with (6.12) imply the existence of a $\bar{\tau} > 0$ such that $\eta(\tau) > 1$ for all $0 < \tau < \tilde{\tau}$ and $\|T_x(x_*, \tau)\| < 1$. Noting Theorem 2.3.4, we conclude that the iterations defined by (6.9) converge linearly.

4. COMPUTATIONAL ASPECTS

In method (2.14) the auxiliary problem (2.3) is to minimize H with respect to x, and iterations are carried out for the dual vectors. Therefore, such methods are often called dual methods.

Conversely, in the methods (6.9), (6.10) the auxiliary problem (6.8) is to maximize H with respect to the dual vector, and the primal vector x is updated iteratively. (Some authors call such methods primal. The terminology has not been firmly established; however, it is convenient to use this term here.)

Within the numerical calculations, most time is spent solving auxiliary problems. Hence, if the total number of constraints m is much greater than the dimension n of the vector x, the dual methods are more appropriate, since in this case, solving (2.3) is much simpler than solving (6.8). If n is much greater than m, then, conversely, one should use primal methods. Whenever the calculation of f, g, h is expensive, it is also better to use primal methods. This is because in solving an auxiliary problem, only the dual vector changes, whereas the primal method is recomputed only on the "exterior" iterations, which occurs far less often.

The region of convergence depends on many factors, in particular, on the form of $\phi(t)$ and on which of the methods, (6.9) or (6.10), has been used. For several test problems, it turned out that the convergence region is greater for method (6.10). The choice of the parameter τ determines the region and rate of convergence. It has been experimentally proved that the convergence rate decreases as τ decreases; hence it is not worthwhile to take τ too close to zero.

If we take ϕ' for ϕ, then (6.9) and (6.10) coincide, and the auxiliary problem (6.8) for $\tau = 1$ is very close to the problem (3.5.7) if $N_k = 2I_n$. Thus, these methods are related to the linearization method.

Chapter 5

RELAXATION METHODS FOR SOLVING NONLINEAR PROGRAMMING PROBLEMS

By relaxation methods we mean iterative numerical methods where the objective function is monotonically decreasing.

1. APPLICATION OF THE REDUCED-GRADIENT METHOD TO SOLVING PROBLEMS WITH EQUALITY-TYPE CONSTRAINTS

1. *THE IDEA OF THE METHOD*

We consider the simplest nonlinear programming problem (1.6.1) involving inequality-type constraints. Suppose we need to find

$$\min_{x\in Y} f(x), \quad Y=\{x\in E^n\colon g(x)=0\}. \tag{1.1}$$

In this chapter we assume that the functions determining the problem are differentiable with respect to x. Let $x(x_0,t)$ denote the solution of the Cauchy problem for the system of ordinary differential equations

$$\frac{dx}{dt}=-[f_x(x)+g_x(x)u], \quad x_0\in Y. \tag{1.2}$$

We will find the feasible values of $u \in E^e$ by the following requirement: the set Y must be invariant with respect to the system (1.2), i.e., if $x_0 \in Y$, then $x(x_0,t) \in Y$ for all

$t \geq 0$, or equivalently $g(x(x_0,t)) \equiv 0$. From this we infer

$$\frac{dg}{dt} = -g_x^T(x)[f_x(x)+g_x(x)u] \equiv 0. \tag{1.3}$$

If the matrix $g_x^T(x)g_x(x)$ is nonsingular, then from (1.3) we find

$$u = -[g_x^T(x)g_x(x)]^{-1}g_x^T(x)f_x(x).$$

Substituting this expression into (1.2), we have

$$\frac{dx}{dt} = -[f_x(x)-g_x(x)[g_x^T(x)g_x(x)]^{-1}g_x^T(x)f_x(x)], \qquad x_0 \in Y. \tag{1.4}$$

Differentiating $f(x)$ using the system (1.4), we obtain

$$\frac{df}{dt} = -f_x^T(x)[f_x(x)+g_x(x)u].$$

After some transformations, noting (1.3), we obtain

$$\frac{df}{dt} = -\|f_x(x)+g_x(x)u\|^2 \leqslant 0. \tag{1.5}$$

Hence the set Y is invariant with respect to the system (1.2), and along any solution $x(x_0,t)$ of the system (1.4) the objective function $f(x(x_0,t))$ monotonically decreases. In the sequel we show that the solutions of (1.4) converge to local solutions of (1.1) as $t \to \infty$. Thus, finding the limit points of solutions of the Cauchy problem (1.4) yields a numerical method of solving the problem (1.1). To implement this method, we have to invert the matrix $\Gamma(x) = g_x^T(x)g_x(x)$. Its determinant is usually called the Gram determinant, for the set of vectors $g_x^1(x), g_x^2(x), \ldots, g_x^e(x)$. It is known (see, e.g., the Gantmacher [1]) that in order that the Gram determinant of the matrix $\Gamma(x)$ be nonzero, it is necessary and sufficient that the vectors $g_x^i(x)$, $i \in [1:e]$, be linearly independent. Hence the right side of (1.4) is defined everywhere on Y if $g(x) = 0$ satisfies the constraint qualification on Y (see Definition 1.7.3).

Let us dwell on a geometric interpretation of the method. We write the system (1.4) in the form:

$$\frac{dx}{dt} = -M(x) f_x(x),$$
$$M(x) = I_n - N(x), \quad N(x) = g_x(x)[g_x^T(x) g_x(x)]^{-1} g_x^T(x). \tag{1.6}$$

The matrices $M(x)$ and $N(x)$ are symmetric. A direct check tells us that $M^2 = MM = M$, $N^2 = NN = N$. Such matrices are called idempotent. The eigenvalues being zero or one, these matrices are nonnegative definite.

If $x \in Y$ and $g(x) = 0$ satisfy the constraint qualification, then the cone of tangential directions (the tangent subspace) to the set Y at the point x is given by

$$K(x) = \{\bar{x} \in E^n : g_x^T(x)\bar{x} = 0\} ,$$

which is the orthogonal complement of the subspace generated by the independent vectors $g_x^1(x), g_x^2(x), \ldots, g_x^e(x)$. An arbitrary vector $z \in E^n$ is representable in the form

$$z = a + b , \tag{1.7}$$

where a is the projection of z on the tangent subspace, b is its projection onto the orthogonal complement. The vector b is a linear combination of the vectors $g_x^i(x)$:

$$b = g_x(x)d , \qquad d \in E^e .$$

Substituting this into (1.7), multiplying (1.7) on the left by $g_x^T(x)$ and noting that $g_x^T(x)a = 0$, we obtain

$$d = [g_x^T(x)g_x(x)]^{-1} g_x^T(x)z ,$$
$$a = M(x)z, \quad b = N(x)z, \quad \langle a,b \rangle = 0 .$$

Hence the matrix $M(x)$ projects each vector $x \in E^n$ onto the tangent subspace and the matrix $N(x)$ projects onto its orthogonal complement. A vector z in the tangent subspace is projected by $M(x)$ onto itself since if $z \in K(x)$ then $g_x^T(x)z = 0$ and $M(x)z = z$; the projection of z onto the orthogonal complement $N(x)z = 0$. If z is any vector in E^n then $M(x)z \in K(x)$ since $g_x^T(x)M(x)z = 0$.

Thus the right side of (1.4) is the orthogonal projection of the antigradient $-f_x(x)$ onto the tangent subspace, so that the $g^i(x)$ are integrals of the system (1.4). Several authors call the vector on the right in (1.4) the "reduced antigradient." Hence we will call the method (1.4) the reduced-gradient method. In a particular case where there are no constraints, the method (1.4) becomes the Cauchy gradient method (2.2.3) of unconstrained minimization of $f(x)$.

Let the feasible point x_* be an equilibrium for the system (1.2). From (1.3) we find the corresponding vector u_*, and noting (1.5) we have the relations

$$g(x_*) = 0 \ , \quad f_x(x_*) + g_x(x_*)u_* = 0 \ .$$

Hence, if the constraints satisfy the constraint qualification, then to each equilibrium point for the system (1.4) there corresponds a Kuhn-Tucker point $[x_*, u_*]$ of the problem (1.1). From (1.5) it follows that at an equilibrium point the projection of f_x onto the tangent subspace is equal to zero, which is a necessary condition for an extremum of the problem (1.1).

2. PROOF OF CONVERGENCE

Introduce the set

$$\Omega = \{x \in E^n: f(x) \leqslant f(x_0), \quad x \in Y\}. \tag{1.8}$$

THEOREM 5.1.1. Let the functions defining the problem (1.1) be differentiable on an open set containing Ω, where $g(x) = 0$ satisfies the constraint qualification; let the local minimum of $f(x)$ be attained on Ω at the unique point x_*. Then the solutions of the system (1.4) converge to the point x_* as $t \to \infty$.

Proof. We use the Lyapunov function $v(x) = f(x) - f(x_*)$. For any $t \geq 0$ we have $x(x_0,t) \subset \Omega$, $v(x(x_0,t)) \geq 0$. By (1.5) $v(x(x_0,t))$ is a monotonically decreasing function of t. Arguing just as in proving Theorem 2.2.3, we arrive at the required assertion. ///

3. COMPUTATIONAL ASPECTS

In implementing this method, the system (1.4) is numerically integrated. Using, say, Euler's method, we obtain the following discrete approximation:

$$x_{k+1} = x_k - \alpha M(x_k) f_x(x_k), \quad x_0 \in Y. \tag{1.9}$$

The accuracy of the integration must be sufficiently high since otherwise the computed trajectory may leave the feasible set. Hence in (1.9) we either need to take small steps α or use more accurate integration schemes.

The simplest case is when the vector function $g(x)$ is linear. Here the matrix M is constant and from (1.9) it follows that

$$g_x^T x_{k+1} = g_x^T x_k - \alpha g_x^T M f_x(x_k) = g_x^T x_k = g_x^T x_0 .$$

Thus $g(x_k) = g(x_0)$ for any value α, so that the choice of the step can be made without caring about the feasibility condition.

Several authors suggest that an additional correction should be made on each kth iteration (1.9), forcing the point x_k back to the feasible set by solving the system of equations $g(x) = 0$. Here the point x_k given by (1.9) is taken as the initial approximation. However, such calculations substantially complicate the the method.

It is interesting that the method (1.6) can be derived from method (4.6.9) as a special case. Indeed, let us take ϕ_1 for ϕ in (4.6.1). We consider the problem of non-linear programming (1.1) with equality-type constraints only. Then the necessary condition of the maximum in (4.6.8) is that $H_u(x,u) = 0$. Determining u from this relation, we have

$$u = [g_x^T g_x]^{-1} [\frac{g}{\tau} - g_x^T f_x] . \tag{1.10}$$

We use the continuous analog of the method (4.6.9):

$$\frac{dx}{dt} = -[f_x + g_x u] ,$$

where we substitute the expression for u from (1.10):

$$\frac{dx}{dt} = -[f_x - g_x [g_x^T g_x]^{-1} [g_x^T f_x - \frac{g}{\tau}]] . \tag{1.11}$$

Differentiating f and g, we obtain

$$\frac{df}{dt} = -\|f_x + g_x u\|^2 - \frac{1}{\tau} f_x^T g_x [g_x^T g_x]^{-1} g$$
$$\frac{dg}{dt} = -\frac{g}{\tau} \quad . \tag{1.12}$$

Integrating the last expression, we find that

$$g(x(t)) = g(x(t_0))\, e^{-t/\tau} \quad . \tag{1.13}$$

The expression (1.11) has a remarkable property: all its trajectories, as $t \to \infty$, approach the feasible set X. Here the function $f(x(t))$ is not monotonic. This is seen from the right side of (1.12) where the first term is always negative and the second term can change sign and can be neglected for small $\|g\|$. The formula (1.13) suggests a new way of handling equality-type constraints. If they are violated, one can integrate (1.11) instead of (1.6), which will guarantee the trajectories to approach X.

From (1.13) it follows that if $g(x(t_0)) = 0$ then $g(x(t)) \equiv 0$ for all t and in this case the equation (1.11) turns into (1.4), and the derivatives (1.12) coincide with (1.5), (1.3), respectively.

Analogously, from (4.6.9) one can obtain computational formulas taking inequality-type constraints into account, but they are too complex to describe them here.

2. A GENERALIZATION OF THE REDUCED-GRADIENT METHOD

1. *PRELIMINARY RESULTS*

The method described above carries over to the general nonlinear programming problem in various ways. We shall examine the approaches, using the notation (1.6.3) and (1.6.4). We say that the points of X_0 are interior points of X and the points of $X\backslash X_0$ are boundary points. We combine the equality- and/or inequality-type constraints in $\Phi = [g,h]$. The vector function $\Phi(x)$ thus defines a mapping $E^n \to E^m$, where $m = e + c$.

Let

$$\gamma(\Phi)=[\gamma(\Phi^1), \gamma(\Phi^2), \ldots, \gamma(\Phi^m)], \sqrt{\gamma(-\Phi)} = [\sqrt{\gamma(-\Phi^1)}, \ldots, \sqrt{\gamma(-\Phi^m)}],$$

where the function $\gamma(z)$ of a scalar argument is defined and continuous for all values $z \geq 0$ and satisfies the following conditions:

$$\gamma(0)=0, \quad \lim_{z \to +0} \frac{\gamma(z)}{z} \geqslant 0; \quad \gamma(z) > 0 \quad \text{if} \quad z > 0. \tag{2.1}$$

As simplest functions γ one can take $\gamma(z) = z$, z^2, $e^z - 1$.

For the numerical solution of problem (1.6.1) we suggest to find the limit (as $t \to \infty$) points of the solution of the Cauchy problem for the system

$$\frac{dx}{dt} = -[f_x(x) + \Phi_x(x)y], \quad x_0 \in X_0. \tag{2.2}$$

Here the vector $y \in E^m$ is determined from solving the following system of m linear equations:

$$\Gamma(x)y + \Phi_x^T(x)f_x(x) = 0, \tag{2.3}$$

where

$$\Gamma(x) = \Phi_x^T(x)\Phi_x(x) + D(\gamma(-\Phi(x))).$$

We find y from (2.3) and substitute it into the right side of (2.2). Then (2.2) can be rewritten in the form

$$\frac{dx}{dt} = -M(x) f_x(x), \tag{2.4}$$

$$M(x) = I_n - N(x), \quad N(x) = \Phi_x(x)\Gamma^{-1}(x)\Phi_x^T(x). \tag{2.5}$$

Introduce the index set:

$$\bar{\sigma}(x) = \{i \in [1:m]: \Phi^i(x) = 0\}.$$

In the particular case where there are no inequality-type constraints, the system (2.4) coincides with (1.4). We show in the sequel that in the general case the system (2.4) conserves the basic properties of the system (1.4): the feasible set X is invariant with respect to (2.4), the function $f(x(x_0,t))$ monotonically decreases on all trajectories of the system (2.4) satisfying the condition $x_0 \in X_0$. One of the ways of deriving the system (2.4) from (1.4) will be described in Subsection 5.2.3 for $\gamma(z) = z$. Let us give sufficient conditions guaranteeing the solvability of the system (2.3).

By Definition (1.7.3), we say that the function $\Phi(x)$ satisfies the constraint qualification at x if all the vectors $\Phi_x^j(x)$, $j \in \bar{\sigma}(x)$, are linearly independent. From this definition it follows that if the constraints satisfy the constraint qualification at x, then the number of coordinates of $\Phi(x)$ which simultaneously vanish does not exceed n.

<u>LEMMA 5.2.1</u>. If at each point $x \in X \backslash X_0$ the function $\phi(x)$ satisfies the constraint qualification, then the matrix $\Gamma(x)$ is nonsingular and nonnegative definite for all $x \in X$.

<u>Proof</u>. We write $\Gamma(x)$ as a product of a rectangular matrix $B(x)$ $(m \times (n+m))$ and its transpose $B^T(x)$, where $B(x)$ consists of

two block matrices:

$$B(x)=[\Phi_x^T(x)\,|\,D(\sqrt{\gamma(-\Phi(x))})],\quad \Gamma(x)=B(x)B^T(x).$$

The Lemma will be proved if we can show that for any $x \in X$ the rank of $B(x)$ is equal to m, since if the rank of $B(x)$ is maximal (equal to m), it then follows that $\Gamma(x)$ is a nonsingular nonnegative definite matrix. If there are no equality-type constraints, then at each interior point $x \in X_0$ the rank of $B(x)$ is equal to m, since in this case for a nonzero minor of $B(x)$ we can take the diagonal matrix $D(\sqrt{\gamma(-\Phi(x))})$. The Lemma is also obvious if $\Phi(x) = 0$. By the linear independence of the vectors $\Phi_x^i(x)$ there is a nonzero minor of order m for $\Phi_x(x)$.

Let s components, $e < s < m$, of the vector function $\Phi(x)$ be zero at $x \in X$. We can assume without loss of generality that they are $\Phi^1(x), \Phi^2(x), \ldots, \Phi^s(x)$. Then the values of the functions $\Phi^{s+1}(x), \ldots, \Phi^m(x)$ are strictly less than zero. In the rectangular matrix

$$V_1(x)=\begin{bmatrix}[\Phi_x^1(x)]^T\\ \cdots\\ [\Phi_x^s(x)]^T\end{bmatrix}$$

with dimensions $s \times n$, we determine a square matrix $c(s)$ of order s such that its determinant being a minor of the matrix $V_1(x)$ of order s is not equal to zero. Such a minor exists by the constraint qualification. The determinant of the matrix

$$V_2(x)=\left[\begin{array}{c|ccc} C(x) & & 0_{s\,(m-s)} & \\ \hline & \sqrt{\gamma(-\Phi^{s+1}(x))} & & 0\\ 0_{(m-s)\,s} & & \ddots & \\ & 0 & & \sqrt{\gamma(-\Phi^m(x))}\end{array}\right]$$

of dimension m is not equal to zero. But the determinant of $V_2(x)$ is simultaneously a minor of the mth order matrix $B(x)$. Hence the rank of $B(x)$ is maximal, i.e., equal to m. ///

From the Lemma it follows that if the constraint qualification is satisfied at each boundary point of the set X, then the right sides of the system (2.4) are defined everywhere on X. In the sequel we will say that the constraint qualification holds everywhere on X if it holds at each boundary point of X.

LEMMA 5.2.2. Let the conditions of Lemma 5.2.1 be satisfied. Then the symmetric matrix $M(x)$ is nonnegative definite for all $x \in X$.

Introduce the matrices

$$A = -\Gamma^{-1}(x)\Phi_x^T(x), \quad P = \left[\frac{M(x)}{D(\sqrt{\gamma(-\Phi(x))})A}\right],$$

having dimension $m \times n$ and $(n+m) \times n$, respectively. The proof of the Lemma follows from the representation $M = P^TP$, which can be checked by direct computations. ///

In what follows we will assume that for each $x_0 \in X$ the system (2.2) determines a unique solution $x(x_0,t)$. Let

$$\Omega = \{x \in E^n\colon\ f(x) \leqslant f(x_0),\ x \in X\}.$$

LEMMA 5.2.3. Let the functions defining the problem (1.6.1) be continuously differentiable on an open set containing the compact set X and let the function $\gamma(z)$ satisfy the conditions (2.1). Then for any $x_0 \in X_0$ the solutions $x(x_0,t)$ of the system (2.4) can be extended as $t \to \infty$, and the sets X, Ω are invariant with respect to (2.4).

Proof. Let us calculate the derivative of the vector function $\Phi(x)$ by the system (2.2):

$$\frac{d\Phi}{dt} = -\Phi_x^T M f_x = -\Phi_x^T f_x - \Phi_x^T \Phi_x y.$$

Using formula (2.3), we obtain

$$\begin{aligned} \Phi_x^T \Phi_x y &= -\Phi_x^T f_x - D(\gamma(-\Phi))\, y, \\ \frac{d\Phi}{dt} &= D(\gamma(-\Phi(x)))\, y. \end{aligned} \tag{2.6}$$

Calculating the square norm of the Lagrangian gradient

$$\begin{aligned} &L(x, y) = f(x) + \langle \Phi(x), y \rangle: \\ &\|L_x(x, y)\|^2 = (f_x^T + y^T \Phi_x^T) M f_x = \\ &\qquad = f_x^T M f_x + y^T \Phi_x^T M f_x = f_x^T M f_x - y^T D(\gamma(-\Phi))\, y, \end{aligned}$$

we obtain

$$\frac{df}{dt} = -\|L_x(x, y)\|^2 - \|D(\sqrt{\gamma(-\Phi(x))})\, y\|^2 \leqslant 0. \tag{2.7}$$

A solution of (2.4) exists at least for t such that $x(x_0,t) \in X$. Let us show that $x(x_0,t)$ does not leave the set X for any $t \geq 0$. Suppose not: let $\Phi^j(x(x_0,t)) > 0$ for some $t > 0$. Then there is a time t_1 such that $\Phi^j(x(x_0,t_1)) = 0$ and $\dot{\Phi}^j(x(x_0,t_1)) > 0$. This contradicts (2.6) since $\gamma(0) = 0$. Hence $x(x_0,t) \in X$ for all $t \geq 0$. Thus the function $\gamma(-\Phi)$ introduced above plays the role of a "barrier," preventing $x(x_0,t)$ from intersecting the hypersurface $\Phi(x) = 0$. The trajectory $x(x_0,t)$ can approach the boundary points only as $t \to \infty$. If the initial point x_0 is on the boundary, the entire trajectory of the system (2.4) belongs to the boundary. The functions $g^i(x)$ defining equality-type constraints are integrals of the system (2.4). Hence, since the set X is bounded, the solutions of the system (2.4) are extendable as $t \to \infty$ and the set X is invariant with respect to (2.4). From this and (2.7), Ω is invariant. ///

Let x_* denote the points in X at which the right sides of (2.2) vanish. We call them stationary points. We will denote the corresponding values $y(x_*)$ by y_*. At the points $x = x_*$ we have

$$L_x(x_*, y_*) = 0, \quad D(\gamma(-\Phi(x_*)))\, y_* = 0,$$
$$\Phi_x^T(x_*)\Phi_x(x_*)\, y_* + \Phi_x^T(x_*)\, f_x(x_*) = 0. \tag{2.8}$$

Each point $\tilde{x}$ being a local solution of the problem (1.6.1) is stationary. Indeed, were this not so, then taking x as an initial point for the system (2.4) we would have that the solution $x(\tilde{x},t) \in X$, $f(x(\tilde{x},t)) < f(\tilde{x})$ for $t > 0$ since $df(\tilde{x})/dt < 0$. But this contradicts the condition for a local minimum of the function $f(x)$.

Let us dwell on a geometric interpretation. Introduce the tangent manifold to the set of active constraints as the point x:

$$K_1(x) = \{\bar{x} \in E^n\colon\ [\Phi_x^j(x)]^T \bar{x} = 0, \quad j \in \bar{\sigma}(x)\}.$$

The vector $\dot{x}$ defined by (2.4) belongs to $K_1(x)$ at each x. Indeed, from (2.6) we have

$$\langle \Phi_x^j(x),\ \dot{x}\rangle = \gamma(-\Phi^j(x))\, y^j.$$

If $j \in \bar{\sigma}(x)$ at the point x, then $\Phi^j(x) = 0$ and the right side of the equality is zero, hence $\dot{x} \in K_1(x)$.

Suppose $j \notin \bar{\sigma}(x)$. By (2.1), if $z \to 0$ then $\gamma(z) \to 0$. Hence the projection of $\dot{x}$ onto the gradient $\Phi_x^j(x)$ of an inactive constraint tends to zero while approaching the hypersurface $\Phi^j(x) = 0$. Owing to this, the trajectories of (2.4) do not intersect the region determined by the inequality-type constraints; the trajectories can approach the boundary arbitrarily close, touching them in the limit. The function $\gamma(-\Phi)$ automatically changes the direction of the vector $\dot{x}(x_0,t)$ near the boundary. Different types of "barrier" functions give rise to different kinds of variation of this rate.

Far away from the hypersurface $h^j(x) = 0$ when $h^j(x) << 0$, one need not fear that the trajectory $x(x_0,t)$ intersects it on a small interval $(t,t+\delta)$, and in the formula for determining $M(x)$ we can omit the function $h^j(x)$ and its derivative, bringing them into view only when $-\varepsilon < h^j(x(x_0,t)) < 0$, where $\varepsilon > 0$ is chosen depending on the step of integrating the system (2.4). This maneuver helps us to lower the order of the system (2.3).

On the other hand, the introduction of the barrier functions $\gamma(-\Phi)$ causes the trajectories to "stick" to the boundary, since if $h^j(x_0) = 0$ then $h^j(x(x_0,t)) \equiv 0$. It was postulated above that all $h^j(x_0) < 0$. To remove this drawback, one can omit in the formulas for M the functions h^j and h^j_x and calculate the derivative $\dot{h}^j = \left\langle h^j_x, \dot{x}\right\rangle$ for the new system. If the derivative turns out to be negative, then we continue the motion along the trajectory of this system. In other words, by removing the "barrier" we check whether this can be done without violating the feasibility condition.

2. PROOF OF CONVERGENCE

<u>THEOREM 5.2.1</u>. Let the conditions of Lemma 5.2.3 be satisfied and let all stationary points of X be isolated. Then for any nonstationary initial points $x_0 \in X_0$ the solution $x(x_0,t)$ of

the system (2.2) and the solution $y(x_0,t)$ found from (2.3) converge as $t \to \infty$ to the Kuhn-Tucker point $[x_*,y_*] = [x_*,u_*,v_*] \in E^{n+m}$.

Proof. Let x_0 be an arbitrary point in X_0 and let $x(x_0,t)$ be the solution of the Cauchy problem (2.2). Since X is compact, the set of ω-limit points ω for the solution $x(x_0,t)$ is non-empty. We show that ω is in the set of feasible stationary points. Since $f(x)$ is bounded below on X and $f(x(x_0,t))$ is a monotonically decreasing function of t, all points of the set ω lie on the same equipotential surface of $f(x)$ (see Barbashin [1]). Let $\tilde{x} \in \omega$. We pass the trajectory $x(x,t)$ through $\tilde{x}$. Any point of it also belongs to ω, hence $\dot{f}(x(x,t)) \equiv 0$ and therefore $\dot{f}(\tilde{x}) = 0$. But we see from (2.7) that this is possible only if x is a stationary point for (2.2). Whence, since all stationary points of X are isolated, we obtain that ω consists of a unique feasible stationary point $\tilde{x}$ which $x(x_0,t)$ approaches as $t \to \infty$.

For each $x = x(x_0,t)$ one can define $y = y(x_0,t)$ from (2.3). Since the function y of t is continuous, the existence of $x_* = \lim_{t\to\infty} (x_0,t)$ implies the existence of $y_* = \lim_{t\to\infty} y(x_0,t)$. Set $u_*^i = y_*^i$ for $i \in [1:e]$ and $v_*^j = y_*^{j+e}$ for $j \in [1:c]$. From the preceding lemma it follows that $x_* \in X$. We show that at the point $[x_*,v_*]$ the complementarity condition is satisfied and $v_* \geq 0$. At the limit point x_* the conditions (2.8) are satisfied. Noting (2.1), we obtain that if $v_*^j \neq 0$ then $\phi^{e+j}(x_*) = h^j(x_*) = 0$, and hence the complementarity condition (1.6.5) holds. From (2.6) we have

$$h^j(x(x_0, t)) = h^j(x_0)\exp(\beta^j(t)), \tag{2.9}$$

where

$$\beta^j(t) = \int_0^t \frac{\gamma(-h^j(x(x_0, t)))}{h^j(x(x_0, t))} v^j(x_0, t)\,dt.$$

If one assumes that $v_*^j < 0$ then $h^j(x_*) = 0$ and there is a $\bar{t}$ such that for all $t > \bar{t}$

$$-v^j(x_0, t) > 0, \quad \frac{\gamma(-h^j(x(x_0, t)))}{-h^j(x(x_0, t))} \geqslant 0, \quad \beta(t) \geqslant \beta(\bar{t}),$$

then $\underline{\lim}_{t\to\infty} \beta(t) \geq \beta(\bar{t})$ and by (2.9) for $h^j(x_0) < 0$ we have $\overline{\lim}_{t\to\infty} h^j(x(x_0,t)) < 0$, which contradicts the complementarity condition by which $h^j(x_*) = 0$. Hence $v_* \geq 0$. The limit point $[x_*,u_*,v_*]$ is thus a Kuhn-Tucker point. ///

It is easy to see that the conditions of the Theorem may be relaxed by requiring that they all hold on the set Ω, rather than on X.

3. ESTIMATION OF THE CONVERGENCE RATE

For simplicity, we consider the case where $\gamma(z) = z$. The method (2.2) and the formula (2.3) do not change, and the formula for $\Gamma(x)$ and the equation (2.6) become

$$\Gamma(x) = \Phi_x^T(x)\Phi_x(x) - D(\Phi(x)), \tag{2.10}$$

$$\frac{d\Phi}{dt} = -D(\Phi(x))\,y. \tag{2.11}$$

For further study, the following elaboration of the method is useful. As in Subsection 1.7.4, we introduce here the additional vector of artificial variables $p \in E^c$ and consider the minimization problem (1.7.17) equivalent to (1.6.1) in the space E^{n+c}

with equality-type constraints only. By the formula (1.7.18), we form the Lagrangian $L^1(x,p,y)$. The reduced-gradient method (1.2) applied to solving this problem is the following:

$$\frac{dx}{dt} = -L_x^1(x, p, y) = -[f_x(x) + g_x(x)u + h_x(x)v], \tag{2.12}$$

$$\frac{dp}{dt} = -L_p^1(x, p, y) = -\frac{1}{2}D(p)v. \tag{2.13}$$

Here $y = [u,v] \in E^m$. Let

$$z = [x, p], \quad R(z) = \Big[g^1(x), \ldots$$
$$\ldots, g^e(x), h^1(x) + \frac{1}{4}(p^1)^2, \ldots, h^c(x) + \frac{1}{4}(p^c)^2\Big].$$

For determining the vector y we have an analog of the equation (1.3):

$$\Phi_x^T(x)\Phi_x(x)y + \Phi_x^T(x)f_x(x) + \left[\begin{array}{c|c} ee & 0_{ec} \\ \hline 0_{ce} & \frac{1}{4}D(p)D(p) \end{array}\right] = 0. \tag{2.14}$$

The system (2.12), (2.13) is solved so that $R(z)$ is its integral. Let the initial point $[x_0, p_0]$ satisfy the conditions

$$g(x_0) = 0, \quad h(x_0) + \frac{1}{4}D(p_0)p_0 = 0. \tag{2.15}$$

Then along the trajectories of (2.12), (2.13) we have

$$h(x) + \frac{1}{4}D(p)p = 0. \tag{2.16}$$

Noting that $R_x(x,p) = \Phi_x(x)$, we conclude that the system (2.2) coincides with the system (2.12), (2.13). We can determine p from (2.16) and then omit (2.13). In the sequel we assume that the conditions (2.15) are satisfied.

Let

$$L^1(z, y) = L^1(x, p, y), \quad R(z) = R(x, p),$$

$$R_z = \left[\begin{array}{c|c} g_x & h_x \\ \hline 0 & \frac{1}{2} D(p) \end{array}\right], \quad L_z^1(z, y) = \left[\begin{array}{c} f_x + \Phi_x y \\ \hline D(\sqrt{-h})\, v \end{array}\right].$$

We can write the system (2.12), (2.13) and the conditions (2.14) in the form

$$\frac{dz}{dt} = -L_z^1(z, y), \quad R_z^T(z) L_z^1(z, y) = 0. \tag{2.17}$$

Assuming that the functions defining the problem are twice differentiable, we differentiate $\phi(x,y) = \frac{1}{2}||L_z^1(z,y)||^2$ by the system (2.17) and obtain

$$\frac{d\phi}{dt} = -(L_z^1)^T L_{zz}^1 L_z^1 - (L_z^1)^T R_z ,$$

$$\frac{dy}{dt} = -(L_z^1)^T L_{zz}^1 L_z^1 .$$

The matrix L_{zz}^1 is given by the formula (1.7.22). We assume that the conditions of Theorem 1.7.5 are satisfied and, furthermore, the matrix $L_{xx}(x,u,v)$ is uniformly positive definite on the cone $K_4(x)$ (see Definition 1.7.7 and the formula (1.7.23)). By (2.17)

$$\frac{d\varphi(t)}{dt} \leqslant -C_1 \| L_z^1 \|^2 = -2C_1 \varphi(t),$$
$$\varphi(t) = \varphi(0) e^{-2C_1 t};$$

passing back to the original notation, we obtain

$$\begin{aligned} \|L_x(x(x_0, t), y(t))\|^2 + \|D(\sqrt{-h(x(x_0, t))})\, v(t)\|^2 \leqslant \\ \leqslant [\|L_x(x_0, y_0)\|^2 + \|D(\sqrt{-h(x_0)})\, v_0\|^2]\, e^{-2C_1 t} \end{aligned} \tag{2.18}$$

implying that the method converges exponentially to a Kuhn-Tucker point.

4. SPECIAL CASES

Suppose we solve the problem (1.6.29) in which the set X is defined by the condition (1.6.2),

$$U = \{x \in E^n:\ x \geqslant 0\}, \quad U_0 = \{x \in E^n:\ x > 0\}.$$

Assume the point $x_0 \in U_0 \cap X_0$ is known. Then, instead of (2.2) the system

$$\frac{dx}{dt} = -D(\gamma(x))\,[f_x(x) + \Phi_x(x)\,y] \tag{2.19}$$

is integrated. The vector $y \in E^m$ is determined from the system

$$\Gamma(x)\,y + \Phi_x^T(x)\,D(\gamma(x))\,f_x(x) = 0,$$
$$\Gamma(x) = \Phi_x^T(x)\,D(\gamma(x))\,\Phi_x(x) + D(\gamma(-\Phi(x))).$$

Eliminating the vector y, we come to a system of the form (2.4), where

$$M(x) = D(\gamma(x))\,[I_n - \Phi_x(x)\,\Gamma^{-1}(x)\,\Phi_x^T(x)\,D(\gamma(x))].$$

If instead of the condition $x \geq 0$ the constraint $x \geq a$ is imposed, then in the formulas we need to write $D(\gamma(x-a))$ in place of $D(\gamma(x))$. If the constraints have the form $a^i \leq x^i \leq b^i$ or $q^j \leq h^j(x) \leq d^j$, then two barrier vector functions $\gamma_1(x)$, $\gamma_2(h(x))$ are introduced whose i^{th} and j^{th} coordinates are, for example $\gamma_1^i(x) = (x^i - a^i)(b^i - x^i)$, $\gamma_2^i(h(x)) = (h^j(x) - q^j)(d^j - h^j(x))$. The systems (2.2) and (2.3) have the form

$$\frac{dx}{dt} = -D(\gamma_1(x))\,(f_x + \Phi_x y),$$
$$[\Phi_x^T D(\gamma_1(x))\,\Phi_x + D(\gamma_2(\Phi(x)))]\,y + \Phi_x^T D(\gamma_1(x))\,f_x = 0_{n1}.$$

Constraints of this kind do not raise the order of the linear system (2.3), which makes the computations much easier.

Consider now the linear programming problem

$$\min_{x \in X} a^T x, \quad X = \{x: Ax = b,\ x \geqslant 0\}, \tag{2.20}$$

where $x, a \in E^n$, $b \in E^e$, A is an $e \times n$ matrix. The dual problem to (2.20) consists in finding

$$\max_{u \in U} b^T u, \quad U = \{u \in E^e: A^T u \leqslant a\}.$$

Setting $\gamma(z) = z$, we obtain that the method (2.19) for solving the primal problem leads to the system

$$\dot{x} = D(x)[A^T u - a], \qquad \text{where } AD(x)A^T u = AD(x)a . \tag{2.21}$$

In this case, $a^T \dot{x} = -\|D(\sqrt{x})(a - A^T u)\|^2 \leq 0$ if $x_0 > 0$ and $Ax_0 = b$. Analogously, for the dual problem

$$\dot{u} = b - Ax, \qquad \text{where } [A^T A - D(A^T u - a)]x = A^T b, \tag{2.22}$$

we have the inequality

$$b^T \dot{u} = \|b - Ax\|^2 + x^T D(x)(a - A^T u) \geq 0$$

if $Au^T \leq a$. The relaxation method (2.21) of solving the primal problem is effective for solving problems of large dimension but with a small number of constraints $(n \gg e)$ since in implementing the method a matrix of low order is inverted. Similarly, the method (2.22) is convenient if $e \gg n$. The methods (2.21) and (2.22) undergo only slight changes for quadratic programming problems.

Suppose the problem

$$\min_{x \in X_1} f(x), \quad X_1 = \{x \in E^n: \sum_{i=1}^{n} x^i = a,\ x \geqslant 0\} \tag{2.23}$$

is to be solved. The method (2.19) leads to the following system:

$$\frac{dx^i}{dt} = x^i \left[\frac{1}{a} \sum_{i=1}^{n} x^i f_{x^i}(x) - f_{x^i}(x) \right]. \tag{2.24}$$

If instead of X_1 we take the set

$$X_2 = \left\{ x \in E^n \colon \sum_{i=1}^{n} x^i \leqslant a, \quad x \geqslant 0 \right\},$$

then the system (2.24) does not change; however as initial data for the Cauchy problem one needs to take an interior point of X_2.

The problem (2.23) is often encountered in applications. Here are two examples.

Let the function $\psi(z)$ be defined on E^S. It is required to find the minimum of $\psi(z)$ on the convex hull of $\{z_1, z_2, \ldots, z_n\}$. The problem reduces to finding

$$\min_{x \in X_3} \psi\left(\sum_{i=1}^{n} x^i z_i \right) \quad X_3 = \left\{ x \in E^n \colon \sum_{i=1}^{n} x^i = 1, \quad x \geqslant 0 \right\}. \tag{2.25}$$

Setting $f(x) = \psi\left(\sum_{i=1}^{n} x^i z_i \right)$, we come to the problem (2.23).

In solving antagonistic two-player games with an infinite number of states we introduce probability measures μ and ν on the σ-algebra of subsets of Z and Y, respectively:

$$\max_{\mu} \min_{\nu} \int_Z \int_Y F(z, y) \mu(dz) \nu(dy).$$

The problem reduces to the following:

$$\max_{\mu} \min_{y \in Y} \int_Y F(z, y) \mu(dz).$$

Approximating the measure μ by an atomic one, we come to a problem close to (2.5):

$$\max_{x \in X_3} \min_{y \in Y} \sum_{i=1}^{n} F(z_i, y) x^i.$$

This argument has been studied in more detail in Evtushenko and Zhadan [1].

Suppose the problem

$$\min_{x \in X_4} f(x), \quad X_4 = \{x \in E^n: a \leqslant x \leqslant b\}$$

is being solved. Treating these constraints as those of the form $h(x) \leq 0$, and using (2.11), (2.10) and (2.3), we easily see that the system (2.3) has an analytic solution. Substituting the vector y found from (2.3) into (2.11), we have

$$\frac{dx^i}{dt} = -\frac{\partial f(x)}{\partial x^i} \varphi_i(x), \quad \varphi_i(x) = \frac{(x^i - a^i)(b^i - x^i)}{b^i - a^i + (x^i - a^i)(b^i - x^i)}.$$

For $x \in X_4$ we have $\phi_i(x) \geq 0$. If $x^i \to a^i$ or $x^i \to b^i$, then $\phi_i(x) \to 0$, due to which the trajectories do not leave the feasible set.

3. A DISCRETE VERSION OF THE REDUCED-GRADIENT METHOD

We integrate the system (2.1) by Euler's formula

$$x_{k+1} = x_k - \alpha_k M(x_k) f_x(x_k). \tag{3.1}$$

It will be shown below that if the equality-type constraints depend linearly on x, then they remain constant for any values of α_k; in this case we can obtain a high convergence rate of the method since one can take relatively large values of α_k. In numerical implementation of the method, the integration step α_k is usually the same on each iteration, but it needs to be checked additionally whether the relaxation as well as feasibility conditions of the point x_{k+1} have been satisfied with respect to the inequality-type constraints. If these constraints are violated, then the step needs to be reduced until these conditions are finally satisfied. As was shown in Section 5.2, near the hypersurface $h^j(x) = 0$ in the method (3.1), the motion in the direction toward this hypersurface automatically becomes faster. Due to this fact the step needs to be reduced relatively seldom.

Numerical computations show that in a number of cases in order to integrate the system (2.11) it is useful to follow implicit methods of integration. If there are no nonlinear constraints among the equality-type constraints, then the process (3.1) slows down considerably since one needs to take sufficiently small integration steps to guarantee the smallness of $\|g(x_k)\|$.

Let us prove the convergence of the process (3.1), considering only the case where the vector function $\Phi(x)$ depends linearly on x, $\gamma(z) = z$, and the step α_k in (3.1) is constant. If the conditions of Lemma 5.2.1 are satisfied, then one can determine the maximum of the norm of the matrix $M(x)$ on X using the relation:

$$\lambda = \max_{x \in X} \max_{\tilde{x} \in E^n} \frac{\tilde{x}^T M(x) \tilde{x}}{\|\tilde{x}\|^2} < \infty.$$

Set

$$\nu = \max_{j \in [1:m]} \max_{x \in X} y^j(x).$$

Here $y(x)$ is determined from (2.3), (2.10). Below we will write $y_k = y(x_k)$.

<u>THEOREM 5.3.1</u>. Let the conditions of Theorem 5.2.1 be satisfied, let the vector function $\Phi(x)$ depend linearly on x, and let the

function $f(x)$ satisfy a Lipschitz condition on X with constant ℓ. Then

● 1. for $0 < \alpha < \min\left[\frac{1}{\nu}, \frac{2}{\lambda\ell}\right]$ and any nonstationary initial points $x_0 \in X_0$, the sequence $\{x_k\}$ converges to a Kuhn-Tucker point, $f(x_{k+1}) \le f(x_k)$ for $k = 0,1,2,\ldots$;

● 2. if, moreover, $B\|z\|^2 \ge z^T L^1_{zz}(x,y(x))z \ge b\|z\|^2$ for all $x \in X$, $z \in E^{n+c}$, then we have the estimate

$$\begin{aligned}&\|L_x(x_k, y_k)\|^2 + \|D(\sqrt{-h(x_k)})\,y_k\|^2 \leqslant \\ &\leqslant [\|L_x(x_0, y_0)\|^2 + \|D(\sqrt{-h(x_0)})\,y_0\|^2]\,[1-\alpha b+\alpha^2 B^2]^k,\end{aligned} \tag{3.2}$$

the sequence $x_k \to x_*$, where x_* is a local solution of the problem (1.6.1).

Here

$$L(x, y) = f(x) + \sum_{i=1}^{m} \Phi^i(x)\,y^i.$$

Proof. From the linearity of $\Phi(x)$ it follows that

$$\Phi(x_{k+1}) = \Phi(x_k) + \left[\frac{d\Phi(x_k)}{dx}\right]^T (x_{k+1} - x_k).$$

Substituting here (3.1), we obtain

$$\Phi(x_{k+1}) = \Phi(x_k) - \alpha\Phi_x^T(x_k)\,M(x_k)\,f_x(x_k).$$

Using the relations found in deriving (2.6), we come to the equality

$$\Phi(x_{k+1}) = \Phi(x_k) + \alpha D(\Phi(x_k))\,y_k.$$

Thus, if $\Phi(x_0) = 0$ then $\Phi(x_k) \equiv 0$ for any α. If $|y^i_k| < \nu$, $\alpha < \frac{1}{\nu}$ and $\Phi^i(x_k) \le 0$, then

$$\Phi^i(x_{k+1}) = \Phi^i(x_k)(1-\alpha y^i_k) \leqslant \Phi^i(x_k)(1-\nu\alpha) \leqslant 0. \tag{3.3}$$

Hence $x_0 \in X$ implies all $x_k \in X$.

Using the Newton-Leibniz formula (see Appendix I), we obtain

$$f(x_{k+1}) \leqslant f(x_k) - \alpha f_x^T(x_k) M(x_k) f_x(x_k) + \frac{l\alpha^2}{2} \| M(x_k) f_x(x_k) \|^2. \qquad (3.4)$$

According to Lemma 5.2.2 the symmetric matrix $M(x_k)$ is positive semi-definite. Let $\sqrt{m}$ denote its square root; $M = \sqrt{m}\,\sqrt{m}$. Introducing the vector $a = \sqrt{M}\, f_x$, we transform the inequality (3.4) to the form

$$f(x_{k+1}) - f(x_k) \leqslant \\ \leqslant \alpha \|a\|^2 \left[-1 + \frac{\alpha l}{2} \frac{a^T M a}{\|a\|^2} \right] \leqslant \alpha \|a\|^2 \left[-1 + \frac{\alpha l}{2} \lambda \right]. \qquad (3.5)$$

Thus, for $\alpha < \frac{2}{\lambda \ell}$ the sequence $f(x_k)$ monotonically decreases. Since $f(x)$ is bounded from below on X, it follows that the limit of the $f(x_k)$ exists. Hence

$$\lim_{k \to \infty} [f(x_{k+1}) - f(x_k)] = 0. \qquad (3.6)$$

From (3.5) we obtain the inequality

$$0 \leqslant f_x^T(x_k) M(x_k) f_x(x_k) \leqslant \frac{2[f(x_k) - f(x_{k+1})]}{\alpha [2 - \alpha \lambda l]}. \qquad (3.7)$$

Using the formulas derived in deducing (2.7), we can write (3.7) as

$$\| L_x(x_k, y_k) \|^2 + \| D(\sqrt{-\Phi(x_k)})\, y_k \|^2 \leqslant \frac{2[f(x_k) - f(x_{k+1})]}{\alpha[2 - \alpha\lambda l]}.$$

Letting k tend to infinity and noting (3.6), we obtain

$$\lim_{k \to \infty} \| f_x(x_k) + \Phi_x^T(x_k) y_k \| = \lim_{k \to \infty} \| D(\sqrt{-\Phi(x_k)})\, y_k \| = 0, \qquad (3.8)$$

i.e., at each limit point of the sequence $\{x_k\}$ the stationary conditions (2.8) are satisfied. Since the stationary points are isolated, the limits exist:

$$\bar{x} = \lim_{k \to \infty} x_k, \quad \bar{y}^j = \lim_{k \to \infty} y_k^j, \quad \bar{\Phi}^j = \lim_{k \to \infty} \Phi^j(x_k).$$

From (3.3) it also follows that

$$\Phi^j(x_{k+1}) \leqslant \Phi^j(x_0) \prod_{s=0}^{k} (1-\alpha y_s^j).$$

If $\bar{\Phi}^j = 0$ then the infinite product

$$\prod_{s=0}^{\infty} (1-\alpha y_s^j)$$

must be zero. For this it is necessary (see Fikhtengol'ts [1]) that

$$\sum_{s=0}^{\infty} \ln[1-\alpha y_s^j] = -\infty,$$

But this is possible only if $\bar{y}^j \geq 0$. By (3.8) for $\bar{\Phi}^j < 0$ one necessarily has $\bar{y}^j = 0$. Thus, the limit point of the sequence $\{x_k, y_k\}$ is a Kuhn-Tucker point. The estimate of the convergence rate of (3.2) is obtained like (2.18). ///

4. THE CONDITIONAL GRADIENT METHOD

1. GENERAL DESCRIPTION OF THE METHOD

We will consider the problem of minimizing a differentiable function on a convex, compact set:

$$\min_{x \in X} f(x), \quad X_* = \operatorname{Arg}\min_{x \in X} f(x). \tag{4.1}$$

Let the point $x_k \in X$ be known. Then the differentiability of $f(x)$ yields the representation

$$\Delta f = f(x) - f(x_k) = \langle f_x(x_k),\ x - x_k\rangle + \|x - x_k\| \beta(x_k,\ x - x_k),$$

where

$$\lim_{x \to x_k} \beta(x_k,\ x - x_k) = 0.$$

We will determine the vector $\bar{x}_k$ yielding the minimum of the linear part of the increment Δf:

$$\bar{x}_k \in \operatorname{Arg} \min_{x \in X} \langle f_x(x_k), x - x_k \rangle = W(x_k). \tag{4.2}$$

Since the set X is closed and bounded, it follows that the point-set mapping W is defined for all $x_k \in X$, with

$$W(x_k) \subset X, \quad R_k = \langle f_x(x_k), \bar{x}_k - x_k \rangle \leqslant 0.$$

If $R_k = 0$ then $x_k, \bar{x}_k \in W(x_k)$ and at $x = x_k$ we have the necessary (and sufficient for convex f) condition of the minimum of a function on X (see Theorems 1.4.1 and 1.4.3), and the computation stops. We now consider the case where $R_k < 0$. As a new point x_{k+1} we take

$$x_{k+1} = x_k + \alpha_k(\bar{x}_k - x_k) \quad . \tag{4.3}$$

Here $0 \leq \alpha_k \leq 1$. By the convexity of X, $x_{k+1} \in X$.

If $W(x_k)$ consists of a single point $\bar{x}_k$, then the vector $\bar{x}_k - x_k$ is called the conditional antigradient of the function $f(x)$ at x_k. It is a vector to which one can move from x_k remaining on X and obtain the greatest projection onto the direction of the antigradient of the objective function $f_k(x_k)$. This is just the direction in which we move from x_k in this method, known as the method of the conditional gradient.

Several versions of this method are possible. The most prevalent is to determine the size of the step α_k from the condition of minimizing the value of $f(x)$ on the straight-line segment joining x_k and $\bar{x}_k$:

$$\alpha_k = \operatorname{Arg} \min_{0 \leqslant \alpha \leqslant 1} f(x_k + \alpha(\bar{x}_k - x_k)).$$

Obviously, in this case the sequence $f(x_k)$ does not increase since

$$f(x_k) \geqslant f(x_k + \alpha_k(\bar{x}_k - x_k)) = f(x_{k+1}).$$

This first version of the method was suggested by Frank and Wolfe. Some authors call it the method of linearization or linear approximation since the basic computational difficulties arise in solving the auxiliary problem (4.2) of finding the minimum of the linearized function on the set X. This problem is not much simpler than the initial problem, and only in particular cases when the set X has a simple structure is its use recommended. For example, if X is defined by linear constraints of equality- and/or inequality-type, then finding $\bar{x}_k$ reduces to an easily solved problem of linear programming.

The method can be somewhat simplified by approximately solving the problem (4.4). In [1], Pshenichnyj and Danilin suggest, for example, to take $\alpha_k = 2^{-i_0}$ where i_0 is the first index $(i = 0,1,2,\ldots)$, for which the inequality

$$f(x_k + 2^{-i}(\bar{x}_k - x_k)) - f(x_k) \leqslant 2^{-1-i}\langle f_x(x_k), \bar{x}_k - x_k\rangle \tag{4.5}$$

is satisfied. If f is twice differentiable on X, then for determining the direction $\bar{x}_k - x_k$ one can use a more accurate approximation of the function

$$f(x) - f(x_k) \approx \langle f_x(x_k), x - x_k\rangle + \frac{1}{2}(x - x_k)^T f_{xx}(x_k)(x - x_k).$$

As $\bar{x}_k - x_k$ we take the vector yielding the minimum of the expression on the right-hand side. After this, the new point is found by the formula (4.3); the size of the step α_k is determined from (4.4) or from (4.5).

2. PROOF OF CONVERGENCE

THEOREM 4.5.1. Let X be a compact convex set in E^n, let the function f be differentiable on X and its gradient satisfy a Lipschitz condition with constant ℓ. Then for any $x_0 \in X$

•1. the sequence $\{x_k\}$ defined from the conditions (4.2) - (4.4) is such that

$$\lim_{k\to\infty} \langle f_x(x_k), \bar{x}_k - x_k\rangle = 0; \tag{4.6}$$

•2. if, moreover, f is convex, then the set of limit points of the sequence $\{x_k\}$ is nonempty and belongs to the set X_*, and we have the estimate

$$0 \leqslant f(x_k) - f(x_*) \leqslant \frac{C}{k}, \tag{4.7}$$

where C is a constant not depending on k, $x_* \in X_*$.

Proof. Let d denote the diameter of the set X:

$$d = \sup_{x,\, y \in X} \|x - y\|.$$

According to the rule (4.4) of choosing the step α, for each $0 \le \alpha \le 1$ the inequality

$$f(x_{k+1}) - f(x_k) \leqslant f(x_k + \alpha(\bar{x}_k - x_k)) - f(x_k)$$

is satisfied. By Theorem 2 in Appendix I the inequality

$$f(x_k + \alpha(\bar{x}_k - x_k)) - f(x_k) \leqslant \leqslant \alpha\langle f_x(x_k), \bar{x}_k - x_k\rangle + \frac{\alpha^2 l}{2}\|\bar{x}_k - x_k\|^2 \tag{4.8}$$

holds, implying in turn that for any $0 \le \alpha \le 1$ we have the estimate

$$|\delta_k| = |\langle f_x(x_k), \bar{x}_k - x_k\rangle| \leqslant \frac{\alpha l}{2} d^2 + \frac{\Delta_k}{\alpha},$$
$$\Delta_k = f(x_k) - f(x_{k+1}).$$

The sequence $\{f(x_k)\}$ is nonincreasing and bounded from below (since all the x_k belong to a compact set). Hence $\{f(x_k)\}$ converges as $k \to \infty$ and $\lim_{k\to\infty} \Delta_k = 0$. Passing to the limit in the preceding inequality as $k \to \infty$, we obtain the estimate

$$0 \leqslant \overline{\lim_{k\to\infty}} |\delta_k| \leqslant \frac{\alpha l}{2} d^2$$

holding for any $0 \leq \alpha \leq 1$. Letting α tend to zero, we get the required property (4.6).

If $f(x)$ is convex, then for any $x \in X$

$$f(x) - f(x_k) \geqslant \langle f_x(x_k),\ x - x_k \rangle \geqslant$$
$$\geqslant \min_{x \in X} \langle f_x(x_k),\ x - x_k \rangle = \langle f_x(x_k),\ \bar{x}_k - x_k \rangle = \delta_k \to 0.$$

Therefore, $\lim_{k\to\infty} f(x_k) \leq f(x)$ $x \in X$, whence we conclude that each limit point of $\{x_k\}$ belongs to X_*.

Note that during the calculations we can adjust the accuracy of solving the initial problem depending on a bound for the error $f(x_k) - f(x_*)$. Indeed, noting the convexity of f, for $x \in X$ and $x_* \in X_*$ we get

$$0 \leqslant f(x_k) - f(x_*) \leqslant \langle f_x(x_k),\ x_k - x_* \rangle$$

while

$$\langle f_x(x_k),\ \bar{x}_k - x_k \rangle \leqslant \langle f_x(x_k),\ x_* - x_k \rangle.$$

Hence

$$0 \leqslant f(x_k) - f(x_*) \leqslant -\langle f_x(x_k),\ \bar{x}_k - x_k \rangle = -\delta_k \geqslant 0. \qquad (4.9)$$

This estimate can be used as a criterion for ending the computations.

To estimate the rate of convergence, we rewrite (4.8) as

$$0 \leqslant f(x_k) - f(x_{k+1}) \geqslant \alpha |\delta_k| - \frac{\alpha^2 l}{2} d^2.$$

The maximum of α on the right side will be attained for

$$\bar{\alpha}_k = \frac{|\delta_k|}{ld^2}.$$

As $k \to \infty$, $\bar{\alpha}_k \to 0$. Hence there is a N such that for all $k > N$ the quantity $\alpha_k \leqslant 1$, and we have the estimate

$$f(x_k) - f(x_{k+1}) \geqslant \frac{1}{2ld^2} |\delta_k|^2.$$

Using (4.9), we obtain

$$f(x_k) - f(x_{k+1}) \geqslant \frac{1}{2ld^2} [f(x_k) - f(x_*)]^2.$$

Setting $a_k = f(x_k) - f(x_*)$, $A = \frac{1}{2} \ell d^2$, we rewrite this inequality in the form

$$a_k - a_{k+1} \geqslant A a_k^2. \tag{4.10}$$

Let all $a_k > 0$. Then from (4.10) it follows that

$$\frac{1}{a_{k+1}} - \frac{1}{a_k} = \frac{a_k - a_{k+1}}{a_k a_{k+1}} \geqslant A \frac{a_k}{a_{k+1}}.$$

Summing this inequality for $k = 1$ to $s - 1$, we obtain

$$\frac{1}{a_s} - \frac{1}{a_1} \geqslant A(s-1).$$

This implies that $a_s < \frac{1}{A(s-1)} \leq \frac{2}{As}$, hence (4.7) holds. ///

One can similarly prove convergence while adjusting the step using the formula (4.5) and choosing the direction from the quadratic approximation.

5. THE GRADIENT-PROJECTION METHOD

1. THE IDEA OF THE METHOD

Consider the problem (4.1) in which we assume $f(x)$ to be convex and differentiable and X to be convex and compact. Then for each point x from the condition (1.1.2) one can define its projection $p(x)$ onto the set X. The gradient-projection method consists in constructing the sequence

$$x_{k+1} = p(x_k - \alpha_k f_x(x_k)). \tag{5.1}$$

There are various ways of changing α_k; depending on this, one can derive various versions of the method. We examine the simplest case.

The auxiliary operation of projection onto the set X necessary for implementing the method is, in general, of the same complexity as the initial problem (4.1). Hence the gradient-projection method as well as the conditional-gradient method are useful when the set X has a simple structure convenient for solving the auxiliary problem (e.g., X is a multidimensional parallelepiped).

2. PROOF OF CONVERGENCE

LEMMA 5.5.1. If the set $X \subset E^n$ is closed and convex, then for all $z \in E^n$, $x \in X$, we have the inequality

$$\langle p(z) - z,\ x - p(z)\rangle \geqslant 0. \tag{5.2}$$

Proof. Consider the strictly convex differentiable function $\phi(x) = \|z - x\|^2$; its derivative is $\phi_x(x) = -2(z - x)$. The minimum of $\phi(x)$ on X is attained at the single point $p(z)$. Hence by Theorem 1.4.1 it is necessary that for any $x \in X$ the inequality (5.2) be satisfied. ///

THEOREM 5.5.1. Let X be a convex, compact set; let the convex differentiable function $f(x)$ be such that its gradient satisfies a Lipschitz condition on X with constant ℓ. Then there exist ε_1, ε_2 such that for any α_k satisfying the condition

$$0 < \varepsilon_1 < \alpha_k < \frac{2}{l+2\varepsilon_2}, \tag{5.3}$$

the method (5.1) converges to X_*, and for $x_* \in X_*$ we have the estimate (4.7).

Proof. Using Theorem 2 in Appendix I, we obtain

$$\Delta_k = f(x_{k+1}) - f(x_k) \leqslant \langle f_x(x_k),\ x_{k+1} - x_k\rangle + \frac{l}{2}\|x_{k+1} - x_k\|^2.$$

We transform this inequality to the form

$$\Delta_k \leqslant \frac{1}{\alpha_k}\langle \alpha_k f_x(x_k) + x_{k+1} - x_k,\ x_{k+1} - x_k\rangle + \\ + \left(\frac{l}{2} - \frac{1}{\alpha_k}\right)\|x_{k+1} - x_k\|^2.$$

Setting $z = x_k - \alpha_k f_x(x_x)$, $x = x_k$ in (5.2), we have

$$\Delta_k \leqslant -\left(\frac{1}{\alpha_k} - \frac{l}{2}\right)\|x_{k+1} - x_k\|^2.$$

Adjusting α_k according to the rule (5.3), we have

$$\begin{gathered} \frac{1}{\alpha_k} - \frac{l}{2} > \frac{l+2\varepsilon_2}{2} - \frac{l}{2} = \varepsilon_2 > 0, \\ \Delta_k \leqslant -\varepsilon_2\|x_{k+1} - x_k\|^2 \leqslant 0. \end{gathered} \tag{5.4}$$

Thus (5.1) is a relaxation method, and $x_{k+1} = x_k$ iff $x_k \in X_*$. Let $x_* \in X_*$. Then, using the formula (1.2.14), we obtain

$$0 \leqslant a_k = f(x_k) - f(x_*) \leqslant \langle f_x(x_k),\ x_k - x_*\rangle.$$

We transform this inequality as follows:

$$a_k \leqslant \langle f_x(x_k),\ x_k - x_{k+1}\rangle + \\ + \frac{1}{\alpha_k}\langle x_k - \alpha_k f_x(x_k) - x_{k+1},\ x_* - x_{k+1}\rangle - \\ - \frac{1}{\alpha_k}\langle x_k - x_{k+1},\ x_* - x_{k+1}\rangle.$$

Again using the formula (5.2) and assuming $z = x_k - \alpha_k f_x(x_k)$, $x = x_*$, we have for a_k

$$a_k \leqslant \| x_{k+1} - x_k \| \left(\| f_x(x_k) \| + \frac{1}{\varepsilon_1} \| x_* - x_{k+1} \| \right).$$

Since X is bounded, so are $\|f_x(x_k)\|$ and $\|x_* - x_k\|$ for any k. Hence there exists λ such that $a_k \leq \alpha \|x_{k+1} - x_k\|$. From (5.4) we obtain

$$\Delta_k = a_{k+1} - a_k \leqslant -\varepsilon_2 \| x_{k+1} - x_k \|^2.$$

Thus

$$a_k - a_{k+1} \geqslant \varepsilon_2 \| x_{k+1} - x_k \|^2 \geqslant \frac{\varepsilon_2 a_k^2}{\lambda^2}.$$

We have arrived at an inequality similar to (4.10), yielding in turn the estimate (4.7). ///

Chapter 6

NUMERICAL METHODS FOR SOLVING OPTIMAL CONTROL PROBLEMS

Intensive efforts to develop numerical methods for solving optimal control problems began in the late 1950's for two reasons: first, at that time high-speed large-memory digital computers were becoming available, opening up the broadest possibilities for using numerical methods and, second, the development of complex engineering systems in rocket and aircraft design, for example, created the need to solve a great multitude of optimal control problems.

Numerical methods draw essentially on the basic results in the general theory of optimal control. A great step forward in this direction was the "maximum principle" of L.S. Pontryagin -- a canonical formulation of necessary conditions for optimality -- providing the basis for development and growth of a new direction in variational calculus and setting the stage for further diverse, prolific studies.

One can distinguish several directions in the development of numerical methods for solving optimal control problems substantially different from each other. First of all are the primal methods based on descent in control space. Then there are methods based on Pontryagin's Maximum Principle, changing the initial problem to a "two-point boundary value" problem. Another approach, Moiseev [2], is based on variations in state space. Yet another direction, developed by Fedorenko [1], involves concepts of the linearization method. The conditional-gradient method and the gradient-projection method have been extended to optimal control problems by Dem'yanov and Rubinov [1]. Much effort has also been expended on numerical methods based on Bellman's dynamic programming (Bellman [1]).

In one of the early monographs on numerical methods of optimal control (Moiseev [2]), a casual mention is made of the possibility of using nonlinear programming methods. This was followed by, and found extensive development, in particular, in Polyak [1], Ermol'ev, Gulenko, and Tsarenko [1], Propoj [2], and Tabak and Kuo [1].

In this book, methods for solving optimal control problems are presented based on the concepts of nonlinear programming theory. This approach turned out to be extraordinarily efficient for many reasons: many earlier heuristic algorithms are now well understood, the possibility of generalizing them has emerged; it allowed the use of an enormous sophisticated arsenal of nonlinear programming methods and of unconstrained minimization methods, and, furthermore, laid the foundation for developing methods of system optimization with high accuracy; the nonlinear programming methods help solve complex problems of optimal control, including those with mixed constraints.

The basic computational formulas to implement the nonlinear programming approach can be found, for example, in Polyak [1]:

however, Polyak's derivation of formulas for the first derivatives cannot be used for more refined integration routines than Euler's method, nor for the second derivatives of the objective functional. This circumstance led us to improve the methodology of deriving formulas. In Section 6.1, the technique for computing derivatives for systems integrated by Euler's method is illustrated by an example. Then similar formulas for systems integrated by the Runge-Kutta method are developed, as well as formulas for computing the second derivatives of an objective function. We can next carry the sufficient conditions for a minimum, so elaborate in nonlinear programming theory, over to discrete-time processes approximating the initial optimal control problems (see Section 6.2). In Section 6.3, we show how the nonlinear programming methods apply to control problems. It is possible to have the structural continuity of all the methods: a change of the methods of integrating the initial systems of differential equations leads only to an algorithmic change of individual blocks for calculating the objective function, and the functions determining the constraints and their derivatives.

In most of the methods of Section 6.3, one constructs a sequence of unconstrained minimization problems with changing auxiliary objective functions of many variables. The local methods used to solve the unconstrained minimization problems yield local solutions of optimal control problems. The auxiliary problems of unconstrained minimization have unique features of interest of their own, and exploit special techniques based on the discrete maximum principle.

The discrete maximum principle has been proved by Propoj [1], Pearson [1], Halkin [1], and others. Among the many Soviet studies of this topic we mention Boltyanskij [1], Gabasov [1], and Yakovlev [1]. In this chapter, for reasons explained in Section 1.8, we call the "discrete maximum principle" the "discrete minimum principle" and give necessary and sufficient conditions for it in Section 6.4. Using these results, in Section 6.5 we show how one can use the conditional-gradient method and the gradient-projection method in order to solve auxiliary unconstrained minimization problems, solving, as well, optimal control problems with mixed constraints. In Section 6.6, these numerical methods are generalized to problems involving control parameters, delays, discontinuous right sides, and the minimal-time problem with simple nondifferential functions. In Section 6.7, some test problems are solved. The application to game problems is illustrated in Section 6.8.

The material of this chapter is a survey of the results obtained during recent years at the Computing Center of the USSR Academy of Sciences and is published partially in Grachev and Evtushenko [5], [6], and Evtushenko [12].

1. BASIC COMPUTATIONAL FORMULAS

1. COMPUTATION OF THE FIRST DERIVATIVES FOR THE EULER SCHEME

In Section 1.8, the necessary conditions for a minimum for optimal control problems were given. In particular, the processes examined are described by the non-autonomous system of ordinary differential equations

$$\frac{dx}{dt}=f(x(t),\ u(t),\ t),\ 0\leqslant t\leqslant T,\ x(0)=x_1, \tag{1.1}$$

where $x(t) \in E^n$, $u(t) \in E^r$.

In constructing numerical methods for solving optimal control problems one usually proceeds from "continuous" systems (1.1) to their discrete approximations. To simplify our discussion, we consider first the case where the system (1.1) is integrated by the Euler scheme.

We decompose $[0,T]$ into $q - 1$ intervals by the points $0 = t_1 < t_2 < \cdots < t_q = T$. We call $[t_i,t_{i+1}]$ the ith interval of integration and denote its length by $h_i = t_{i+1} - t_i$. Furthermore, we set

$$t_i=\sum_{s=1}^{i-1} h_s,\ t_q=\sum_{s=1}^{q-1} h_s=T,\ x_i=x(t_i),\ u_i=u(t_i),$$
$$z_i=[x_i,\ u_i,\ t_i],\ z_q=[x_q,\ u_q,\ t_q],\ f(z_i)=f(x_i,\ u_i,\ t_i).$$

In the remainder of this chapter, i takes on all possible integer values in the interval $[1:q-1]$. Integrating the system (1.1) by the Euler scheme, we have

$$x_{i+1}=x_i+h_if(z_i).$$

It is convenient to write this system in the form

$$x_{i+1}=F(z_i),\qquad F(z_i)=x_i+h_if(z_i). \tag{1.2}$$

When we considered the "continuous" system (1.1), the control $u(t)$ was a vector function. For a given discrete approximation the control is a finite-dimensional vector $w = [u_1,\ldots,u_q] \in E^{rq}$ which we call the complete control vector.

Substituting the components of the vector w into the system (1.2) (1.2), we determine sequentially the components of the vector

$$x = [x_1, x_2, \dots, x_q] \in E^{nq}$$

and call this vector the complete state vector. We next consider the function of the two complete vectors x and w:

$$R(x, w) = b(z_q) + \sum_{i=1}^{q-1} h_i B(z_i). \tag{1.3}$$

Given the vector w, with the aid of (1.2) we determine uniquely the vector x and denote this dependence by writing: x = x(w), R = R(x(w), w). The objective of this section is to derive formulas for computing the first and second derivatives of a composite function with respect to the components of the vector w. We need these formulas to implement various numerical methods of approximate solution of optimal control problems. In deriving the formulas, no constraints will be imposed on the complete control vector w; and it may be unfeasible and nonoptimal. Accounting for constraints and numerical methods will be described in Section 6.3.

Introduce the auxiliary n-dimensional vector

$$p_i = \frac{dR(x, w)}{dx_i}. \tag{1.4}$$

Let us explain the meaning of a derivative. Let Δ denote the n-dimensional vector of increments. We introduce a new complete state vector

$$\tilde{x} = [x_1, x_2, \dots, x_{i-1}, \tilde{x}_i, \tilde{x}_{i+1}, \dots, \tilde{x}_q] \in E^{nq},$$

whose first $i-1$ components correspond to the components of the vector s; $\tilde{x}_i = x_i + \Delta$, all subsequent components are obtained from the recurrence relation (1.2) if we take $\tilde{x}_i$ as x_i and then increase i to the value $i = q-1$. The vector p_i, called the derivative of R with respect to x_i at the point $[x,w]$, is defined by

$$\lim_{\|\Delta\|\to 0} \frac{1}{\|\Delta\|}[R(\tilde{x}, w) - R(x, w) - \langle p_i, \Delta\rangle] = 0.$$

It should be emphasized that in computing p_i, the complete control vector w remains constant. If the functions R, F are continuously differentiable with respect to the components of the vector x_i, then the vector p_i exists. For $i = q$ the vector p_i is computed simply:

$$p_q = \frac{dR(x, w)}{dx_q} = \frac{\partial R(x, w)}{\partial x_q} = \frac{\partial b(z_q)}{\partial x_q}, \tag{1.5}$$

since by the recurrence relations (1.2) none of the vectors $x_1, x_2, \ldots, x_{q-1}$ depends on x_q. Let

$$R(x, w) = b(z_q) + \sum_{i=1}^{q-1} h_i B(z_i) = b(z_q) + \sum_{i=1}^{q-1} C(z_i),$$
$$H(z_i, p_{i+1}) = C(z_i) + \langle F(z_i), p_{i+1}\rangle. \tag{1.6}$$

Noting the formula (1.2) expressing the dependence of x_{i+1} on the "preceding" vector x_i, one can write out a more detailed formula for computing p_i:

$$p_i = \frac{\partial R}{\partial x_i} + \frac{\partial x_{i+1}}{\partial x_i}\frac{dR}{dx_{i+1}}.$$

By the formulas (1.2) - (1.6) this expression can be made concise:

$$p_i = C_x(z_i) + F_x(z_i)p_{i+1} = H_x(z_i, p_{i+1}). \tag{1.7}$$

In what follows the subscripts x and u denote the partial derivatives with respect to explicit components of the vectors x and u, respectively. In particular,

$$H_x(z_i, p_{i+1}) = h_i \frac{\partial B(z_i)}{\partial x_i} + \frac{\partial F(z_i)}{\partial x_i} p_{i+1},$$
$$H_u(z_i, p_{i+1}) = h_i \frac{\partial B(z_i)}{\partial u_i} + \frac{\partial F(z_i)}{\partial u_i} p_{i+1}.$$

For the derivative of R at the point [x,w] with respect to the ith component $u_i \in E^r$ of the vector w we write

$$y_i = \frac{dR(x(w), w)}{du_i}.$$

The meaning of this formula is distinct from the definition of the vector p_i; here we mean the differentiation of R as of a composite function w. More precisely, let the complete vectors x and w be given. Introduce two new auxiliary complete vectors

$$\tilde{x} = [x_1, x_2, \ldots, x_i, \tilde{x}_{i+1}, \ldots, \tilde{x}_q] \in E^{nq},$$
$$\tilde{w} = [u_1, u_2, \ldots, u_{i-1}, u_i + \Delta, u_{i+1}, \ldots, u_q] \in E^{rq}.$$

All components of $\tilde{w}$ coincide with the corresponding components of w except the ith component equal to $u_i + \Delta$, where $\Delta \in E^r$ is the increment vector. The first i components of the vector $\tilde{x}$ coincide with the first i components of the vector x. The complete vector $\tilde{x}$ obtains from (1.2) if one takes the vector $\tilde{w}$ as a complete control vector. We define the vector y_i from the condition

$$\lim_{\|\Delta\| \to 0} \frac{1}{\|\Delta\|} [R(\tilde{x}, \tilde{w}) - R(x, w) - \langle y_i, \Delta \rangle] = 0.$$

Using the rule for differentiating composite functions, we obtain

$$y_i = \frac{dR}{du_i} = C_u(z_i) + F_u(z_i)\,p_{i+1} = H_u(z_i, p_{i+1}),$$
$$y_q = \frac{dR}{du_q} = \frac{\partial R}{\partial u_q} = \frac{\partial b(z_q)}{\partial u_q}. \tag{1.8}$$

If in the formulas obtained, all the h_i tend to zero and the number of steps q tends to infinity, then the difference equation (1.7) becomes the following ordinary differential equation:

$$\frac{dp(t)}{dt} = -[f_x(x(t), u(t), t)\,p(t) + B_x(x(t), u(t), t)], \tag{1.9}$$

which in Mayer's problem (for $B \equiv 0$) coincides with equation (1.8.3) of Chapter 1, which describes the change of impulses (adjoint multiplier) in optimal control theory. Hence in the sequel we will call the vector p_i an impulse.

Given the complete control vector w, using (1.2) one can determine in sequence the complete state vector x and compute the value of R. For a continuous system, this means integrating the system (1.1) "from left to right." Next, using (1.5), (1.7) we compute the sequence p_i -- in optimal control this is called "integrating the impulse equations." After this we determine the derivatives by the formula (1.8).

It is worthwhile to compare the formulas (1.4) and (1.7) for the same vector p_i. One can use (1.4) for numerical computations without introducing the impulse equations (1.7). Fixing w and giving x_i increments, we integrate (1.2) from the ith step to the qth, calculate the changes of R, and approximately find the vector p_i. These computations are more cumbersome than those used in (1.7) since calculating only once "from right to left" in (1.7), one can immediately obtain all the vectors p_i. Nevertheless, it is convenient to use (1.4) for adjusting the programs since it helps verify the correctness of the programmed formulas (1.7).

In this section, using the finite-difference approximation (1.2) of the system of differential equations (1.1) we derive recurrence relations for computing the impulses (1.7), after which by passing to the limit we find the differential equation (1.9). One might try to argue backward, starting from the "continuous" systems (1.1) and (1.9) and going to the discrete ones (1.2) and (1.7) -- but a few difficulties arise on the way. Indeed, formal integration of the system (1.9) with the same error of order h_i^2 leads to different results. One can write, for example, the following two formulas equivalent in the sense of accuracy of integrating (1.1):

$$p_{i+1} = p_i - h_i[f_x(z_i)p_i + B_x(z_x)] , \tag{1.10}$$

$$p_{i+1} = p_i - h_i[f_x(z_i)p_{i+1} + B_x(z_i)] . \tag{1.11}$$

However, only the latter is equivalent to (1.7). At the same time, using (1.7) and also (1.8), one can obtain an exact formula for computing the derivative of R. Hence it is preferable to use the formula (1.11) rather than (1.10), although the difference between them is small, that is, of the same order as the error of integration. In schemes of integration of the system (1.1) of higher order of accuracy this difference becomes more crucial. Hence instead of formal integration of the differential impulse equation, exact formulas need to be used. Formulas for computing the derivatives of R are usually employed to implement various numerical methods of unconstrained minimization. The differentiation errors may substantially complicate the computations. This is especially serious if refined minimization methods are used, e.g., the conjugate-gradient method.

It is often imperative to study systems of the form (1.2) in solving optimization problems for systems the behavior of which is described by recurrence relations only; their "continuous" notation (1.1) is not quite sufficient, the coefficients h_i need not be small quantities. In analyzing such systems, exact differentiation formulas are particularly appropriate.

In making discrete approximations of the initial system (1.1), one can adopt special hypotheses concerning the behavior of the control within the integration interval. In (1.2) it is assumed that the control is constant within the integration interval. One can postulate, for example, that within each ith integration interval, the control is a linear function of t; for $t_i \leqslant t \leqslant t_i + h_i$ we have

$$u(t) = u_i + \frac{t - t_i}{h_i}(u_{i+1} - u_i) .$$

For the Euler scheme, the computational formulas do not change, but for more exact integration schemes new terms appear.

2. COMPUTATION OF THE SECOND DERIVATIVES

Suppose that the functions $R(x,w)$ and $F(x_i,u_i,t_i)$ are twice continuously differentiable with respect to the components of the vectors x and w. Let us find the matrix of the second derivatives of the composite function $R(x(w), w)$ with respect to w. Introduce the square symmetric matrices

$$P_i = \frac{d^2R(x,w)}{dx_i\,dx_i}, \quad P_q = \frac{d^2R(x,w)}{dx_q\,dx_q} = \frac{\partial^2 b(z_q)}{\partial x_q^2},$$
$$\frac{dy_i}{du_j} = \frac{d^2R}{du_j du_i}, \quad \frac{dy_q}{du_q} = \frac{\partial^2 b(z_q)}{\partial u_q^2}.$$

The interpretation of these formulas is analogous to the definitions of the first derivatives of R with respect to the components of the vectors x and w. In differentiating with respect to the components of the vectors x the control vector w is assumed to be constant.

Differentiating p_i and y_i with respect to x_i and u_i and noting (1.7), we obtain

$$\begin{aligned} P_i &= \frac{dp_i}{dx_i} = \frac{d^2R(x,w)}{dx_i\,dx_i} = H_{xx}(z_i, p_{i+1}) + F_x(z_i)P_{i+1}F_x^T(z_i), \\ \frac{dp_i}{du_i} &= \frac{d^2R(x(w),w)}{du_i\,dx_i} = H_{ux}(z_i, p_{i+1}) + F_u(z_i)P_{i+1}F_x^T(z_i), \\ \frac{d^2R(x(w),w)}{du_i\,du_i} &= H_{uu}(z_i, p_{i+1}) + F_u(z_i)P_{i+1}F_u^T(z_i). \end{aligned} \tag{1.12}$$

If $1 \le s < i \le q-1$, then we easily derive the following formulas:

$$\frac{d^2R}{du_s du_i} = \frac{\partial x_{s+1}}{\partial u_s}\frac{d^2R}{dx_{s+1}du_i} = F_u(z_s)\frac{d^2R}{dx_{s+1}du_i},$$
$$\frac{d^2R}{dx_s du_i} = \frac{\partial x_{s+1}}{\partial x_s}\frac{d^2R}{dx_{s+1}du_i} = F_x(z_s)\frac{d^2R}{dx_{s+1}du_i}.$$

Similarly for $q > s > i$ we have

$$\frac{d^2R}{du_s du_i} = \frac{d^2R}{du_s\,dx_{i+1}}F_u^T(z_i),$$
$$\frac{d^2R}{du_s dx_i} = \frac{d^2R}{du_s\,dx_{i+1}}F_x^T(z).$$

From these formulas one can determine sequentially the elements of the matrix $d^2R(x(w),w)/dw^2$. We give only part of this matrix, writing the block matrices occupying the block columns with subscripts from $s-1$ to $s+1$ and block rows with subscripts from $s-1$ to $s+1$:

	$s-1$	s	$s+1$
$s-1$	$\dfrac{d^2R}{du_{s-1}\,du_{s-1}}$	$F_u(z_{s-1})\dfrac{d^2R}{dx_s du_s}$	$F_u(z_{s-1})\dfrac{d^2R}{dx_s du_{s+1}}$
s	$\dfrac{d^2R}{du_s dx_s}F_u^T(z_{s-1})$	$\dfrac{d^2R}{du_s\,du_s}$	$F_u(z_s)\dfrac{d^2R}{dx_{s+1}du_{s+1}}$
$s+1$	$\dfrac{d^2R}{du_{s+1}dx_s}F_u^T(z_{s-1})$	$\dfrac{d^2R}{du_{s+1}\,dx_{s+1}}F_u^T(z_s)$	$\dfrac{d^2R}{du_{s+1}du_{s+1}}$

Thus, in computing the second derivatives it is necessary "to integrate from right to left" not only the equation (1.7) for the impulse vectors p_j but also to recalculate the impulse matrices p_i and the matrices $d^2R/dx_s dx_i$. Cases are possible where for some i the vector $p_i = 0$, which corresponds to "singular" regimes in optimal control. Obviously, these formulas still hold and are useful in theoretical studies of these cases.

Although these formulas are cumbersome, they are expected to be of considerable importance, since they open broad possibilities of using the approach described in Chapter 4 for constructing rapidly convergent computational procedures analogous to Newton's method. Some examples of such computations are given in Section 6.7.

3. THE RUNGE-KUTTA SCHEMES

For a number of practical problems, it is required to guarantee high accuracy in integrating the initial system (1.1). The simplest way to meet this condition is to reduce the integration step in size. In that case, however, the dimension of the control vector w increases making the optimization process much more complicated. Another way of improving the integration accuracy is to use more accurate integration formulas. We shall consider the use of Runge-Kutta schemes. As before, the problem will be to generate formulas for computing the derivatives of the composite function $R(x(w),w)$ with respect to the components of the vector w.

We integrate the system (1.1), using

$$x_{i+1} = x_i + h_i \sum_{j=1}^{\rho} g_j f(z_i^j), \tag{1.13}$$

where

$$z_i^j = [x_i^j, u_i^j, t_i^j], \quad u_i^j = u(t_i^j),$$

and

$$x_i^j = x_i + \beta_{j-1} h_i f(z_i^{j-1}), \quad t_i^j = t_i + \beta_{j-1} h_i, \tag{1.14}$$

where g_j, β_{j-1} is a set of numbers, all $0 \le \beta_{j-1} \le 1$ and $\beta_0 = 0$, hence the values $f(z_i^0)$ are inessential. In (1.13), (1.14) and later in this section the indices i and j take on integer values in the intervals $[1:q-1]$ and $[1:\rho]$, respectively.

To different (parameters) g_j and β_{j-1} there correspond different integration schemes. Hence the formula (1.13) determines a set of methods of numerical integration of the system (1.1) usually called the family of Runge-Kutta methods. The error of integration of the system (1.1) on the ith step is estimated by the difference $\eta(h_i) = x(t_{i+1}) - x_{i+1}$, where $x(t)$ is the solution of (1.1) with initial condition $x(t_i) = x_i$. The quantity η is a function of the integration step h_i. If $f(z)$ is a sufficiently smooth function of its arguments, then the function $\eta(h_i)$ is representable by a Taylor series:

$$\eta(h_i) = \sum_{k=0}^{s} \frac{\eta^{(k)}(0)}{k!}(h_i)^k + \frac{\eta^{(k+1)}(\theta h_i)}{(k+1)!}(h_i)^{k+1},$$

where $0 < \theta < 1$. The parameters of the integration method g_j, β_{j-1} are such that

$$\eta(0) = \eta^{(1)}(0) = \cdots = \eta^{(s)}(0) = 0$$

for arbitrary sufficiently smooth functions $f(z)$. Here if $\eta^{(s+1)}(0) \neq 0$, then s is called the order of the error of the integration method on one step.

A detailed analysis of the various schemes for integrating ordinary differential equations may be found in many texts on numerical methods. Here we limit ourselves to few sets of possible parameters from the family of Runge-Kutta methods.

If we set $\rho = 1$ in (1.13), we obtain the Euler scheme considered above, which has a first-order error of integration.

If we set $\rho = 2$, $g_1 = 0$, $g_2 = 1$, $\beta = \frac{1}{2}$, we obtain the so-called Euler scheme with recalculation, with second-order error of integration; and the computations are made by the formulas:

$$\begin{aligned} &x_{i+1} = x_i + h_i f(z_i^2), \quad z_i^2 = [x_i^2, u_i^2, t_i^2], \\ &x_i^2 = x_i + \frac{h_i}{2} f(z_i^1), \quad z_i^1 = z_i, \quad t_i^2 = t_i + \frac{1}{2} h_i. \end{aligned} \tag{1.15}$$

A modified Euler scheme in which $\rho = 2$, $g_1 = g_2 = \frac{1}{2}$, $\beta_1 = 1$ also has second-order error

$$\begin{gathered} x_{i+1} = x_i + \frac{h_i}{2}[f(z_i^1) + f(z_i^2)], \\ x_i^1 = x_i, \quad x_i^2 = x_i + h_i f(z_i), \\ t_i^1 = t_i, \quad t_i^2 = t_i + h_i. \end{gathered}$$

Among the schemes having fourth-order error, the most common is one with

$$\rho = 4, \quad g_1 = g_4 = 1/6, \quad g_2 = g_3 = 1/3,$$
$$\beta_1 = \beta_2 = 1/2, \ \beta_3 = 1, \quad \beta_4 = 0.$$

Let us go over to the derivatives of the composite function R. By the formula (1.13) we write the expression for R in the form

$$R(x, w) = b(x_q, u_q, t_q) + \sum_{i=1}^{q-1} h_i \sum_{j=1}^{\rho} g_j B(z_i^j),$$

where x is a complete state vector and w is a complete control vector:

$$x = [x_1, x_1^1, \ldots, x_1^\rho, x_2, x_2^1, \ldots, x_2^\rho, \ldots, x_q],$$
$$w = [u_1^1, \ldots, u_1^\rho, u_2^1, \ldots, u_2^\rho, \ldots, u_q].$$

Introduce the auxiliary n-dimensional vectors

$$p_q = \frac{dR}{dx_q} = \frac{\partial b(z_q)}{\partial x_q}, \quad p_i = \frac{dR(x, w)}{dx_i}, \quad p_i^j = \frac{dR(x, w)}{dx_i^j}. \tag{1.16}$$

Noting the relations (1.13), (1.14) determining the vectors x_{i+1}, x_i^{j+1} as differentiable functions of the "preceding" vectors x_i, x_i^j, we obtain

$$p_i = \frac{\partial R}{\partial x_i} + \sum_{j=1}^{\rho} \frac{\partial x_i^j}{\partial x_i} \frac{dR}{dx_i^j} + \frac{\partial x_{i+1}}{\partial x_i} \frac{dR}{dx_{i+1}},$$
$$p_i^e = \frac{\partial R}{\partial x_i^e} + \frac{\partial x_i^{e+1}}{\partial x_i^e} \frac{dR}{dx_i^{e+1}} + \frac{\partial x_{i+1}}{\partial x_i^e} \frac{dR}{dx_{i+1}},$$
$$p_i^\rho = \frac{\partial R}{\partial x_i^\rho} + \frac{\partial x_{i+1}}{\partial x_i^\rho} \frac{dR}{dx_{i+1}}, \qquad e \in [1:\rho-1].$$

Noting (1.13), (1.14), (1.16) we rewrite these expressions in the form of recurrence relations

$$p_i = p_{i+1} + \sum_{j=1}^{\rho} p_i^j, \qquad p_q = \frac{\partial b(z_q)}{\partial x_q},$$

$$p_i^e = h_i [g_e B_x(z_i^e) + f_x(z_i^e) [g_e p_{i+1} + \beta_e p_i^{e+1}]], \tag{1.17}$$

$$p_i = h_i g_\rho [B_x(z_i^\rho) + f_x(z_i^\rho) p_{i+1}] , \qquad e \in [1:\rho-1] .$$

To simplify the formulas, it is convenient to add one additional coefficient β_ρ to the coefficients β_j, assuming $\beta_\rho = 0$. In this case, the vector $p_i^{\rho+1}$ will be multiplied everywhere by β_ρ, hence its value is not essential. Define the function

$$H(z_i^j, p_{i+1}, p_i^{j+1}) = h_i [g_j B(z_i^j) + \langle f(z_i^j), g_j p_{i+1} + \beta_j p_i^{j+1} \rangle].$$

We rewrite the relations (1.17) compactly:

$$p_i^j = H_x(z_i^i, p_{i+1}, p_i^{j+1}) .$$

We derive similarly the next, basic formula for computing the components of the vector of composite derivatives of R with respect to w:

$$\frac{dR(x(w), w)}{du_i^j} = \frac{\partial R}{\partial u_i^j} + \frac{\partial x_i^{j+1}}{\partial u_i^j} \frac{dR}{dx_i^{j+1}} + \frac{\partial x_{i+1}}{\partial u_i^j} \frac{dR}{dx_{i+1}} = H_u(z_i^j, p_{i+1}, p_i^{j+1}). \tag{1.18}$$

These formulas do not differ by much from those found for the Euler scheme. The recalculation of the "impulses" needs more complex formulas.

From the general relations one can easily obtain formulas for computing the derivatives for each particular integrating scheme. For the Euler scheme, with recalculation (1.15), we have, for example,

$$p_i^1 = \frac{1}{2} h_i f_x(z_i) p_i^2, \quad p_i^2 = h_i [B_x(z_i^2) + f_x(z_i^2) p_{i+1}],$$

$$\frac{dR}{du_i^1} = \frac{1}{2} h_i f_u(z_i) p_i^2, \quad \frac{dR}{du_i^2} = h_i [B_u(z_i^2) + f_u(z_i^2) p_{i+1}],$$

$$p_i = p_{i+1} + p_i^1 + p_i^2.$$

Compared to the Euler scheme, the number of points at which the control vector is sought has doubled. One can set

$u_i = u_i^1 = u_i^2$ everywhere and note that in using the Euler scheme with recalculation, quantities of order h_1^3 have been discarded on each integration step. We determine the vectors p_i and the derivatives of R with the same error. We thus obtain

$$p_i = p_{i+1} + h_i [f_x(z_i^2)\bar{p}_{i+1} + B_x(z_i^2)],$$
$$\bar{p}_{i+1} = p_{i+1} + \frac{1}{2} h_i [f_x(z_i^2) p_{i+1} + B_x(z_i^2)],$$
$$\frac{dR}{du_i} = h_i [f_u(z_i^2)\bar{p}_{i+1} + B_u(z_i^2)].$$

Here all the partial derivatives f_x, f_u, B_x, B_u have to be computed with error of order h_i^2 in order to ensure error of order h_i^3 in determining dR/du_i.

For the Runge-Kutta scheme (1.13), when the control is constant on the integration interval, we have

$$p_i = p_{i+1} + \sum_{s=1}^{4} M_x(z_i^s) + h_i \sum_{s=2}^{4} \beta_{s-1} f_x(z_i^{s-1}) M_x(z_i^s) +$$
$$+ h_i^2 \sum_{s=3}^{4} \beta_{s-2}\beta_{s-1} f_x(z_i^{s-2}) f_x(z_i^{s-1}) M_x(z_i^s) +$$
$$+ h_i^3 \beta_1\beta_2\beta_3 f_x(z_i^1) f_x(z_i^2) f_x(z_i^3) M_x(z_i^4),$$
$$\frac{dR}{du_i} = \sum_{s=1}^{4} M_u(z_i^s) + h_i \sum_{s=2}^{4} \beta_{s-1} f_u(z_i^{s-1}) M_x(z_i^s) +$$
$$+ h_i^2 \sum_{s=3}^{4} \beta_{s-2}\beta_{s-1} f_u(z_i^{s-2}) f_x(z_i^{s-1}) M_x(z_i^s) +$$
$$+ h_i^3 \beta_1\beta_2\beta_3 f_u(z_i^1) f_x(z_i^2) f_x(z_i^3) M_x(z_i^4),$$
$$M(z_i^s) = g_s h_i [B(z_i^s) + \langle f(z_i^s),\ p_{i+1}\rangle].$$

One can similarly obtain formulas for other integration schemes.

Using integration schemes of the form (1.13) for $\rho \geqslant 2$, one presumes that the function $f(x,u,t)$ has bounded derivatives in all the arguments on each integration interval. If within the integration interval the control $u(t)$ changes sharply (by quantities of order 1), the accuracy of the computations deteriorates.

Hence, in finding the optimal control, one needs to check the position of grid points with respect to t and, when needed, either change their positions or limit the size of variation of the control on each integration interval. The simplest of all things to do is to assume that the control is constant within each interval. This can be done in solving many practical problems, when the system (1.1) has to be integrated with high accuracy; whereas the optimal control need be computed only coarsely, because usually the optimal control can be implemented only approximately and therefore the sampling interval for the control need not be too small.

If the control is fixed on the integration interval, then $u_i = u_i^1 = \dots = u_i^\rho$, and the vector w may be viewed as the set $u_1, u_2, \dots, u_q$. For computing the derivatives with respect to u_i one needs the following formula instead of (1.18):

$$\frac{dR(x(w), w)}{du_i} = \sum_{j=1}^{\rho} H_u(z_i^j,\ p_{i+1},\ p_i^{j+1}).$$

This approach may be carried further, assuming the control is constant on several integration steps. This enables us to lower the dimension of the vector w, lowering at the same time the accuracy of solving the optimization problem. In particular, if we assume that the control is constant everywhere, then

$u_1 = u_2 = \dots = u_q = u$ and

$$\frac{dR(x(w), w)}{du} = \sum_{i=1}^{q-1} \sum_{j=1}^{\rho} H_u(z_i^j,\ p_{i+1},\ p_i^{j+1}) + b_u(z_q).$$

Actually, in this case the vector u becomes a control parameter, rather than a control. Analogous formulas for the Euler scheme will be given in Section 6.6.

Formulas for computing the derivatives of the function R with respect to the components of the vector u_i are needed in the sequel for minimizing R numerically with respect to w. Hence the accuracy of determining the derivatives ought to be coordinated with the accuracy of the minimization process. In particular, in carrying out the coarse preliminary computations, in the formulas for determining the gradients and the impulses one can discard small terms proportional to high powers of h_i, thus easing the computations. Other simplifying techniques are also feasible. For example, one can assume that the control $u(t)$ is a specified function of t within the integration interval. It is not hard to obtain formulas for differentiation in this case as well. Much more complex are the formulas for the second derivatives of R; they can be found in Grachev and Evtushenko [5].

2. NECESSARY AND SUFFICIENT CONDITIONS FOR A MINIMUM

1. *THE STATEMENT OF THE PROBLEM*

We will consider the problem of optimal control with "mixed" constraints on state and controls. For the sake of simplicity, we study the case where the system (1.1) is integrated by the Euler scheme (1.2) and control must satisfy the mixed constraints along the trajectory

$$\Gamma^1(x_i,u_i,t_i) = 0, \quad \Gamma^2(x_i,u_i,t_i) \leqslant 0 \tag{2.1}$$

and at the end of the trajectory

$$\Gamma^3(x_q, u_q, t_q) = 0, \qquad \Gamma^4(x_q, u_q, t_q) \leqslant 0 . \tag{2.2}$$

By (1.2), to each complete control vector $w = [u_1, u_2, \ldots, u_q] \in E^{rq}$ there corresponds a unique complete state vector $x = [x_1, x_2, \ldots, x_q] \in E^{nq}$. Hence we write $x = x(w)$. We write the conditions (2.1) and (2.2) in concise form:

$$g(x(w), w) = 0, \qquad h(x(w), w) \leq 0, \tag{2.3}$$

where the vector functions g and h are a union of constraints of equality- and/or inequality-type along as well as at the end of the trajectory, respectively:

$$g(x, w) = [\Gamma^1(z_1), \Gamma^1(z_2), \ldots, \Gamma^1(z_{q-1}), \Gamma^3(z_q)],$$
$$h(x, w) = [\Gamma^2(z_1), \Gamma^2(z_2), \ldots, \Gamma^2(z_{q-1}), \Gamma^4(z_q)],$$
$$z_i = [x_i, u_i, t_i].$$

One can assume without loss of generality that these functions define mappings $g\colon E^{rq} \to E^e$, $h\colon E^{rq} \to E^c$.

We say that the control vector w is feasible if the vectors w and $x(w)$ are such that the conditions (2.3) are satisfied. The feasible set W of complete control vectors can be defined in standard fashion:

$$W = \{w \in E^{rq}\colon g(x(w), w) = 0, \ h(x(w), w) \leqslant 0\}$$

emphasizing that g and h are composite functions of w. The partial derivatives of g and h with respect to x_i and u_i are simply expressed in terms of the original functions:

$$\frac{\partial g}{\partial x_i} = \frac{\partial \Gamma^1(z_i)}{\partial x_i} = \Gamma^1_x(z_i), \qquad \frac{\partial h}{\partial u_i} = \frac{\partial \Gamma^2(z_i)}{\partial u_i} = \Gamma^2_u(z_i),$$
$$\frac{\partial g}{\partial x_q} = \frac{\partial \Gamma^3(z_q)}{\partial x_q} = \Gamma^{3'}_x(z_q), \qquad \frac{\partial h}{\partial u_q} = \frac{\partial \Gamma^4(z_q)}{\partial u_q} = \Gamma^4_u(z_q).$$

Moreover, in the sequel the new fact (not included in (2.3)) that the functions defining the constraints for $t = t_i$ depend only on x_i, u_i, t_i, will be exploited to significantly simplify the calculations.

The problem of discrete optimal control of (1.2) consists in finding the complete control vector w and the complete state vector x such that the conditions (2.1) and (2.2) are satisfied and the objective function

$$R_1(x, w) = b_1(z_q) + \sum_{i=1}^{q-1} h_i B_1(z_i) \tag{2.4}$$

takes on the smallest possible value.

The vector functions F, g, h and the functions b_1, B_1 will be referred to as the functions "determining" the discrete optimal control problem. We always assume that a solution of the problem exists. An analogous problem with state constraints along a trajectory can be formulated as well for the system (1.1). Intuitively one would expect that for a large class of systems (1.1) the solutions of both problems will be close if the integration steps for (1.1) are sufficiently small. We will not do a rigorous study of this property since it has already been done, for instance, by Ermol'ev, Gulenko, and Tsarenko [1], Fedorenko [1], Budak, Berkovich, and Solov'eva [1], and by many others.

2. NECESSARY AND SUFFICIENT CONDITIONS FOR A MINIMUM

The discrete optimal control problem is a special nonlinear programming problem. Hence one can obtain extremality conditions as well as numerical methods by using the well-known results of nonlinear programming theory. Let us use the Lagrangian

$$R(x, w, \tilde{u}, v) = b_1(z_q) + \sum_{i=1}^{q-1} h_i B_1(z_i) + \\ + \langle \tilde{u}, g(x, w) \rangle + \langle v, h(x, w) \rangle \tag{2.5}$$

with the Lagrange multipliers $\tilde{u} \in E^e$, $v \in E_+^c$.

This Lagrangian is of the same form as (1.3); hence if the functions defining the problem are differentiable with respect to the components of the vectors x and w, then we can apply to (2.5) the differentiation formulas obtained in the preceding section. The necessary and sufficient conditions for minima in nonlinear programming problems given in Sections 1.6 and 1.7 carry over almost verbatim to our problem here. Hence we limit ourselves to recalling a few results only.

Let there exist a complete control vector w_*, the corresponding complete state vector $x_* = x(w_*)$ and dual vectors $\tilde{u}_*$, $v_* \geqslant 0$ such that for any w_*, $\tilde{u}$, $v \geqslant 0$ we have the saddle-point conditions:

$$R(x_*, w_*, \tilde{u}, v) \leq R(x_*, w_*, \tilde{u}_*, v_*) \leq R(x(w), w, \tilde{u}_*, v_*) .$$

Then, by Theorem 1.6.1, the vector w_* is a solution of the discrete optimal control problem.

By Theorem 1.6.7, if the discrete optimal control problem is a convex programming problem and Slater's or Karlin's constraint qualifications hold, then the Lagrangian R has saddle points.

By Theorem 1.7.6, if, in addition, the functions defining the problem are differentiable with respect to the components of the vectors x and w and the Arrow-Hurwicz-Uzawa condition is satisfied, then in order that the vector w_* be a solution of the discrete optimal control problem, it is necessary that there exist Lagrange multipliers $\tilde{u}_*$, $v_* \geqslant 0$ such that the triplet $[w_*, \tilde{u}_*, v_*]$ is a Kuhn-Tucker point, i.e.,

$$
\begin{gathered}
\frac{dR(x(w_*), w_*, \tilde{u}_*, v_*)}{dw} = 0, \quad g(x(w_*), w_*) = 0, \\
h(x(w_*), w_*) \leqslant 0, \quad v_*^j h^j(x(w_*), w_*) = 0, \quad j \in [1:c].
\end{gathered} \tag{2.6}
$$

Here the components of the derivative dR/dw are found by the formulas derived in the preceding section:

$$
\begin{gathered}
\frac{dR(x(w_*), w_*, \tilde{u}_*, v_*)}{du_i^*} = H_u(z_i^*, p_{i+1}, \tilde{u}_i^*, v_i^*), \\
z_i^* = [x_i^*, u_i^*, t_i], \\
H(z_i, p_{i+1}, \tilde{u}_i\ v_i) = h_i B_1(z_i) + \langle F(z_i), p_{i+1}\rangle + \\
+ \langle \tilde{u}_i, \Gamma^1(z_i)\rangle + \langle v_i, \Gamma^2(z_i)\rangle, \\
p_i = \frac{dR(x_*, w_*, \tilde{u}_*, v_*)}{dx_i} = H_x(z_i^*, p_{i+1}, \tilde{u}_i^*, v_i^*), \\
p_q = \frac{\partial b_1(z_q^*)}{\partial x_q} + \langle \tilde{u}_q^*, \Gamma_x^3(z_q^*)\rangle + \langle v_q^*, \Gamma_x^4(z_q^*)\rangle,
\end{gathered} \tag{2.7}
$$

where the vectors $\tilde{u}_i$, v_i are the components of the vectors $\tilde{u}$ and v; their dimensions coincide respectively with those of the vector functions Γ^1 and Γ^2 for $1 \leq i \leq q-1$ and Γ^3 and Γ^4 for $i = q$.

Introduce the cone

$$
\begin{aligned}
K(w_*, v_*) = \Big\{ w\colon\ & w^T \frac{dg(x(w_*), w_*)}{dw} = 0, \\
& w^T \frac{dh^j(x(w_*), w_*)}{dw} = 0, \quad w^T \frac{dh^s(x(w_*), w_*)}{dw} \leqslant 0 \Big\},
\end{aligned}
$$

where

$$
\begin{gathered}
j \in \theta(w_*, v_*), \quad s \in \sigma(w^*) \backslash \theta(w_*, v_*), \\
\sigma(w_*) = \{k \in [1:c]\colon\ h^k(x(w_*), w_*) = 0\}, \\
\theta(w_*, v_*) = \{k \in [1:c]\colon\ v_*^k > 0,\ k \in \sigma(w_*)\}.
\end{gathered}
$$

We use McCormick's Theorem 1.7.2. Suppose that the functions defining the problem are twice differentiable with respect to the components of the vectors x and w, and there exist vectors w_*, $x_* = w(x_*)$, $\tilde{u}_*$, $v_* \geqslant 0$ such that the conditions (2.6) are satisfied; the matrix of the second derivatives

$$\frac{d^2R(x(w_*), w_*, \tilde{u}_*, v_*)}{dw^2}$$

defined by (1.12) is positive definite on the cone $K(w_*,v_*)$. Then w_* is an isolated local solution of the discrete optimal control problem.

In these assertions, the first and second derivatives of the functions defining the problem are computed with the aid of auxiliary variables (impulses), which is somewhat unusual in nonlinear programming. However, this is only a technicality and has no effect on the matter. Various schemes of integrating (1.1) yield diverse formulas for computing the derivatives without changing the form of necessary and sufficient conditions for an extremum. A similar situation arises in describing and proving numerical methods of nonlinear programming, used to solve discrete optimal control problems. In the next section we shall dwell only briefly on basic numerical methods, without proving the convergence, since they have been extensively treated in the preceding chapters.

3. NUMERICAL METHODS BASED ON THE REDUCTION TO NONLINEAR PROGRAMMING PROBLEMS

The discrete optimal control problem stated in Section 6.2 is a particular case of a nonlinear programming problem. It involves relatively simple computations of the derivatives of the function R with respect to the components of w. This property suggests many ways of effectively using the nonlinear programming methods, based on the computation of first derivatives.

Most of the methods follow a common pattern: given a vector w, the values of the objective function and of the constraints are computed and the auxiliary function R of the form (1.3) is minimized with respect to w by unconstrained minimization methods. Numerical methods differ from each other with respect to the construction of the functions R and the rules for change during the iteration. The choice of method for integrating the system (1.1) affects only the formulas for computing R, g, h and their derivatives. In computer programs, it is convenient to construct the procedures for computing R, g, h and finding the derivatives as separate modules. As the schemes of integrating the system (1.1) change, only these modules change in the numeriical algorithms; other blocks of the program remain intact. The choice of integration scheme is divorced from that of method. Schemes of high-order accuracy are employed in integration, whereas relatively coarse methods are used in optimization, and vice versa.

We shall describe several most commonly used methods for solving discrete optimal control problems. Numerical computations of test problems are given in Section 6.7. To simplify the references, each method is designated OPTS, S denoting the number of the computer program realizing the method. In solving complex optimal control problems, several distinct optimization methods are usually used. Methods having a relatively large region of convergence are used first, followed by rapidly converging methods upon narrowing to a sufficiently small neighborhood of the solution. According to this procedure, each operating method "prepares" necessary data for a subsequent method to operate.

In each numerical method, an iterative process of solving the problem of Section 6.2 is used. The number of the iteration is labeled k. We show how the complete control vector w changes on the kth iteration, and specify the rule for changing the vector of dual variables $\{\tilde{u}, v\}$ in some methods.

<u>OPT41. EXTERIOR PENALTY-FUNCTION METHOD</u> (see Section 3.1). We compose the auxiliary function

$$R(x, w, \tau) = R_1(x, w) + \tau\left[\sum_{i=1}^{e}[g^i(x, w)]^2 + \sum_{j=1}^{c}\psi(h^j(x, w))\right], \tag{3.1}$$

where the objective function R_1 is given by (2.4) and the penalty function ψ is defined by (3.1.6).

For any monotonic increasing sequence $\tau_0 < \tau_1 < \tau_2 < \cdots$, one constructs the sequence of vectors $w(\tau_0)$, $w(\tau_1)$, ... defined by the approximate solution of the unconstrained minimization problem

$$w_k = w(\tau_k) \in \operatorname{Arg}\min_{w} R(x(w), w, \tau_k). \tag{3.2}$$

If the functions defining the problem are differentiable with respect to the components of the vector x, then one can approximately determine the dual vectors

$$\tilde{u}_k^i = \frac{R(x(w_k), w_k, \tau_k)}{\partial g^i}, \quad v_k^j = \frac{\partial R(x(w_k), w_k, \tau_k)}{\partial h^j},$$

useful for further computations via the methods based on modified Lagrangians.

If the conditions of Theorem 3.1.1 are satisfied, the method converges to a solution of the problem. The computing process depends to a great extent on the policy used to increase the penalty coefficient τ, on the unconstrained minimization method employed, on the accuracy of solving the auxiliary problems (3.2), and on many other factors. The accuracy of solving the problems (3.2) has different interpretations in unconstrained minimization methods. In particular, in methods involving derivatives of the objective function, the computations are interrupted upon finding the control w_k satisfying

$$\left\| \frac{dR(x(w_k),\ w_k,\ \tau_k)}{dw} \right\| \leqslant \varepsilon_k.$$

Thus, the second simplified version of the penalty function method will be used (see Section 3.1).

If the functions defining the problem are nondifferentiable in x and u, then as the auxiliary function R it is appropriate to take a function depending on a nondifferentiable penalty (see Section 3.2).

<u>OPT45. THE FIRST VERSION OF OBJECTIVE-FUNCTION PARAMETRIZATION METHOD</u> (see Section 3.3). We compose the auxiliary function

$$R(x,\ w,\ \eta) = [R_1(x,\ w) - \eta]^2 + \sum_{i=1}^{e} [g^i(x,\ w)]^2 + \sum_{j=1}^{c} \psi(h^j(x,\ w)) , \tag{3.3}$$

where the function ψ is the same as in (3.1), η is the lower bound of the optimal value R_1^* of the objective function. According to the results of Section 3.2, the value of η can be obtained after at least one iteration by the method OPT41.

Let the control w_{k-1} and the value η_k be known, and let the new control w_k be determined from the solution of the auxiliary problem

$$w_k \in \operatorname{Arg}\min_{w} R(x(w), w, \eta_k). \tag{3.4}$$

The control w_{k-1} is taken as an initial approximation. Using (3.3.6), we set

$$\eta_{k+1} = \eta_k + \sqrt{R(x(w_k), w_k, \eta_k)}.$$

The computing process stops if at least one of the following three conditions is violated:

$$k \leqslant d, \quad \sum_{i=1}^{e} [g^i(x(w_k), w_k)]^2 + \sum_{j=1}^{c} \psi(h^j(x(w_k), w_k)) \leqslant \varepsilon,$$
$$\eta_{k+1} - \eta_k < \varepsilon(1 + |\eta_k|).$$

Here d is the prescribed maximal number of iterations and ε is the accuracy of solving the unconstrained minimization problem. The quantities d and ε are assigned by the user.

If the functions defining the problem are differentiable with respect to the components of the vector x, one can find the dual vectors

$$\tilde{u}^i = \frac{1}{a} \frac{\partial R(x(w_k), w_k, \eta_k)}{\partial g^i}, \quad v^j = \frac{1}{a} \frac{\partial R(x(w_k), w_k, \eta_k)}{\partial h^j},$$
$$a = 2[R_1(x(w_k), w_k) - \eta_k]. \tag{3.5}$$

OPT46. THE SECOND VERSION OF THE OBJECTIVE-FUNCTION PARAMETRIZATION METHOD. The auxiliary function (3.3) is constructed, problem (3.4) is solved in the kth iteration, the parameter η_k is changed according to (3.3.10), which in this case has the form

$$\eta_{k+1} = \eta_k + \frac{R(x(w_k), w_k, \eta_k)}{R_1(x_k, w_k) - \eta_k}.$$

The dual vectors are defined by (3.5).

The essential drawback of these two versions of the objective-function parametrization method is the requirement that the unconstrained minimization be performed with high accuracy. Only in this case will the necessary condition $\eta_k \leqslant R_1^*$ be satisfied. In the second version, due to the fact that η changes more drastically, this requirement is really crucial. Hence, in solving the auxiliary unconstrained minimization problems, a higher accuracy for the computation is required than with the other method.

The penalty function method and the objective-function parametrization method are most useful when the initial approximation is known to be coarse. Usually, the computations start with these methods and the initial values of the dual variables are determined. It is, however, difficult to solve the problem with high accuracy, using these methods.

OPT53. THE METHOD OF MODIFIED LAGRANGIANS. The simple iteration method (4.3.21) is used. The modified Lagrangian has the form

$$R(x, w, \tilde{u}, v) = R_1(x, w) + \sum_{i=1}^{e} \left[\tilde{u}^i + \frac{\tau}{2} g^i(x, w)\right] g^i(x, w) + \\ + \frac{1}{2\tau} \sum_{j=1}^{c} [(v^j + \tau h^j(x, w))_+]^2.$$

In the kth iteration the following is performed:

$$w_k \in \operatorname{Arg\,min}_w R(x(w), w, \tilde{u}_k, v_k),$$
$$\tilde{u}_{k+1} = \tilde{u}_k + \tau g(x_k, w_k), \quad v_{k+1}^j = (v_k^j + \tau h^j(x_k, w_k))_+,$$
$$x_k = x(w_k).$$

A drawback of this method is that R has no continuous second derivatives in w. This cuts down the number of unconstrained minimization methods usable for finding w_k. In this respect, the next method is most advantageous.

OPT55. THE FIRST VERSION OF THE SIMPLE ITERATION METHOD. The method (4.3.20) is used. The modified Lagrangian has the form

$$R(x, w, \tilde{u}, v) = R_1(x, w) + \sum_{i=1}^{e}\left[\tilde{u}^i + \frac{\tau}{2} g^i(x, w)\right] g^i(x, w) + \sum_{j=1}^{c}\left[\tau^3 [h_+^j(x, w)]^4 + \frac{v^j}{\tau}\begin{cases} 1 + \tau h^j(x, w) + [\tau h^j(x, w)]^2 + [\tau h^j(x, w)]^3 & \text{if } h^j \geqslant 0, \\ [1 - \tau h^j(x, w)]^{-1} & \text{if } h^j < 0. \end{cases}\right]$$

The method consists in the following:

$$w_k \in \operatorname{Arg\,min}_w R(x(w), w, \tilde{u}_k, v_k),$$

$$\tilde{u}_{k+1} = \tilde{u}_k + \tau g(x(w_k), w_k),$$

$$v_{k+1}^j = 4[\tau (h_k^j)_+]^3 + v_k^j \begin{cases} 1 + 2\tau h_k^j + 3\tau^2 [h_k^j]^2 & \text{if } h_k^j \geqslant 0, \\ [1 - \tau h_k^j]^{-2} & \text{if } h_k^j < 0. \end{cases}$$

Here $h_k = h(x(w_k), w_k)$.

To implement this and the preceding methods, it is necessary to know approximate values for the components of the dual vectors. These methods are especially efficient in computations in the neighborhood of the solution. As was shown in Chapter 4, they are in fact versions of the simple iteration method. A higher rate of convergence can be attained by using a modification of Newton's method.

OPT8. NEWTON'S METHOD. The simplest version (4.1.11) is implemented. The modified Lagrangian has the form

$$R(x,\ w,\ \tilde{u},\ \tilde{v}) = R_1(x,\ w) + \sum_{i=1}^{e} \tilde{u}^i g^i(x,\ w) + \sum_{j=1}^{c} (\tilde{v}^j)^2 h^j(x,\ w).$$

The iterative process involves solving the following system of linear equations

$$\begin{gathered}
\frac{d^2R(x_k,\ w_k,\ \tilde{u}_k,\ \tilde{v}_k)}{dw\,dw}(w_{k+1}-w_k) + \frac{dg(x_k,\ w_k)}{dw}(\tilde{u}_{k+1}-\tilde{u}_k) + \\
+ 2\sum_{j=1}^{c} \tilde{v}_k^j \frac{dh^j(x_k,\ w_k)}{dw}(\tilde{v}_{k+1}^j - \tilde{v}_k^j) = -\frac{dR(x_k,\ w_k,\ \tilde{u}_k,\ \tilde{v}_k)}{dw}, \\
\left[\frac{dg(x_k,\ w_k)}{dw}\right]^T (w_{k+1}-w_k) = -g(x_k,\ w_k), \\
\tilde{v}^j\left\langle \frac{dh^j(x_k,\ w_k)}{dw},\ w_{k+1}-w_k \right\rangle + h^j(x_k,\ w_k)(\tilde{v}_{k+1}^j - \tilde{v}_k^j) = \\
= -\tilde{v}^j h^j(x_k,\ w_k),
\end{gathered} \tag{3.6}$$

(3.6)

where $j \in [1:c]$, $x_k = x(w_k)$, and the matrix of the second derivatives of R with respect to w is defined from the recurrence formulas (1.12). In implementing this method, it is necessary to store a symmetric matrix of dimension $(rq)^2$, which limits the size of the discrete optimal control problem being solved.

Some difficulties arise in solving the linear system (3.6) when the constraints (2.3) do not depend explicitly on the components of the vector w, since in this case the determinant of the system (3.6) may be zero. For example, if the inequality constraints has the form $h(x) \leqslant 0$, then for the discrete approximation (1.2) we have $h(x_1) \leqslant 0$ on the first step. Therefore, the value $h(x_1)$ does not depend on the complete vector w, the gradient $dh(x_1)/dw = 0$, and if $h(x_1) = 0$ then one column in the matrix of the second derivatives of R with respect to $[w,\tilde{u},v]$ is equal to zero, which makes the matrix singular. In this case, the constraint $h(x_1) \leqslant 0$ should be dropped and all the sequential constraints be represented as

$$\Gamma_i = h(x_i + h_i f(x_i, u_i, t_i)), \quad i \in [1:q-2] \tag{3.7}$$

and then use the usual computational formulas.

OPT1. THE LINEARIZATION METHOD. In the kth iteration, the cost function (2.4) and the constraints are linearized:

$$R_1(x(w), w) \approx R_1(x(w_k), w_k) + \left\langle \frac{dR_1(x(w_k), w_k)}{dw}, \delta w \right\rangle,$$
$$g(x(w), w) \approx g(x(w_k), w_k) + \left[\frac{dg(x(w_k), w_k)}{dw}\right]^T \delta w,$$
$$h(x(w), w) \approx h(x(w_k), w_k) + \left[\frac{dh(x(w_k), w_k)}{dw}\right]^T \delta w,$$

where $\delta w = w - w_k$, and one formulates the following quadratic programming problem of finding the minimum of the function with respect to δw:

$$\left\langle \frac{dR_1(x(w_k), w_k)}{dw}, \delta w \right\rangle + a\langle \delta w, \delta w\rangle \tag{3.8}$$

satisfying the conditions

$$\begin{aligned} g(x(w_k), w_k) + \left[\frac{dg(x(w_k), w_k)}{dw}\right]^T \delta w = 0,\\ h(x(w_k), w_k) + \left[\frac{h(x(w_k), w_k)}{dw}\right]^T \delta w \leqslant 0, \end{aligned} \tag{3.9}$$

where a is a positive coefficient. Upon finding the optimal value δw, we set: $w_{k+1} = w_k + \alpha\delta w$, while the step α is obtained by minimizing the nondifferentiable penalty function,

$$P = R_1(x(w_{k+1}), w_{k+1}) + \\ + \tau\left[\sum_{i=1}^{e} |g^i(x(w_{k+1}), w_{k+1})| + \sum_{j=1}^{c} h_+^j(x(w_{k+1}), w_{k+1})\right],$$

where τ is sufficiently large.

If one sets $a = 0$, then (3.8), (3.9) becomes a linear programming problem and in this case one should require in addition that the components of δw be bounded. One can use all the other versions of the method described in Section 3.5. A special version of the linearization method oriented toward solving optimal control problems has been suggested by Fedorenko [1].

OPT7. MODIFICATION OF THE ARROW-HURWICZ METHOD. Following the arguments of Section 4.1, we form the simplest modification of the Lagrangian, setting

$$R(x, w, \tilde{u}, \tilde{v}) = R_1(x, w) + \langle \tilde{u}, g(x, w)\rangle + \sum_{j=1}^{c} (\tilde{v}^j)^2 h^j(x, w).$$

The method consists in constructing the sequence

$$\begin{aligned} w_{k+1} &= w_k - \alpha \frac{dR(x(w_k), w_k, \tilde{u}_k, \tilde{v}_k)}{dw}, \\ \tilde{u}_{k+1} &= \tilde{u}_k + \varepsilon\alpha g(x(w_k), w_k), \\ \tilde{v}^j_{k+1} &= \tilde{v}^j_k + 2\varepsilon\alpha\tilde{v}^j_k h^j(x(w_k), w_k) , \end{aligned}$$

where the step α must be sufficiently small, ε is either equal to one or sufficiently small. The method has a rather low rate of convergence.

This will end our description of the basic methods. In similar fashion, numerous other algorithms of nonlinear programming can be carried over to solving discrete optimal control problems. As mentioned earlier, programs implementing these methods can be modified by changing individual blocks as as to make them suitable for computing diverse schemes of integrating the system (1.1.). In the programs OPT41, OPT45, OPT46, OPT53, OPT55, OPT1, OPT8, OPT7, the system (1.1) is integrated by the Euler scheme. In Section 6.7, we shall be referring to the programs OPT413, OPT553, using the methods described in OPT41 and OPT55, respectively; but the Euler method with recalculation (1.15) will be used to integrate the system (1.1).

4. DISCRETE MINIMUM PRINCIPLES

1. BASIC DEFINITIONS

As seen from the numerical methods described in Section 6.3, in solving discrete optimal control problems one often needs to find the unconstrained minimum of the auxiliary function $R(x(w),w)$. The vector w usually has large size, thus minimization of $R(x(w),w)$ with respect to w is a rather complex problem. One can use an extensive library of standard programs implementing diverse methods of unconstrained minimization of multivariable functions. However, the problem we are considering is of peculiar nature, extrinsic in the general unconstrained minimization problem, and therefore it is not accounted for in these methods. The fact that the minimization problem is connected with the solution of a discrete optimal control problem makes it possible in a number of cases to use special properties which, for continuous systems, are embodied in Pontryagin's maximum principle. Their analogs in nonlinear programming are the results presented in Section 1.4. The objective of this section is to derive for the system (1.2) necessary conditions for a minimum of the function R that are analogous to Pontryagin's maximum principle. Numerical methods based on these results will be examined in the next section.

We assume that the multistep process is described by the relation (1.2). The problem consists in finding the ith component

u_i of the vector w, yielding the minimum of the composite function

$$R(x(w), w) = b(z_q), \quad z_q = [x_q, u_q, t_q]$$

over all possible values u_i belonging to some compact set U. All the other components of w will be assumed to be fixed. The subscript i is arbitrary and less than q. Let W_* denote the solution set of our problem:

$$W_* = \operatorname*{Arg\,min}_{u_i \in U} b(z_q) . \tag{4.1}$$

We assume in the sequel that this set is nonempty.

If the mapping $F(z_i)$ and the function $b(z_q)$ depend continuously on all their arguments, then each vector $u_i \in U$ (1.2) yields the unique sequence $x_{i+1}, x_{i+2}, \ldots, x_q$. This operation defines a continuous function b of the vector x_{i+1} and a mapping x_{i+1} as a function of the vector u_i. To simplify the formulas, we denote the vector x_{i+1} by the letter a and the composite function b of a by $B(a)$. Then

$$R(x(w), w) = B(a) = B(a(u_i)), \quad a(u_i) = F(x_i, u_i, t_i) .$$

The set W_* is representable in the form

$$W_* = \operatorname*{Arg\,min}_{u_i \in U} B(a(u_i)).$$

Instead of minimizing $B(a(u_i))$ with respect to the control vector $u_i \in U$, we consider the problem of minimizing $B(a)$ with respect to the state vector $a \in \Omega = a(U)$ (in the state space) and define the set

$$\Omega_* = \operatorname{Arg}\min_{a\in\Omega} B(a).$$

Obviously, Ω_* is the image of the set W_* under the mapping $a(u_i)$: $\Omega_* = a(W_*)$.

If the vector function F and the function b are differentiable with respect to the components of the complete state vector, then by the formulas of Section 6.1 we have

$$p_s = \frac{dR}{dx_s} = H_x(z_s, p_{s+1}), \quad i \leqslant s \leqslant q-1, \quad p_q = \frac{\partial b\ (z_q)}{\partial x_q}. \tag{4.2}$$

With the usual notation $H(z_s, p_{s+1}) = \langle F(z_s), p_{s+1}\rangle$.

Now for each vector u_i U one can compute the vectors $x_{i+1}, x_{i+2}, \ldots, x_q$ and, also, from (4.2) find $p_q, p_{q-1}, \ldots, p_{i+1}$. This operation defines the single-valued mappings

$$p_{i+1} = p_{i+1}(a) = p_{i+1}(a(u_i)).$$

If the function $B(a)$ is defined on an open set containing Ω and is differentiable at the point $\bar{a} \in \Omega$, then one can introduce the point-set mapping

$$W_1(\bar{a}) = \operatorname{Arg}\min_{a\in\Omega} \langle p_{i+1}(\bar{a}),\ a - \bar{a}\rangle$$

or, passing to the control space, define for $\bar{u}_i \in U$ the multi-valued mapping

$$W_2(\bar{u}_i) = \operatorname{Arg}\min_{u_i\in U} \langle p_{i+1}(a(\bar{u}_i)),\ a(u_i) - a(\bar{u}_i)\rangle.$$

Let $\bar{a} = a(\bar{u}_i)$. Then the set $W_1(\bar{a})$ is the image of the set $W_2(\bar{u}_i)$ under the mapping $a(u_i)$:

$$W_1(\bar{a}) = a(W_2(\bar{u}_i)) .$$

If we assume that the mapping F is differentiable with respect to the components of the complete state vector as well as the components of the vector w, then we can use the formula (1.8) obtained in Section 6.1, which can be rewritten in the form

$$\frac{dB(a(u_i))}{du_i} = F_u(x_i, u_i, t_i)\, p_{i+1}(a(u_i)) =$$
$$= H_u(x_i, u_i, t_i, p_{i+1}(a(u_i))) = h_i f_u(x_i, u_i, t_i)\, p_{i+1}(a(u_i)).$$

For each vector $\bar{u}_i \in U$ we define the point-set mapping

$$W_3(\bar{u}_i) = \operatorname{Arg}\min_{u_i \in U} \left\langle \frac{dB(a(\bar{u}_i))}{du_i},\ u_i - \bar{u}_i \right\rangle.$$

Similarly to Section 1.4, one can pose the problem of finding the fixed points of the multivalued mappings $W_1(a)$, $W_2(u_i)$, $W_3(u_i)$ or the problem of solving the corresponding variational inequalities, i.e., the points satisfying the conditions

$$\bar{a} \in W_1(\bar{a}),\quad \langle p_{i+1}(\bar{a}),\ a - \bar{a}\rangle \geqslant 0 \quad \forall a \in \Omega, \tag{4.3}$$

$$\bar{u}_i \in W_2(\bar{u}_i),\ \langle p_{i+1}(a(\bar{u}_i)),\ f(x_i, u_i, t_i) - f(x_i, \bar{u}_i, t_i)\rangle \geqslant 0 \quad \forall u_i \in U, \tag{4.4}$$

$$\bar{u}_i \in W_3(\bar{u}_i),\ \langle f_u(x_i, \bar{u}_i, t_i)\, p_{i+1}(a(\bar{u}_i)),\ u_i - \bar{u}_i\rangle \geqslant 0 \qquad \forall u_i \in U. \tag{4.5}$$

If $\bar{u}_i \in W_*$, then (4.4) is a discrete analog of Pontryagin's minimum principle (see Section 1.8). The condition (4.3) implies the same property, however, in the state space. The condition (4.5) is a discrete analog of the linearized minimum principle.

We replace the initial problem (4.1) of computing the minimum of the composite function $b(z_q)$ with respect to $u_i \in U$ by that of finding vectors satisfying any of the conditions (4.3)-(4.5). First, however, we formulate conditions under which such a reduction is appropriate.

2. NECESSARY AND SUFFICIENT CONDITIONS FOR A MINIMUM

Let

$$u_i^* \in W_*, \quad a_* = a(u_i^*) \in \Omega_*, \quad \bar{u}_i \in U, \quad \bar{a} = a(\bar{u}_i) \in \Omega.$$

Also, let

$$\Delta(\bar{a}) = R(x(\bar{w}), \bar{w}) - R(x(w_*), w_*) = B(\bar{a}) - B(a_*),$$

where all the components of the complete control vectors $\bar{w}$ and w_* coincide except the ith component: for $\bar{w}$ the vector $\bar{u}_i$ is taken as u_i, for w_* the vector u_i^* is taken as u_{i*}. Thus the quantity Δ represents an increment of the cost function R when the control u_i^* is replaced by $\bar{u}_i$.

<u>THEOREM 6.4.1</u>. Let the function B(a) be defined on an open set containing Ω and suppose there exists a point $\bar{u}_i \in U$ such that the function B(a) is differentiable at $\bar{a} = a(\bar{u}_i)$. Then:

•1. if $\bar{u}_i \in W_*$ and the set Ω is convex, then the conditions (4.3) and (4.4) are satisfied;

•2. if B(a) is pseudoconvex at $\bar{a}$ with respect to Ω and either at the point $\bar{a}$ the condition (4.3) is satisfied or at the point $\bar{u}_i$ the condition (4.4) is satisfied, then $\bar{a} \in \Omega_*$, $\bar{u}_i \in W_*$;

•3. if the function B(a) is convex on Ω, then for any $a \in W_1(\bar{a})$, $u_i \in W_2(u_i)$ we have the inequalities

$$0 \leqslant \Delta(\bar{a}) \leqslant \langle p_{i+1}(\bar{a}), \ \bar{a} - \bar{\bar{a}} \rangle, \tag{4.6}$$

$$0 \leqslant \Delta(\bar{a}) \leqslant H(x_i, \bar{u}_i, t_i, p_{i+1}(a(\bar{\bar{u}}_i))) - H(x_i, \bar{\bar{u}}_i, t_i, p_{i+1}(a(\bar{u}_i))). \tag{4.7}$$

Assertion 1 is a necessary condition for a minimum of problem (4.1). Its proof is the same as that of Theorem 1.4.1. Assertion 2 yielding sufficient conditions follows from a similar assertion stated in Theorem 1.4.3. From the differentiability of $B(a)$ at $\bar{a}$ we have

$$B(a)-B(\bar{a})=\langle p_{i+1}(\bar{a}),\ a-\bar{a}\rangle+\|a-\bar{a}\|\alpha(\bar{a},\ a-\bar{a}), \tag{4.8}$$

where $\lim_{\alpha\to\bar{a}} (\bar{a}, a-\bar{a}) = 0$.

If $B(a)$ is convex, then by Theorem 1.2.5 for any $a \in \Omega$ we have

$$\langle p_{i+1}(\bar{a}),\ a-\bar{a}\rangle \leqslant B(a)-B(\bar{a}), \quad p_{i+1}(\bar{a})=\frac{dB(\bar{a})}{da}.$$

Taking for a the vector $a_* \in \Omega_*$, we obtain that for any $\tilde{a} \in W_1(\bar{a})$:

$$\Delta(\bar{a}) = B(\bar{a})-B(a_*) \leq \langle p_{i+1}(\bar{a}),\ \bar{a}-a_*\rangle \leq \langle p_{i+1}(\bar{a}),\ \bar{a}-\tilde{a}\rangle .$$

We have thus arrived at (4.6); if the right side of this formula is expressed in terms of the function H, then we obtain (4.7).///

The necessary condition (4.3) can be called a discrete "state" minimum principle, and (4.4) a discrete "control" minimum principle. To obtain analogous assertions concerning the conditions (4.5), we define the function

$$\psi(\bar{u}_i,\ u_i)=\langle p_{i+1}(a(\bar{u}_i)),\ F(x_i,\ u_i,\ t_i)\rangle = \\ = H(x_i,\ u_i,\ t_i,\ p_{i+1}(\bar{u}_i)).$$

THEOREM 6.4.2. Let the composite function $B(a(u_i))$ of u_i be defined on an open set containing U and suppose there exists a point $\bar{u}_i \in U$ such that the function $B(a(u_i))$ is differentiable at $\bar{u}_i$.

•1. if $\bar{u}_i \in W_*$ and the set U is convex, then the condition (4.5) holds;

•2. if $\bar{u}_i \in W_*$, the set U is convex, $\psi(\bar{u}_i, u_i)$ is a pseudoconvex function of u_i at the point $\bar{u}_i$ with respect to U, then at the point $\bar{u}_i$ the condition (4.4) is satisfied and at the point $\bar{a} = a(\bar{u}_i)$ the condition (4.3) is satisfied;

•3. if at the point $\bar{u}_i$ either the condition (4.4) is satisfied and U is convex or the condition (4.5) is satisfied, and in addition, at the point $\bar{u}_i$ the function $B(x(u_i))$ is pseudoconvex in u_i with respect to U, then $\bar{u}_i \in W_*$.

•4. if $B(a(u_i))$ is a convex function of u_i on U, then for any $\tilde{u}_i \in W_3(\bar{u}_i)$ we have the estimate

$$0 \leqslant \Delta(\bar{a}) \leqslant h_i \langle f_u(x_i, \bar{u}_i, t_i) p_{i+1}(a(\bar{u}_i)), \bar{u}_i - \tilde{u}_i \rangle. \tag{4.9}$$

Let us prove assertion 1. From the differentiability of the composite function $B(a(u_i))$ it follows that

$$\begin{aligned} B(a(u_i)) - B(a(\bar{u}_i)) = \\ = \langle H_u(x_i, \bar{u}_i, t_i, p_{i+1}(a(\bar{u}_i))), u_i - \bar{u}_i \rangle + \\ + \|u_i - \bar{u}_i\| \beta(\bar{u}_i, u_i - \bar{u}_i), \end{aligned}$$

where $\lim_{u_i \to \bar{u}_i} \beta(\bar{u}_i, u_i - \bar{u}_i) = 0$.

The condition $\bar{u}_i \in W_*$ implies that $B(a(u_i))$ attains its minimum in U at $\bar{u}_i$. Noting the convexity of U and using Theorem 1.4.1, we conclude that for any $u_i \in U$ we have the inequality

$$\left\langle \frac{dB(a(\bar{u}_i))}{du_i}, u_i - \bar{u}_i \right\rangle \geqslant 0, \tag{4.10}$$

i.e., the condition (4.5) is satisfied.

Let us prove assertion 2. Using the differentiability of $\psi(\bar{u}_i, u_i)$ in u_i at $u_i = \bar{u}_i$, we rewrite (4.10) as follows:

$$\left\langle \frac{d\psi(\bar{u}_i, \bar{u}_i)}{du_i}, u_i - \bar{u}_i \right\rangle \geqslant 0.$$

Here $\psi(\bar{u}_i, u_i)$ is differentiated with respect to the second argument. But from the pseudoconvexity of $\psi(\bar{u}_i, u_i)$ it then follows that

$$\langle F(x_i, \bar{u}_i, t_i), p_{i+1}(a(\bar{u}_i))\rangle = \psi(\bar{u}_i, \bar{u}_i) \leqslant \psi(\bar{u}_i, u_i). \tag{4.11}$$

Since the vector $u_i \in U$ is arbitrary, we conclude that the conditions (4.3) and (4.4) hold.

Let us show assertion 3. Let $\bar{u}_i$ satisfy (4.4). Then for any $u_i \in U$, (4.11) is satisfied. From the condition that $\bar{u}_i$ is a minimum of $\psi(\bar{u}_i, u_i)$ with respect to u_i on the convex set U, we obtain that for each $u_i \in U$ we have the inequality

$$0 \leqslant \left\langle \frac{d\psi(\bar{u}_i, \bar{u}_i)}{du_i}, u_i - \bar{u}_i \right\rangle = \\ = \langle H_u(x_i, \bar{u}_i, t_i, p_{i+1}(a(\bar{u}_i))), u_i - \bar{u}_i\rangle,$$

i.e., (4.5) is satisfied. From the pseudoconvexity of $B(a(u_i))$ in u_i we obtain that $\bar{u}_i \in W_*$.

The inequality (4.9) is obtained exactly as (4.6). ///

Omitting conditions of continuity and differentiability, we can summarize the assertions of the last two theorems as follows:

○ if $\bar{u}_i \in W_*$, then for (4.4) to be satisfied at this point it suffices that either Ω be convex, or $H(x_i, u_i, t_i, p_{i+1}(a(\bar{u}_i)))$ be pseudoconvex in u_i at the point $u_i = \bar{u}_i$ with respect to the convex set U;

○ if at $\bar{u}_i \in U$ the condition (4.4) holds, then for the condition $\bar{u}_i \in W_*$ to be satisfied it suffices that either the function $B(a)$ be pseudoconvex at $\bar{a} = a(\bar{u}_i)$ with respect to Ω or the function $B(a(u_i))$ be pseudoconvex in u_i at $\bar{u}_i$ with respect to the convex set U.

In proving the necessary conditions for a minimum, the convexity of Ω can be dispensed with if the variational inequalities are treated locally. For example, instead of (4.4) we require that the inequality

$$\langle p_{i+1}(a(\bar{u}_i)), \quad f(x_i,u_i,t_i) - f(x_i,\bar{u}_i,t_i)\rangle \geqslant 0$$

be satisfied for all vectors u_i belonging to U and lying in a sufficiently small neighborhood of $\bar{u}_i$. The assertion that for the differentiable function $B(a)$ the last inequality follows from the condition $\bar{u}_i \in W_*$ is called the local control minimum principle. We shall not formulate theorems on the local minimum principles, since they are similar to the statement of Theorem 1.4.2.

3. THE QUASIMINIMUM PRINCIPLE

We shall compare the minimum principle formulated in Section 1.8 with the results of the preceding section. The justification of the minimum principle for the system of differential equations (1.1) does not require the convexity of Ω, whereas for discrete optimal control problems this condition is essential. It is not hard to give examples in which Ω is not convex and in the space of controls and states the discrete minimum principles do not hold. Nevertheless, it is clear that taking sufficiently small integration steps in the numerical schemes for integrating the system (1.1), it is possible to obtain arbitrarily close approximations of the solution of the initial system (1.1) under very general assumptions, properties of the solutions obtained differing only slightly. This apparent contradiction is easily dispelled if the discrete principles are given a different interpretation delineated by Gabasov and Kirillova in [1].

Let us show that approximate discrete minimum principles hold without the assumption of convexity, and the smaller the integration step of the initial system (1.1) the closer the approximation. Here we are talking about the minimum principle in the space of controls and states. Since the linearized minimum principle for the system (1.1) holds only for convex U.

The sets $W_1(\bar{a})$, $W_2(\bar{u}_i)$ introduced in the preceding subsection can be represented in the equivalent form

$$W_1(\bar{a}) = \{a \in \Omega:\ \langle p_{i+1}(\bar{a}),\ a\rangle = \min_{\hat{a}\in\Omega} \langle p_{i+1}(\bar{a}),\ \hat{a}\rangle\},$$
$$W_2(\bar{u}_i) = \{u_i \in U:\ H(x_i,\ u_i,\ t_i,\ p_{i+1}(a(\bar{u}_i))) = \\ = \min_{\hat{u}_i\in U} H(x_i,\ \hat{u}_i,\ t_i,\ p_{i+1}(a(\bar{u}_i)))\}.$$

Instead of these sets we shall introduce sets obtainable by solving similar minimization problems in which, however, only an approximate minimum is sought, with an error not exceeding εh_i:

$$W_1^{\varepsilon}(\bar{a}) = \{a \in \Omega:\ \langle p_{i+1}(\bar{a}),\ a\rangle \leqslant \min_{\hat{a}\in\Omega} \langle p_{i+1}(\bar{a}),\ \hat{a}\rangle + \varepsilon h_i\} \tag{4.12}$$

$$W_2^{\varepsilon}(\bar{u}_i) = \{u_i \in U:\ H(x_i,\ u_i,\ t_i,\ p_{i+1}(a(\bar{u}_i))) \leqslant \\ \leqslant \min_{\hat{u}_i\in U} H(x_i,\ \hat{u}_i,\ t_i,\ p_{i+1}(a(\bar{u}_i))) + \varepsilon h_i\}. \tag{4.13}$$

Obviously, for any $\varepsilon > 0$ we have the inclusions

$$W_1(\bar{a}) \subset W_1^{\varepsilon}(\bar{a}), \quad W_2(\bar{u}_i) \subset W_2^{\varepsilon}(\bar{u}_i).$$

From (4.13) it follows that if $\tilde{u}_i \in W_2^{\varepsilon}(\bar{u}_i)$, then for any $\tilde{u}_i \in U$ we have the inequality

$$H(x_i,\ \tilde{u}_i,\ t_i,\ p_{i+1}(a(\bar{u}_i))) \leqslant H(x_i,\ u_i,\ t_i,\ p_{i+1}(a(\bar{u}_i))) + \varepsilon h_i.$$

The converse is also true: if this inequality holds for any $u_i \in U_1$, then $\tilde{u}_i \in W_2(\bar{u}_i)$. An analogous property follows from the definition (4.12).

If, say, we rewrite the inequality in (4.13) in terms of the system (1.2), εh in (4.12), (4.13) becomes

$$h_i \langle f(x_i, u_i, t_i), p_{i+1}(\bar{u}_i)\rangle \leqslant \\ \leqslant h_i \min_{\hat{u}_i \in U} \langle f(x_i, \hat{u}_i, t_i), p_{i+1}(a(\bar{u}_i))\rangle + \varepsilon h_i ,$$

whence, cancelling h_i, we obtain that $W_2^{\varepsilon}(\hat{u}_i)$ is the set of ε-optimal points in the problem of the minimum of the scalar product $\langle f(x_i, u_i, t_i), p_{i+1}(a(\bar{u}_i))\rangle$, i.e.,

$$W_2^{\varepsilon}(\hat{u}_i) = \{u_i \in U: \langle f(x_i, u_i, t_i), p_{i+1}(a(\bar{u}_i))\rangle \leqslant \\ \leqslant \min_{\hat{u}_i \in U} \langle f(x_i, \hat{u}_i, t_i), p_{i+1}(a(\bar{u}_i))\rangle + \varepsilon\}.$$

THEOREM 6.4.3. Let the set $\tilde{\Omega} = f(x_i, U, t_i) \subset E^n$ be bounded, let the function $B(a)$ be defined on an open set containing the set $\tilde{\Omega}$ and suppose there exists a point $\bar{u}_i \in W_*$ such that $B(a)$ is differentiable at $\bar{a} = a(\bar{u}_i)$. Then for any $\varepsilon > 0$ there exists $\bar{h}_i > 0$ such that for any integration step satisfying the condition $0 < h_i < \bar{h}_i$ the following assertions hold:

$$\bar{a} \in W_1^{\varepsilon}(\bar{a}), \quad -\varepsilon h_i \leqslant \langle p_{i+1}(\bar{a}), a - \bar{a}\rangle \quad \forall a \in \Omega, \tag{4.14}$$

$$\bar{u}_i \in W_2^{\varepsilon}(\bar{u}_i), \quad -\varepsilon \leqslant \langle p_{i+1}(a(\bar{u}_i)), f(x_i u_i, t_i) - \\ - f(x_i, \bar{u}_i, t_i)\rangle \quad \forall u_i \in U. \tag{4.15}$$

Proof. From the boundedness of the set $\tilde{\Omega}$ there exists a number d (the diameter of $\tilde{\Omega}$) such that for any $\hat{u}_i, \tilde{u}_i \in U$ the condition

$$\|f(x_i, \hat{u}_i, t_i) - f(x_i, \tilde{u}_i, t_i)\| \leqslant d$$

is satisfied.

To prove the theorem we need to estimate the difference $B(a) - B(\bar{a})$. The Newton-Leibniz and Lagrange formulas are not suitable for this purpose since they hold only if Ω is convex. To get rid of this condition, we use the formula (4.8) for differtiating $B(a)$. From the condition $a \in \Omega_*$ it follows that for any $a \in \Omega$, $B(\bar{a}) \in B(a)$, which, noting (4.8), can be written

$$\langle p_{i+1}(\bar{a}) \quad a-\bar{a}\rangle \geqslant -\|a-\bar{a}\| \alpha(\bar{a}, a-\bar{a}) \geqslant \\ \geqslant -\|a-\bar{a}\| \cdot |\alpha(\bar{a}, a--\bar{a})|.$$

From the property of the limit of the function $\alpha(\bar{a}, a-\bar{a})$ as $a \to \bar{a}$ we obtain that for ratio $\frac{\varepsilon}{d}$ there exists a $\delta > 0$ such that $|\alpha(\bar{a}, a-\bar{a})| < \frac{\varepsilon}{d}$, only if

$$\|a-\bar{a}\| = h_i \|f(x_i, u_i, t_i) - f(x_i, \bar{u}_i, t_i)\| < \delta.$$

This inequality holds for any $u_i \in U$ if the integration step h_i is such that

$$0 < h_i < \bar{h}_i = \delta/d. \tag{4.16}$$

In this case,

$$\|a-\bar{a}\| \cdot |\alpha(\bar{a}, a-\bar{a})| \leqslant \varepsilon h_i.$$

When we conclude that for any h_i satisfying (4.16) and for all $a \in \Omega$ we have the inequality

$$-\varepsilon h_i \leqslant \langle p_{i+1}(\bar{a}), a-\bar{a}\rangle,$$

i.e., (4.14) is satisfied at $\bar{a} \in \Omega$; passing to the control space we obtain (4.15). ///

The requirement (4.14) can be called the discrete state principle for a quasiminimum and (4.15) the discrete control principle.

This Theorem makes clear that in solving many practical optimal control problems, it is possible to use a sufficiently accurate discrete approximation of the initial system (1.1) and ignore completely the discrete minimum principle, using, instead, the minimum principle for the system (1.1). Fedorenko [1], for example, notes that based on computational experience, "the author has not been able to draw any practical recommendations that would follow from distinguishing between the discrete maximum principle and the maximum principle for differential equations and which would be efficient in computations."

Necessary conditions for a minimum are very useful for solving practical problems, since they "screen" the nonoptimal points. Most advantageous are the necessary conditions or sets of conditions which can discard a large number of nonoptimal points. In this regard, the theorems of the preceding subsection are more essential than Theorem 6.4.3. The practical value of Theorem 6.4.3, which is, however, applicable to a wider class of problems, can even be enhanced when used together with the local discrete minimum principles mentioned in Subsection 6.4.2. For instance, the points $u_i \in U$ yielding the minimum of the function H with error εh_i are likely to be optimal.

4. SECOND-ORDER NECESSARY CONDITIONS

In some problems, the condition

$$H(x_i, u_i, t_i, p_{i+1}(a(\bar{u}_i))) \equiv H(x_i, \bar{u}_i, t_i, p_{i+1}(a(\bar{u}_i))) \qquad (4.17)$$

holds for any $u_i \in U$, $\bar{u}_i \in W_*$. These situations are called singular regimes in optimal control theory. As in Section 1.4, one can derive necessary conditions for a minimum in this case as well. If $B(a)$ is twice differentiable at $\bar{a}$, then

$$B(a)-B(\bar{a})= \\ =\left\langle \frac{dB(\bar{a})}{da},\ a-\bar{a} \right\rangle + \frac{1}{2}(a-\bar{a})^T \frac{d^2B(\bar{a})}{da^2}(a-\bar{a})+ \\ +\|a-a\|^2 \beta(\bar{a},\ a-\bar{a}), \tag{4.18}$$

where

$$\lim_{a \to \bar{a}} \beta(\bar{a},\ a-\bar{a}) = 0 .$$

We use the notation introduced in Section 6.1:

$$p_{i+1}(\bar{a}) = \frac{db(x_q(\bar{a}))}{da} = \frac{dB(\bar{a})}{da}, \quad p_q = \frac{\partial b(x_q)}{\partial x_q},$$

$$P_{i+1}(\bar{a}) = \frac{d^2b(x_q(\bar{a}))}{da^2} = \frac{d^2B(\bar{a})}{da^2}, \quad P_q = \frac{\partial^2 b(x_q)}{\partial x_q^2}.$$

The vectors p_i and the matrices p_i are defined by the recurrence relations of Section 6.1. The singularity condition (4.17) can be rewritten in the state space as follows:

$$\left\langle \frac{dB(\bar{a})}{da},\ a \right\rangle \equiv \left\langle \frac{dB(\bar{a})}{da},\ \bar{a} \right\rangle \quad \forall a \in \Omega ,$$

which together with (4.18) implies that if $\bar{a} \in \Omega_*$ and Ω is convex, then we have the inequality

$$(a-\bar{a})^T P_{i+1}(\bar{a})(a-\bar{a}) \geqslant 0 \quad \forall a \in \Omega.$$

Just as in defining the linearized minimum principle, we take the function $B(a(u_i))$ of u_* and call the control singular if

$$\langle H_u(x_i,\ \bar{u}_i,\ t_i,\ p_{i+1}(\bar{u}_i)),\ u_i \rangle \equiv \\ \equiv \langle H_u(x_i,\ \bar{u}_i,\ t_i,\ p_{i+1}(\bar{u}_i)),\ \bar{u}_i \rangle \tag{4.19}$$

for any $u_i \in U$, $\bar{u}_i \in W_*$. If $B(a(u_i))$ is a twice differentiable function of u_i at $u_i = \bar{u}_i$ and the condition (4.19) is satisfied, then for any $u_i \in U$ we need to have the inequality

$$(u_i - \bar{u}_i)^T \frac{d^2 B(a(\bar{u}_i))}{du_i^2} (u_i - \bar{u}_i) \geqslant 0,$$

where the matrix of the second derivatives is computed by the formula (1.12). Similarly one can derive necessary conditions of higher orders.

5. ACCOUNTING FOR INTEGRAL FUNCTIONALS

All the above results easily extend to the case where the function R depends explicitly not only on the terminal state vector x_q but also on intermediate vectors x_i. Let

$$R(x(w), w) = b(x_q) + \sum_{i=1}^{q-1} C(z_i). \qquad (4.20)$$

We extend the state space by introducing an additional state variable x^{n+1}. Let

$$\begin{gathered} x_{i+1}^{n+1} = x_i^{n+1} + C(z_i), \quad x_1^{n+1} = 0, \\ \hat{x}_i = [x_i, x_i^{n+1}] \in E^{n+1}, \quad R(x, w) = x_q^{n+1} + b(x_q). \end{gathered} \qquad (4.21)$$

To the extended state space there corresponds an extended impulse vector $\hat{p}_i = [p_i, p_i^{n+1}]$ for which p_i is definable by the formulas (1.4) and (1.7) and the last coordinate is

$$p_i^{n+1} = \frac{dR(x, w)}{dx_i^{n+1}} = p_{i+1}^{n+1} = \dots = \frac{\partial R(x, w)}{\partial x_q^{n+1}} = 1. \qquad (4.22)$$

The derivative of the function R with respect to $\hat{x}_i$ is given by

$$\hat{p}_i = \frac{dR}{d\hat{x}_i} = \begin{bmatrix} \dfrac{dR}{dx_i} \\ \dfrac{dR}{dx_i^{n+1}} \end{bmatrix} = \begin{bmatrix} C_x(z_i) + F_x(z_i)\, p_{i+1} \\ 1 \end{bmatrix}. \qquad (4.23)$$

The derivative of the function R with respect to u_i is given by

$$\frac{dR}{du_i} = C_u(z_i) + F_u(z_i)p_{i+1}$$

and therefore coincides with (1.8).

We state the problem of minimizing $R(x(w),w)$ with respect to the component u_i of the vector w. To this end, we write R as a composite function of the extended vector

$$R(x(w),\ w) = B(\hat{a}) = B(\hat{a}(u_i)) \ , \tag{4.24}$$

$$\begin{aligned} \hat{a}(u_i) &= [x_{i+1},\ x_{i+1}^{n+1}] = [a(u_i),\ x_{i+1}^{n+1}(u_i)] \\ &= [F(z_i),\ x_i^{n+1} + C(z_i)] \ . \end{aligned} \tag{4.25}$$

We assume that the functions defining the problem are differentiable in x. Then for each vector $u_i \in U$, (1.5), (1.7) and (4.22) define uniquely the sequence of extended impulses $\hat{p}_q, \hat{p}_{q-1}, \ldots, \hat{p}_{i+1}$. This enables us to write

$$\hat{p}_{i+1} = \hat{p}_{i+1}(\hat{a}) = \hat{p}_{i+1}(\hat{a}(u_i)) = [p_{i+1}(a(u_i)),\ 1].$$

Let us introduce some notation:

$$W_* = \operatorname{Arg}\min_{u_i \in U} B(\hat{a}(u_i)),$$

$$W_1(\bar{\hat{a}}) = \operatorname{Arg}\min_{\hat{a} \in \hat{\Omega}} \langle \hat{p}_{i+1}(\bar{\hat{a}}),\ \hat{a} - \bar{\hat{a}} \rangle,$$

$$W_2(\bar{u}_i) = \operatorname{Arg}\min_{u_i \in U} \langle \hat{p}_{i+1}(\hat{a}(\bar{u}_i)),\ \hat{a}(u_i) - \hat{a}(\bar{u}_i) \rangle.$$

Here $\hat{\Omega} = \hat{a}(U)$ is the image of the set U.

Further arguments and formulations of theorems repeat almost verbatim those given above. The condition that the set Ω be convex is replaced by the condition that $\hat{\Omega}$ be convex. In particular, the necessary condition of a minimum takes the following form in the control space (4.4): if $u_i^* = W_*$ then for any $u_i \in U$ we have

$$\langle p_{i+1}(F(x_i, u_i^*, t_i)), \; F(x_i, u_i, t_i) - F(x_i, u_i^*, t_i)\rangle + \\ + C(x_i, u_i, t_i) - C(x_i, u_i^*, t_i) \geqslant 0.$$

An analogous generalization can be given to the conditions (4.3) and (4.5).

6. ACCOUNTING FOR MIXED CONSTRAINTS

We assume that the vector w is fixed as before, with the exception of its ith component u_i which should be such that the control u_i belongs to U, the conditions (2.3) hold, and the function $R_1(x(w), w) = b_1(z_q)$ takes on the smallest possible value. We also assume that the set of solutions of this problem W_* is nonempty. We denote by W_* the complete control vector w for which one takes $u_i^* \in W_*$ as u_i; all the components of w coincide with the components of w_*, except for the ith component equal to u_i. We now introduce the auxiliary function R analogous to (1.6.25):

$$R(x, w) = \mu b_1(z_q) + \langle \tilde{u}, g(x, w)\rangle + \langle v, h(x, w)\rangle. \tag{4.26}$$

Here $\mu \in E_+^1$, $\tilde{u} \in E^e$, $v \in E_+^c$. Set

$$C(z_i) = \langle \tilde{u}_i, \Gamma^1(z_i)\rangle + \langle v_i, \Gamma^2(z_i)\rangle,$$
$$b(z_q) = \mu b_1(z_q) + \langle \tilde{u}_q, \Gamma^3(z_q)\rangle + \langle v_q, \Gamma^4(z_q)\rangle.$$

Then the function $R(x,w)$ coincides with the function (4.20), enabling us to use the formulas and notation (4.21)-(4.25).

From minimization in the control space we proceed to minimization in the state space. Thus the vector $\hat{x}_{i+1}$ has to be chosen from the set $\hat{\Omega} = \hat{a}(U)$ such that $R(x(w),w)$ takes on the smallest value and the conditions (2.3) are satisfied in which only these are essential:

$$\begin{gathered}\Gamma^1(a, u_{i+1}, t_{i+1}) = \Gamma^1(z_{i+2}) = \ldots = \Gamma^1(z_{q-1}) = 0,\\ \Gamma^3(z_q) = 0, \quad \Gamma^2(a, u_{i+1}, t_{i+1}) \leqslant 0,\\ \Gamma^2(z_{i+2}) \leqslant 0, \ \ldots, \ \Gamma^2(z_{q-1}) \leqslant 0, \quad \Gamma^4(z_q) \leqslant 0,\end{gathered}$$

where $x_{i+2}, x_{i+3}, \ldots, x_q$ are composite functions of the vector $\hat{a} = \hat{x}_{i+1}$. Nevertheless, we keep the form of writing the auxiliary function (4.26), noting, however, nonessential constraints too. We shall use the general Theorem 1.7.6. By the formulas (4.22) - (4.25) the condition (1.7.24) becomes

$$\begin{aligned}0 \leqslant \left\langle \frac{dR(x(w_*), w_*)}{d\hat{x}_{i+1}}, \ \hat{x}_{i+1}(u_i) - \hat{x}_{i+1}(u_i^*) \right\rangle = \\ = \langle \hat{p}_{i+1}(\hat{a}_*), \ \hat{a} - \hat{a}_* \rangle \quad \forall u_i \in U.\end{aligned}$$

Or, passing to the control space, we obtain

$$\begin{aligned}0 \leqslant \langle p_{i+1}(F(x_i, u_i^*, t_i)), \ F(x_i, u_i, t_i) - F(x_i, u_i^*, t_i)\rangle + \\ + \langle \tilde{u}_i, \ \Gamma^1(x_i, u_i, t_i) - \Gamma^1(x_i, u_i^*, t_i)\rangle + \\ + \langle v_i, \ \Gamma^2(x_i, u_i, t_i) - \Gamma^2(x_i, u_i^*, t_i)\rangle \quad \forall u_i \in U.\end{aligned} \tag{4.27}$$

Using the linearized principle of the minimum, we have

$$\begin{aligned}\langle F_u(x_i, u_i^*, t_i)\, p_{i+1}(F(x_i, u_i^*, t_i)) + \Gamma_u^1(x_i, u_i^*, t_i)\, \tilde{u}_i + \\ + \Gamma_u^2(x_i, u_i^*, t_i)\, v_i, \ u_i - u_i^* \rangle \geqslant 0 \quad \forall u_i \in U.\end{aligned}$$

Analogously, one can extend the remaining assertion of Theorems 6.4.1 and 6.4.2 to the case in question.

Theorem 1.7.6 can be reformulated as follows:

THEOREM 6.4.4. Let $\hat{\Omega}$ be a convex set whose interior is nonempty. The minimum of the composite function $b_1(z_q)$ with respect to $u_i \in U$, taking into account the constraints (2.3), is attained at the point $u_i^* = W_*$; the functions defining the problem are continuously differentiable with respect to the components of the vector x. Then there exist $\mu \in E_+^1$, $\tilde{u}_i, v_i \geq 0$, not all equal to zero, such that for any $u_i \in U$ the inequality (4.27) holds and the complementarity condition

$$v_i^s h^s(x_i, u_i^*, t_i) = 0 \tag{4.28}$$

is satisfied.

Noting the feasibility of the vector u_i^* and the condition (4.28), we write the inequality (4.27) in the form

$$0 \leqslant \langle p_{i+1}(F(x_i, u_i^*, t_i)),\ F(x_i, u_i, t_i) - \\ - F(x_i, u_i^*, t_i)\rangle + \langle \bar{u}_i,\ \Gamma^1(x_i, u_i, t_i)\rangle + \langle v_i, \Gamma^2(x_i, u_i, t_i)\rangle.$$

If the set U is open and the functions determining the problem are differentiable in u_i, then by Theorem 1.3.1, from (4.27) it follows that

$$F_u(x_i, u_i^*, t_i)\, p_{i+1}(F(x_i, u_i^*, t_i)) + \\ + \Gamma_u^1(x_i, u_i^*, t_i)\tilde{u}_i + \Gamma_u^2(x_i, u_i^*, t_i)\, v_i = 0.$$

If the problem is one of convex programming, then these conditions are simultaneously sufficient for the minimum also. ///

7. SOME GENERALIZATIONS

Throughout this subsection, all components of the vector w are fixed except for u_i, which essentially simplifies our discussion. All results are easily extendable to the case where the complete vector w changes. We make the notation as close as possible to to that introduced in Subsection 6.4.1. The feasibility condition of the control vector $w = [u_1, u_2, \ldots, u_q]$ is that all $u_i \in U$.

By (1.2), to each vector w there corresponds a complete state vector $a = [x_1, x_2, \ldots, x_q]$. We denote this dependence by $a = a(w)$ and assume that Ω is the union of all possible complete state vectors corresponding to all possible feasible controls. The function b to be minimized depends explicitly on x_q and implicitly on the other components of the vector $a \in \Omega$; hence we can use the earlier representation $b(x_q) = B(a)$, however, with a different meaning. If the vector function f is differentiable in x_i and the function b in x_q, then by the formulas of Section 6.1, the function $B(a)$ is differentiable in a and we have

$$B(a) - B(\bar{a}) = \left\langle \frac{dB(\bar{a})}{da},\ a - \bar{a} \right\rangle + \|a - \bar{a}\| \alpha(\bar{a}, a - \bar{a}) =$$

$$= \sum_{i=1}^{q} \langle p_i(\bar{a}),\ x_i - \bar{x}_i \rangle + \|a - \bar{a}\| \alpha(\bar{a}, a - \bar{a}).$$

Set $\Omega_* = \arg\min_{a \in \Omega} B(a)$. If Ω is convex and $\bar{a} \in \Omega_*$, then it is necessary that for all $a \in \Omega$

$$\left\langle \frac{dB(\bar{a})}{da},\ a - \bar{a} \right\rangle \geq 0 .$$

Analogously one can extend the results derived above to the complete vectors w and a.

It is not hard to show that the results of this section are also extendable to the Runge-Kutta integration of the system (1.1). In this case the formulas are more cumbersome and hence we omit them.

5. NUMERICAL METHODS BASED ON DISCRETE MINIMUM PRINCIPLES

Discrete minimum principles make it possible to construct several numerical methods for finding the points of the set W_* defined by (4.1). This is an auxiliary problem of implementing the methods described in Section 6.3 for solving discrete optimal control problems.

We use the notation introduced in Subsection 6.4.1. The problem of minimizing $b(z_q)$ with respect to $u_i \in U$ can be replaced by that of seeking the fixed points of the multivalued mappings $W_1(a)$, $W_2(u_i)$, $W_3(u_i)$, i.e., the points satisfying respectively condition (4.3), (4.4), (4.5). It is not hard to formulate requirements in terms of the initial problem that are sufficient to invoke Kakutani's theorem, providing sufficient conditions for the existence of fixed points (see Appendix III). For simplicity we will consider the point-set mapping $W_1(a)$.

LEMMA 6.5.1. Let the function $B(a)$ be continuously differentiable and convex on an open set containing the convex compact set Ω. Then the multivalued mapping $W_1(a)$ has a fixed point a_* at which (4.3) holds and the set of preimages of a_* lies in W_*.

Indeed, for each $a \in \Omega$ the set $W_1(a)$ is nonempty, convex, compact and contained in Ω, hence by Kakutani's theorem W_1 has a fixed point.

The vector p_{i+1} is the gradient of the function B with respect to x_{i+1}, hence the minimization (used in defining the set $W_1(a)$) of the scalar product $\langle p_{i+1}, x_{i+1} \rangle$ with respect to x_{i+1} for the fixed vector p_{i+1} is, in essence, a minimization of the linearized function B. This is reminiscent of the conditional gradient method described in Section 5.4.

We shall assume that B is defferentiable in x_{i+1} and the discrete minimum principle is satisfied for all $u_i^* \in W_*$. In the kth iteration for minimizing B with respect to u_i let the vector u_i^k and the corresponding vectors p_{i+1}^k, x_{i+1}^k be known. We shall describe several computational procedures for the conditional gradient method.

We determine the auxiliary vector $\tilde{x}_{i+1}$ by

$$\tilde{x}_{i+1} \in \operatorname{Arg} \min_{x_{i+1} \in \Omega} \langle p_{i+1}^k, x_{i+1} \rangle. \tag{5.1}$$

Set $d_k = \tilde{x}_{i+1} - x_{i+1}^k$, $\delta(x_{i+1}^k) = \langle p_{i+1}^k, d_k \rangle$. By (5.1), $\delta(x_{i+1}^k) \leq 0$. We assume that Ω is convex and closed.

We find the minimum of $B(x_{i+1})$ on the segment joining x_{i+1}^k and $\tilde{x}_{i+1}$:

$$\tau_k \in \operatorname{Arg} \min_{0 \leqslant \tau \leqslant 1} B(x_{i+1}^k + \tau d_k). \tag{5.2}$$

Next we determine the new vector

$$x_{i+1}^{k+1} = x_{i+1}^k + \tau_k d_k. \tag{5.3}$$

We find the corresponding gradient p_{i+1}^{k+1}, etc. The sequence $\{B(x_{i+1}^k)\}$ decreases monotonically. Indeed, if $\delta(x_{i+1}^k) < 0$ then we have

$$B(x_{i+1}^k + \tau d_k) = B(x_{i+1}^k) + \tau\delta(x_{i+1}^k) + O(\tau^2).$$

Hence at least for sufficiently small $\tau > 0$

$$B(x_{i+1}^k) > B(x_{i+1}^k + \tau d_k) \geqslant B(x_{i+1}^{k+1}).$$

If $\delta(x_{i+1}^k) = 0$ and $B(x_{i+1})$ is pseudoconvex, then for all $x_{i+1} \in \Omega$, $B(x_{i+1}^k) \leq B(x_{i+1})$. Thus the minimum of B on Ω is attained at x_{i+1}^k and the minimization process stops at this point.

On these lines, one needs to solve the problem (5.1) of minimizing a linear function on the set Ω and then to determine the control u_i ensuring transition of the system (1.2) from x_i to x_{i+1}^{k+1}. The structure of the set Ω may be very complex, which makes the solution of (5.1) difficult. It is simpler to pass from the minimization in the state space to minimization in the control space. Instead of (5.1) we seek

$$\tilde{u}_i \in \operatorname{Arg}\min_{u_i \in U} \langle p_{i+1}^k,\ x_{i+1}\rangle = \operatorname{Arg}\min_{u_i \in U} H(x_i,\ u_i,\ t_i,\ p_{i+1}^k). \tag{5.4}$$

Next we set $\tilde{x}_{i+1} = F(x_i, \tilde{u}_i, t_i)$. By (5.2) and (5.3) we find x_{i+1}^{k+1}, the τ_k and the control u_i^{k+1}. Both versions of the method generate the same sequence $\{x_{i+1}^k\}$ (under the condition of unique solvability of the minimization problems); but numerically, to implement the version in the control space is, as a rule, easier. Conversely, the state space is more convenient for proofs.

A third version of the method can be devised , using a linearized discrete minimum principle. We define the control u_i^{k+1} by

$$\tilde{u}_i \in \operatorname{Arg}\min_{u_i \in U} \left\langle \frac{\partial H(x_i, u_i, t_i, p_{i+1}^k)}{\partial u},\ u_i - u_i^k \right\rangle, \tag{5.5}$$

$$\tau_k \in \operatorname{Arg}\min_{0 \leqslant \tau \leqslant 1} B(a(u_i^k + \tau(\tilde{u}_i - u_i^k)), \tag{5.6}$$

$$u_i^{k+1} = u_i^k + \tau_k(\tilde{u}_i - u_i^k). \tag{5.7}$$

Obviously, in this case the values B also decrease monotonically on each iteration.

Using the results of Section 5.4, it is easy to formulate and prove the convergence of all these schemes. Hence we limit ourselves to a convergence theorem for the first version.

THEOREM 6.5.1. Let $B(a)$ be continuously differentiable and pseudoconvex (with respect to Ω) on an open set containing the convex, compact set Ω and let the gradient of $B(a)$ satisfy a Lipschitz condition on Ω. Then the set of limit points of the sequence $\{x_{i+1}^k\}$ is nonempty and consists of points $\bar{x}_{i+1}$ at which (4.3) holds, the corresponding controls being such that $\bar{u}_i \in W_*$.

If we simplify the conditional gradient method by dropping the line search and, in addition, always assuming that $\tau = 1$, we obtain the following methods:

$$x_{i+1}^{k+1} \in W_1(x_{i+1}^k) \,, \tag{5.8}$$

$$u_i^{k+1} \in W_2(u_i^k) \,, \tag{5.9}$$

$$u_i^{k+1} \in W_3(u_i^k) \,. \tag{5.10}$$

These modifications do not ensure that the function to be minimized decrease monotonically. Rather, one seeks the fixed points of the point-set mappings W_1, W_2 and W_3 respectively, using the simple iteration method. For such problems this method is frequently divergent.

If the discrete minimum principle holds for any i, then these methods can be used, changing the control simultaneously for all $i \in [1:q-1]$. Obviously, all the results extend to this case. Such schemes of minimizing R with respect to w make it possible to "decompose" the initial problem into many minimization problems with respect to vectors u_i of small dimension and, next, choose the steps τ_k. These schemes are alluring because the computations are simple; however, if the initial control is known only approximately, they converge slowly--in fact, the methods (5.8)-(5.10) frequently diverge. Hence, first the function R is minimized with respect to w by some unconstrained minimization method, and only then one switches to these procedures.

Applied to the system (1.1), the process (5.9) coincides with the method suggested by Krylov and Chernoous'ko [1]. To eliminate the frequently observed divergence, Chernoous'ko and Banichuk [1] suggest one should determine u_i from (5.4) and τ_k and u_i^{k+1} from (5.6) and (5.7). The convergence of this procedure, if the u_i obtained from (5.4) are different from those obtained according to the rule (5.5), needs special justification.

We note that the "decomposition" of the problem of minimizing the function R with respect to w can be solved without the discrete minimum principle. Indeed, we can perform component-by-component minimization of $b(z_q)$, solving sequentially the problems of minimizing $b(z_q)$ with respect to u_i for $i = 1,2,\ldots, q-1$. We apply, for example, the gradient-projection method in the state space and control space, thus obtaining the following numerical schemes:

$$x_{i+1}^{k+1} = \pi_\Omega (x_{i+1}^k - \tau_k p_{i+1}(x_{i+1}^k)),$$
$$u_i^{k+1} = \pi_U [u_i^k - \tau_k H_u (x_i, u_i^k, t_i, p_{i+1}(a(u_i^k)))].$$

Here $\pi_X(x)$ denotes the projection of x on the set X. Proof of convergence and properties of the method follow from the results of Section 5.5.

6. SOME GENERALIZATIONS

The foregoing approach can be applied to more general problems than described. Let us delineate basic possible generalizations, invoking, for the sake of simplicity, discrete approximations of the system (1.1) using the Euler scheme.

1. OPTIMIZATION OF CONTROL PARAMETERS

The statement of this problem for systems described by ordinary differential equations with control parameters is given in Section 1.8. We use discrete approximation of the system (1.8.10):

$$x_{i+1}=x_i+h_i f(x_i, u_i, t_i, \xi)=F(x_i, u_i, t_i, \xi). \tag{6.1}$$

The function

$$R(x(w),w,\xi) = b(w_q,u_q,t_q,\xi) + \sum_{i=1}^{q-1} h_i B(x_i,u_i,t_i,\xi)$$

is the control performance criterion. One must choose the complete control vector w and the control parameter vector ξ so that the function R has the smallest possible value. Assuming that all the functions determining the problem are ξ-differentiable and using the arguments of Section 6.1, we obtain the following formula for computing the derivatives of R with respect to ξ:

$$\frac{dR}{d\xi}=b_\xi(z_q, \xi)+\sum_{i=1}^{q-1} H_\xi(z_i, p_{i+1}, \xi), \tag{6.2}$$

$$z_i=[x_i, u_i, t_i];$$

where

$$H(z_i, p_{i+1}, \xi)=h_i B(z_i, \xi)+\langle F(z_i, \xi), p_{i+1}\rangle.$$

A necessary condition of a minimum is given by

$$b_\xi(z_q, \xi) + \sum_{i=1}^{q-1} H_\xi(z_i, p_{i+1}, \xi) = 0 ,$$

which can also be obtained directly from (1.8.11).

The formula (6.2) makes it possible to calculate the gradient of R with respect to ξ and apply, in turn, the approach used earlier in determining the vector w. Let us combine the control vector and the control parameter vector, setting $\tilde{w} = [w, \xi]$. Then we minimize R with respect to the extended vector $\tilde{w}$. Therefore, there is no fundamental distinction between the optimization with respect to the complete control vector and that of the control parameter vector. At the same time, the "incommensurate scaling" of w and ξ sometimes aggravates the numerical implementation of the process. These vectors differ in their meaning and the extent of their effect on R. Hence it is essential to make a special scaling of the vector ξ. In many cases it is also useful to decompose the process of minimizing R with respect to w and ξ. First R is minimized with respect ξ and next with respect to the complete control vector $\tilde{w}$.

The control parameter vector can be introduced into the system artificially. For instance, when the interval of motion T is large, or high accuracy for numerical integration of the system (1.1) is required, there is a need for a small integration step, which leads to a high dimension of the vector w. In such cases the control is sought in the "feedback" (or synthesis) form:

$$u(t) = \gamma_1(\xi, t) , \quad u(t) = \gamma_2(x(t), \xi, t) .$$

The functions γ_1 and γ_2 depend on t as well as on the control parameter vector. Substituting these formulas into the right side of the system (6.1), we reduce the initial problem to an optimization problem with respect to the control parameter vector. In passing to discrete approximation, this enables us to lower the dimension of the auxiliary problems of unconstrained minimization.

2. A PROBLEM WITH INCOMPLETE INITIAL CONDITIONS

For the system (1.2), some coordinates of the initial vector x_1 need not be specified. To define (1.2) more precisely, we introduce an additional fictitious step $x_1 = h_0\xi$; using the formula (1.4), we obtain

$$\frac{dR}{d\xi} = \frac{dx_1}{d\xi}\frac{dR}{dx_1} = h_0 p_1.$$

Once the gradient is found, we can carry out the computations as usual, changing the missing components of the vector x_1 and improving the complete control vector w.

This formula can be used to solve boundary value problems by Newton's method. Indeed, let the complete control vector w be fixed. Each of the vectors x_1 and x_q consists of two vectors: $x_1 = [\bar{x}_1, \tilde{x}_1]$, $x_1 = [\bar{x}_q, \tilde{x}_q]$, where $\bar{x}_1, \bar{x}_q \in E^\nu$, $\tilde{x}_1, \tilde{x}_q \in E^{n-\nu}$, and the vectors $\tilde{x}_1$ and $\bar{x}_q$ are given but $\bar{x}_1$ and $\tilde{x}_q$ are unknown. Given an arbitrary vector $\bar{x}_1$ we integrate (1.1) by some scheme and determine the corresponding vector $\hat{\bar{x}}_q$. Thus, we have the functional dependence $\hat{\bar{x}}_q = (\bar{x}_1)$, where ϕ is a mapping from E^ν in E^ν. The problem consists in finding a vector $\bar{x}_1^*$ such that

$$\phi(\bar{x}_1^*) = \bar{x}_q \quad .$$

Newton's method yields the following computation scheme:

$$\bar{x}_1^{k+1} = \bar{x}_1^k - A_k^{-1}(\phi(\bar{x}_1^k) - \bar{x}_q) . \tag{6.3}$$

Here A_k is the $\nu \times \nu$ matrix of the first derivative of the vector function $\phi(\bar{x}_1^k)$ whose j^{th} row consists of the components of the vector

$$\frac{d\varphi^j(\bar{x}_1{}^k)}{d\bar{x}_1} = \left[\frac{d\varphi^j(\bar{x}_1{}^k)}{d\bar{x}_1^1}, \quad \frac{d\varphi^j(\bar{x}_1{}^k)}{d\bar{x}_1^2}, \quad \ldots, \quad \frac{d\varphi^j(\bar{x}_1{}^k)}{d\bar{x}_1^\nu}\right].$$

One can compute the coefficients of A_k, using the results obtained in Section 6.1. To do this, we take the jth components of $\hat{\bar{x}}_q^{jk}$ as R. Then, by virtue of (1.5), the impulse vector p_q will have all components equal to zero except the jth component equal to one. We determine the vector p_1 from the recurrence relation (1.7). We combine its first ν components corresponding to $\bar{x}_1$ and write $\bar{p}_1 \in E^\nu$. From the results of Section 6.1 it follows that

$$\bar{p}_1 = \frac{d\phi^j(\bar{x}_1^k)}{d\bar{x}_1} .$$

After the computations for $j = 1,2,\ldots,\nu$, we compute all the elements of A_k. To this end, the recurrence relation (1.7) has to be calculated ν times, upon which we can make one step of Newton's method (6.3). In the numerical implementation of this process, certain complications associated with divergence may develop. This occurs if the initial approximation for $\bar{x}_1$ is known to be bad. Hence, at first a gradient scheme is used; only after that is Newton's method used.

3. A VARIABLE TIME PROBLEM

For the system (1.2) it is required to find the complete control vector w and the interval of motion $[0, t_q]$ so that a function R of the form (1.3) at the final time T have the smallest possible value.

We set and fix the sequence of integration steps $h_1, h_2, \ldots, h_{q-1}$. The interval of motion is changed with the aid of the scale factor ξ. Just as in Section 1.8, by introducing an additional state variable we convert the system into an autonomous one and substitute the independent variable $t = \tau\xi$. The variable τ varies in the interval $[0, T_0]$, where $T_0 = \sum_{i=1}^{q-1} h_i$; the variable t varies in the interval $[0, T]$, where $T = T_0\xi$. A discrete approximation of the system (1.8.18) by the Euler scheme yields

$$\begin{aligned} x_{i+1} &= x_i + \xi h_i f(x_i, u_i, x_i^{n+1}), \\ x_{i+1}^{n+1} &= x_i^{n+1} + \xi h_i, \quad x_1^{n+1} = 0. \end{aligned} \tag{6.4}$$

The function (1.3) to be minimized is given by

$$R = b(x_q, u_q, x_q^{n+1}) + \xi \sum_{i=1}^{q-1} h_i B_{\bullet}(x_i, u_i, x_i^{n+1}). \tag{6.5}$$

For the system (6.4) in standard form (6.1) one seeks the complete control vector w and the control parameter ξ such that the function R assumes the smallest possible value. Applying the formula (6.2) to the system (6.4), we obtain

$$\begin{aligned} \frac{dR}{d\xi} &= \sum_{i=1}^{q-1} h_i [B(x_i, u_i, x_i^{n+1}) + \langle f(x_i, u_i, x_i^{n+1}), p_{i+1}\rangle + p_{i+1}^{n+1}], \\ p_i &= p_{i+1} + \xi h_i [B_x(x_i, u_i, x_i^{n+1}) + f_x(x_i, u_i, x_i^{n+1}) p_{i+1}], \\ p_i^{n+1} &= p_{i+1}^{n+1} + \xi h_i [B_t(x_i, u_i, x_i^{n+1}) + \langle f_t(x_i, u_i, x_i^{n+1}), p_{i+1}\rangle]. \end{aligned} \tag{6.6}$$

The conditions at the right endpoint are:

$$p_q = \frac{\partial b(x_q, u_q, t_q)}{\partial x_q}, \quad p_q^{n+1} = \frac{\partial b(x_q, u_q, t_q)}{\partial t_q}$$

implying

$$p_i^{n+1} = \frac{\partial b(x_q, u_q, t_q)}{\partial t_q} + \xi \sum_{s=i}^{q-1} h_s [B_t(x_s, u_s, x_s^{n+1}) + \\ + \langle f_t(x_s, u_s, x_s^{n+1}), p_{s+1}\rangle], \quad 1 \leqslant i \leqslant q-1.$$

The formula (6.6) is simplified if the initial system (1.1) is autonomous and the function B does not depend explicitly on t:

$$\frac{dR}{d\xi} = \sum_{i=1}^{q-1} h_i [B(x_i, u_i) + \langle f(x_i, u_i), p_{i+1}\rangle] + \\ + T_0 b_{t_q}(x_q, u_q, t_q).$$

The connection of this result with (1.8.20) is obvious. These formulas make it possible to construct gradient methods for the minimization with respect to the parameter ξ, finding thereby the optimal size of the interval of motion.

4. A MINIMAL TIME PROBLEM

It is required to find the complete control vector w and the corresponding solution of the system (1.2) so that the conditions (2.3) be satisfied and the interval of motion be smallest.

We use the representation (6.4) but, in contrast to (6.5), the function to be minimized $R_1 = x_q^{n+1}$. Following (2.5), we have the Lagrangian

$$R = x_q^{n+1} + \langle \tilde{u}, g(x, w)\rangle + \langle v, h(x, w)\rangle.$$

The condition (6.2) for this problem is

$$\frac{dR}{d\xi} = \sum_{i=1}^{q-1} h_i [\langle f(x_i, u_i, t_i), p_{i+1}\rangle + p_{i+1}^{n+1}].$$

The conditions at the right endpoint are:

$$p_q = g_x(x_q, u_q)\tilde{u}_q + h_x(x_q, u_q)v_q, \quad p_q^{n+1} = 1.$$

For an autonomous system, in particular, we obtain

$$\frac{dR}{d\xi} = \sum_{i=1}^{q-1} h_i[1 + \langle f(x_i, u_i), p_{i+1}\rangle].$$

5. ACCOUNTING FOR NONDIFFERENTIABILITY OF COST FUNCTIONALS

Among practical problems, it is often possible that the cost functionals are not differentiable, for example:

$$R_1 = \max_{0 \leqslant t \leqslant T} \varphi(x(t), u(t), t), \quad R_2 = \int_0^T |\varphi(x(t), u(t), t)|\,dt.$$

In spite of the fact that ϕ is a differentiable function of its arguments, the functionals R_1 and R_2 are differentiable only directionally. For R_1 we introduce an additional control parameter ξ with respect to which we make minimization, and also a new constraint, setting

$$R_1 = \xi, \quad \phi(x(t), u(t), t) \leqslant \xi .$$

For R_2 we introduce a new control $\bar{u}(t)$ and two additional constraints, setting

$$R_2 = \int_0^T \bar{u}(t)dt, \quad -\phi(x(t), u(t), t) \geqslant \phi(x(t), u(t), t) .$$

The minimization is with respect to $u(t)$ and $\bar{u}(t)$. In both cases, the problems are reduced to standard form. The nondifferentiability of the functionals is removed and all the methods can be used without any additional modifications. One can treat the simplest minimax problems (see, e.g., Davydov [1]) in a similar way.

It makes sense to use this transformation if all the functions determining the problem, except R, are differentiable with respect to the components of the vectors x and u. Otherwise, there is no need to get rid of the nondifferentiable cost functions, and the computations need to follow some less efficient technique avoiding differentiation.

6. ACCOUNTING FOR BOX CONSTRAINTS

Many optimal control problems involve constraints of the form

$$a_i \leqslant u_i \leqslant b_i \ . \qquad (6.7)$$

If solution methods are used which are based on auxiliary procedures of unconstrained minimization, it is not advisable to treat the constraints (6.7) as general-type constraints (2.3). The former are simpler to account for in solving unconstrained minimization problems. Indeed, one can show that most of the methods given in Section 6.3 remain effective if instead of the auxiliary unconstrained minimization of the function R with respect to w, one solves the problem of minimizing R with respect to w on the set (6.7). Hence the library of unconstrained minimization problems has to contain two kinds of programs: one that accounts for constraints of the form (6.7) and one that does not.

7. ACCOUNTING FOR CONTROL "CONTINUITY"

In some optimal control problems, an additional constraint is imposed on the control variation rate:

$$\left|\frac{du(t)}{dt}\right| \leqslant c \ . \qquad (6.8)$$

If t_i and t_{i+1} are neighboring points in the t-integration grid, then in the discrete version this constraint becomes

$$-ch_i \leqslant u(t_{i+1}) - u(t_i) \leqslant ch_i \ .$$

These two constraints can be considered as constraints of the type $\Gamma^2 \leqslant 0$. In using the formulas of Section 6.1 for computing the derivatives, one needs to take into account the additional summands, since the constraints on the ith integration step depend on both u_i and u_{i+1}. We note that the constraints (6.8) can be introduced artificially for regularization.

8. PROBLEMS WITH DISCONTINUOUS RIGHT SIDES

If in the system (1.1) the vector function $f(x,u,t)$ is differentiable in x and u everywhere at a finite number of points $\{t_j\}$, where there is a discontinuity of the first kind, then the sizes of steps h_i need to be such that all the points $\{t_j\}$ are nodes of the principal grid in t. The formulas given in Section 6.1 do not change. Similarly, if at some points t_i of the principal grid in t the state trajectory has a preset discontinuity not depending on x, u, then the formulas of Section 6.1 for computing the derivatives do not change either. If the discontinuity depends on x, u, i.e.,

$$x(t_+) = (x(t_{i-}), u(t_i)) \ ,$$

where

$$x(t_{i+}) = \lim_{t \to t_i + 0} x(t), \quad x(t_{i-}) = \lim_{t \to t_i - 0} x(t) \ ,$$

then the impulses need to be recalculated at these points. In determining p_{i-1}, instead of p_i we need to take the vector

$\psi_x(x(t_{i-}), u(t_i))p_i$ and in the formula for computing the derivative dR/du_i add the summand $\psi_u(x(t_{i-}), u(t_i))p_i$. It is not hard to account for discontinuities of more complex form (see, e.g., Velichenko [2] and Chentsov [1]).

9. ACCOUNTING FOR DELAY

Let the control process be described by the relations

$$x_{i+1}=F(\bar{z}_i),\quad \bar{z}_i=[z_i,\ x_{i-s},\ u_{i-s}],\quad z_i=[x_i,\ u_i,\ t_i],$$
$$R=b(z_q)+\sum_{i=1}^{q=1} h_i B(\bar{z}_i),$$

where s is a positive integer and the $x_{1-s}, x_{2-s}, \ldots, x_1$ are given and fixed; the complete control vector

$$w = [u_{1-s},\ u_{2-s},\ \ldots,\ u_0,\ u_1,\ \ldots,\ u_q]$$

is sought. Following the arguments of Section 6.1, we obtain that for $i = 1,2,\ldots,q-1$

$$p_i=\frac{dR}{dx_i}=\frac{\partial R}{\partial x_i}+F_x(\bar{z}_i)\,p_{i+1}+\tilde{F}_x(\bar{z}_{i+s})\,p_{i+1+s}\theta(i+1+s,\ q),$$

where $\theta(a,c)=1$ if $a \le c$ and $\theta(a,c)=0$ otherwise. Moreover,

$$F_x(\bar{z}_i)=\frac{\partial}{\partial x_i}F(x_i,\ u_i,\ t_i,\ x_{i-s},\ u_{i-s}),$$
$$\tilde{F}_x(\bar{z}_{i+s})=\frac{\partial}{\partial x_i}F(x_{i+s},\ u_{i+s},\ t_{i+s},\ x_i,\ u_i).$$

Setting $p_q=\partial R/\partial x_q$, we determine $p_{q-1}, p_{q-2}, \ldots, p_1$. Next we compute the required derivatives of R with respect to u_i:

$$\frac{dR}{du_i}=\frac{\partial R}{\partial u_i}+F_u(\bar{z}_i)\,p_{i+1}\theta(1,\ i)+$$
$$+\tilde{F}_u(\bar{z}_{i+s})\,p_{i+1+s}\,\theta(i+1+s,\ q).$$

Here $i \in [1-s;q-1]$, with

$$F_u(\bar{z}_i) = \frac{\partial}{\partial u_i} F(x_i, u_i, t_i, x_{i-s}, u_{i-s}),$$
$$\tilde{F}_u(\bar{z}_{i+s}) = \frac{\partial}{\partial u_i} F(x_{i+s}, u_{i+s}, t_{i+s}, x_i, u_i).$$

Likewise, one can take into account more complex forms of delay. In all these cases we can compute the derivatives of the function R, reducing thus the problem to a form amenable to nonlinear programming methods.

10. A SPECIAL TECHNIQUE

Some problems involve contraints

$$\phi(x(t), t) \leqslant u(t) \leqslant \psi(x(t), t) , \tag{6.9}$$

where ϕ and ψ are r-dimensional vector functions. Such constraints are a special case of (2.3) and may be accounted for by the methods described earlier. However, it is simpler to introduce a new control $v \in E^r$, setting

$$u^i(t) = v^i(t)\phi^i(x(t),t) + (1 - v^i(t))\psi^i(x(t),t) , \tag{6.10}$$

where $i \in [1:r]$. We impose the constraints

$$0 \leqslant v^i(t) \leqslant 1 . \tag{6.11}$$

We insert (6.10) for $u(t)$ in the right sides of (1.1) and replace the initial problem by an optimization problem with respect to $v(t)$ under the simple constraints (6.11). The conditions (6.9) are thereby guaranteed. This simple technique makes it possible to account for (6.9) without introducing additional state constraints.

7. EXAMPLES OF NUMERICAL COMPUTATIONS

Many of the methods described above were tried out for solving a variety of test and practical problems. We shall list some results obtained on a BESM-6 computer. In order to follow the convergence of the process closely, the same conjugate-gradient method was used to solve auxiliary unconstrained minimization problems.

1. THE SIMPLEST MINIMAL TIME PROBLEM (ZERMELO'S PROBLEM)

The statement of this problem is given, for example, in the article of Powers and Shich [1]. The control process is described by the following system of differential equations:

$$\frac{dx^1}{dt} = \cos x^3, \quad \frac{dx^2}{dt} = \sin x^3, \quad \frac{dx^3}{dt} = u,$$
$$x^1(0) = x^2(0) = x^3(0) = 0.$$

The constraint $|u(t)| \leqslant 0.5$ and two terminal constraints $x^1(T) = 4$, $x^2(T) = 3$ are imposed. The functional is the time of motion, i.e., $R = T$. For definiteness, time is measured in seconds.

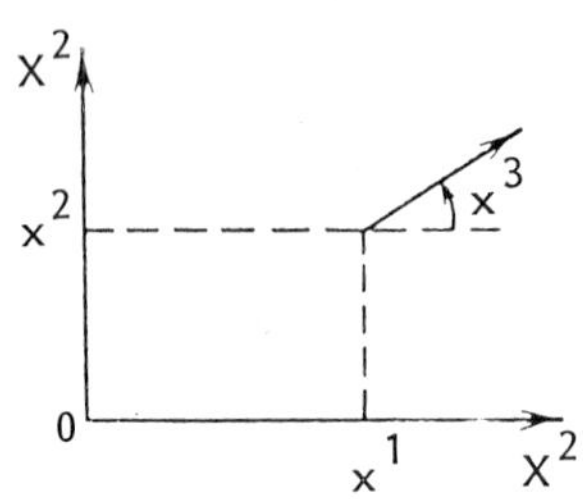

Figure 3

The problem has an obvious physical interpretation. Let x^1, x^2 be the Cartesian coordinates of a material point moving in the plane. Its velocity vector is equal to one in modulus and forms an angle x^3 with the positive direction of the axis OX^1 (Figure 3). At the initial moment $t = 0$ the velocity vector is directed along the axis OX^1, i.e., $x^3(0) = 0$. It is required to move the point from the origin to the point $[4,3]$ in the shortest time. If by a material point we mean, say, a car, then we can regard the turn angle of the drive wheels as a control, and the constraint $|u(t)| \leqslant 0.5$ has thus a simple physical meaning.

Proceeding to numerical solution of the problem, we make a change of independent variable, setting $t = \tau T$. Then the system takes the form

$$\frac{dx^1}{d\tau} = T\cos(x^3),\quad \frac{dx^2}{d\tau} = T\sin(x^3),\quad \frac{dx^3}{d\tau} = Tu.$$

It is integrated for $0 \leqslant \tau \leqslant 1$ by the Euler scheme with constant step $h = 0.02$, which corresponds to the number of discretization points $q = 51$. The total time of motion T becomes a control parameter.

The initial approximation is $u_0(\tau) \equiv 0$, $T_0 = 3$ sec. The constraint $|u(t)| \leqslant 0.5$ is treated as a "box constraint": $-0.5 \leqslant u(t) \leqslant 0.5$ (see Section 6.6) and is accounted for in the conjugate-gradient method. All computations are carried out in the dialogue (interactive) mode with the aid of the DISO system.

Solution of the problem starts with the first version of the cost-function parametrization method OPT45. The lower estimate of the optimal value of the functional in this problem is easily obtainable. The shortest distance between the points [0,0] and [4,3] is 5, and is covered in 5 sec. at unit velocity. Since the initial direction of the velocity vector coincides with the axis OX^1, the trajectory and motion time will be greater than these values. The lower estimate of the optimal value of the functional $T_0 = 3$ sec. satisfies the conditions for OPT45. After the first two iterations, the computation continues by OPT53 (the modified Lagrangian method).

In Table 1 the values of the functional are shown in the respective iterations of each method. When the computation ends, the results are: $R = T_* \approx 5.12277$ sec., $x^1(T_*) = 3.99997$, $x^2(T_*) = 2.99999$, which one may regard optimal with sufficiently high degree of accuracy. The corresponding control and trajectory are shown in Figures 4 and 5.

Table 1

Method	OPT45		
Number of Iterations	0	1	2
Value of the Functional	3.00000	3.05383	4.51062

Method	OPT53				
Number of Iterations	1	2	3	4	5
Value of the Functional	5.12532	5.12416	5.12297	5.12284	5.12277

The problem was also solved by another combination of algorithms: first by OPT41 (2 iterations) and next by OPT53 (5 iterations). The results are given in Table 2.

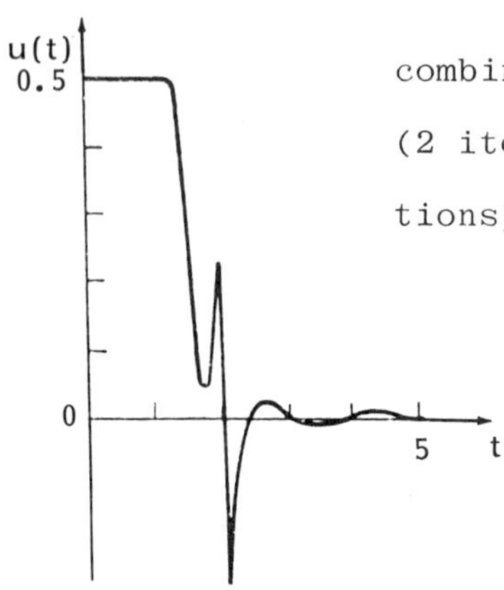

Figure 4

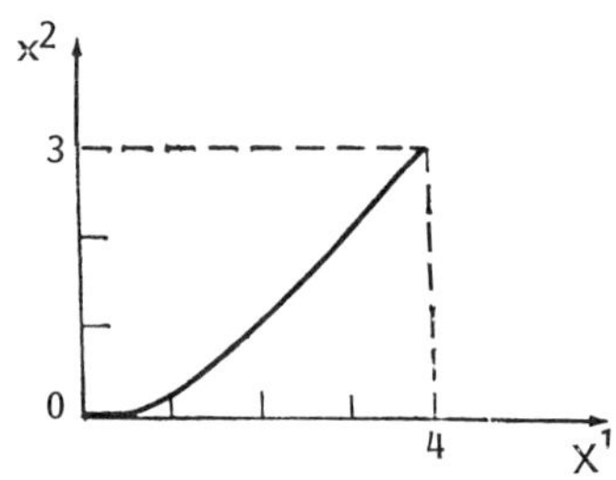

Figure 5

Table 2

Method	OPT41		
Number of Iterations	0	1	2
Value of the Functional	3.00000	4.61421	4.78960

Method	OPT53				
Number of Iterations	1	2	3	4	5
Value of the Functional	5.12230	5.12339	5.12277	5.12266	5.12265

The final values of the functional and the state coordinates were close to those given above: $R = T_* \approx 5.12265$ sec., $x^1(T_*) = 3.99998$, $x^2(T_*) = 2.99999$. The corresponding control is shown in Figure 6; the explicit "switching" form should be noted: at first, an abrupt turn in the direction of the terminal point [4,3] with the greatest possible angular velocity and next, motion on the line. The switching occurs at the moment when the velocity vector is first directed toward the required point [4,3]. Although the end results in both versions of the problem are very close, the graphs of the controls differ after the switching. The method OPT45 introduced these "defects" in the graph in Figure 4. Subsequent computations by OPT53 bridged the gap between the "peaks" only slightly but did not

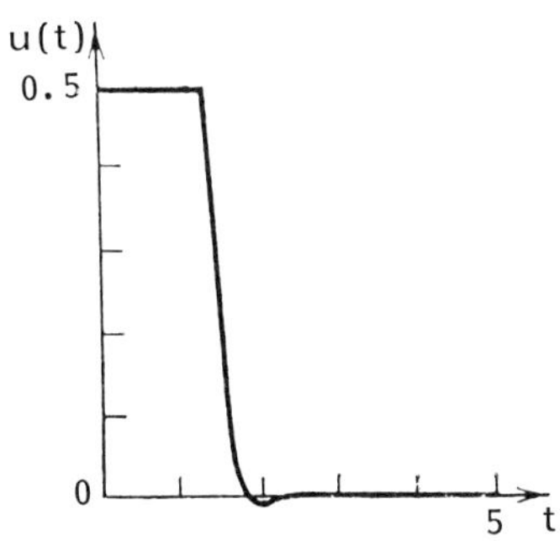

Figure 6

eliminate it. The control shown in Figure 6 seems to be more physically correct than others. One may draw two conclusions: first, the well-known fact that convergence of the functional does not imply at all convergence of controls; second, differing algorithms are useful since it then becomes possible to analyze constructively and choose the easiest implementable controls.

2. A SIMPLE PROBLEM WITH STATE CONSTRAINTS

The statement of the problem has been taken from the article of Mehra and Davis [1]. The numerical results are given therein. The control process is described by the equations

$$\frac{dx^1}{dt} = x^2 , \quad \frac{dx^2}{dt} = u - x^2 , \qquad\qquad (7.1)$$
$$x^1(0) = 0 , \quad x^2(0) = -1 , \quad 0 \leqslant t \leqslant 1 .$$

The problem consists in minimizing

$$R = \int_0^1 [(x^1)^2 + (x^2)^2 + 0.005u^2]dt$$

under the constraints along the trajectory

$$\Gamma(x,t) = x^2(t) - 8(t-0.5)^2 + 0.5 \leqslant 0 .$$

The system (7.1) was integrated by the Euler scheme with recalculation with constant step $h = 0.02$ $(q = 51)$. The control was assumed constant within the integration interval. The initial approximation was $u_0(t) \equiv 0$. The approximation problem of nonlinear programming consists in minimizing the function of 50 variables under 50 inequality constraints: $\Gamma_i = \Gamma(x_i,t_i) \leqslant 0$.

The strategy of seeking the solution is similar to the one used in the previous problem, i.e., after the first three iterations by the penalty-function method OPT413, the computation continues by the first version of the simple iteration method OPT533. In Table 3 one can see how the functional R changes depending on the number of the iteration. The optimal trajectory $x^2(x)$ and the corresponding control law U(t) are represented in Figures 7 and 8. In Figure 7 the lower curve is the trajectory $x^2(t)$ under the initial control $u_0(t)$.

Table 3

Method	OPT413			
Number of Iterations	0	1	2	3
Value of the Functional	0.600368	0.163748	0.166120	0.167267

Method	OPT533				
Number of Iterations	1	2	3	4	5
Value of the Functional	0.169152	0.169468	0.169487	0.169480	0.169480

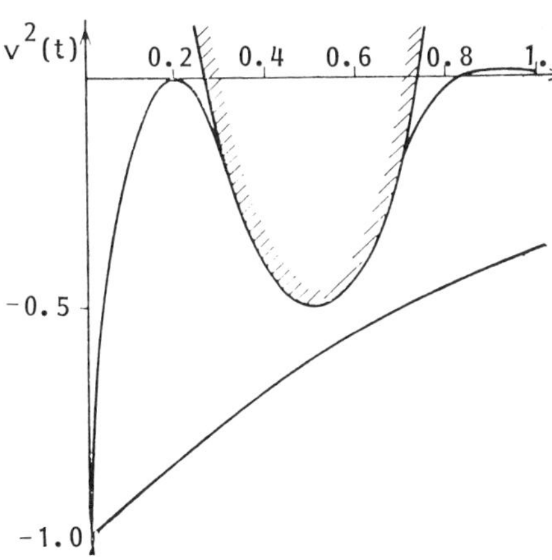

Figure 7

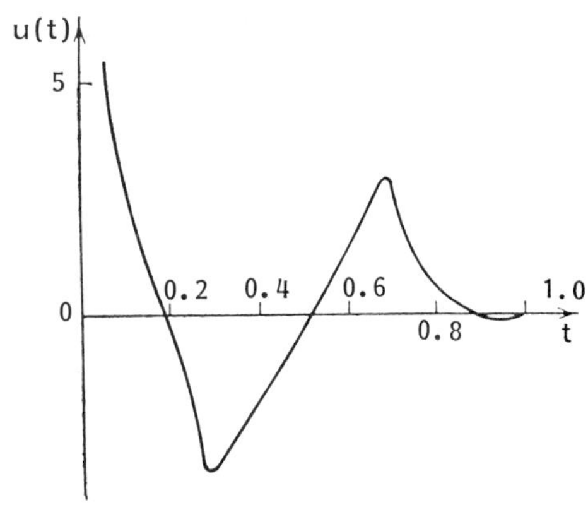

Figure 8

The "jump-on" time ($t_1 = 0.3$, $x^2(t_1) = -0.179077$) and "jump-off" time ($t_1 = 0.7$, $x^2(t_2) = -0.179997$) with respect to the state constraint yield oscillations (break points) on the graph of the optimal control law $u(t)$. In this solution of the problem, some constraints are not satisfied, the maximum violation being $\max_i \Gamma_i = 3 \cdot 10^{-4}$.

This problem was solved using also the programs OPT41, OPT53 and OPT8 in which the system (7.1) was integrated by the Euler scheme. In the first series of computations, the first three iterations were made by OPT41 and next eight iterations by OPT53, obtaining: $R_* = 0.181378$ and $\max_i \Gamma_i = 3 \cdot 10^{-5}$. In the second series of computations, one iteration first by OPT41 yielded approximate values of the dual variables, obtaining: $R_* = 0.187064$, $\max_i \Gamma_i = 3.3 \cdot 10^{-2}$, upon which the program OPT8 was used via Newton's method and in eight iterations the results showed high accuracy of the constraints: $\max_i \Gamma_i = 1.4 \cdot 10^{-12}$, $R_* = 0.181378$.

The controls differ from those in Figure 8 by a quantity of order 10^{-5} and it is impossible to observe this difference graphically. That is why we itemize in Table 4 the values of optimal control and of the constraints which have been computed by the formula (3.7). Note that in this problem we again encounter the situation mentioned in Section 6.3 concerning the program OPT8, when the function does not depend on the controls explicitly. Using the program OPT8 with the number of integration steps $q = 70$, we obtain $R_* = 0.17790$. Thus the value of R_* approaches the values upon integration of the system (7.1) by the Euler scheme with recalculation. The most exact solution can be found by Newton's

Table 4

i	u_i	Γ_i
1	1.300398_{10+01}	-2.1_{10+00}
2	9.756003_{10+00}	-1.7_{10+00}
3	7.299726_{10+00}	-1.4_{10+00}
4	5.434525_{10+00}	-1.1_{10+00}
5	4.008026_{10+00}	-9.3_{10+01}
6	2.903673_{10+00}	-7.5_{10-01}
7	2.031207_{10+00}	-5.9_{10-01}
8	1.319297_{10+00}	-4.5_{10-01}
9	7.097200_{10-01}	-3.3_{10-01}
10	1.526038_{10-01}	-2.2_{10-01}
11	-3.976422_{10-01}	-1.4_{10-01}
12	-9.860504_{10-01}	-7.3_{10-02}
13	-1.660771_{10+00}	-2.5_{10-02}
14	-2.477003_{10+00}	-2.1_{10-08}
15	-3.472799_{10+00}	-4.5_{10-13}
16	-3.220000_{10+00}	-2.3_{10-13}
17	-2.960800_{10+00}	0.0_{10+00}
18	-2.695200_{10+00}	0.0_{10+00}
19	-2.423200_{10+00}	9.1_{10-13}
20	-2.144800_{10+00}	9.1_{10-13}
21	-1.860000_{10+00}	9.1_{10-13}
22	-1.568800_{10+00}	9.1_{10-13}
23	-1.271200_{10+00}	9.1_{10-13}
24	-9.672000_{10-01}	9.1_{10-13}
25	-6.568000_{10-01}	1.4_{10-12}
26	-3.400000_{10-01}	9.1_{10-13}
27	-1.680000_{10-02}	9.1_{10-13}
28	3.128000_{10-01}	9.1_{10-13}
29	6.488000_{10-01}	9.1_{10-13}
30	9.912000_{10-01}	9.1_{10-13}
31	1.340000_{10+00}	9.1_{10-13}
32	1.695200_{10+00}	0.0_{10+00}
33	2.056800_{10+00}	9.1_{10-13}
34	2.424800_{10+00}	6.8_{10-13}
35	2.799200_{10+00}	1.1_{10-12}
36	2.636892_{10+00}	-1.1_{10-02}
37	1.937987_{10+00}	-4.3_{10-02}
38	1.411949_{10+00}	-9.3_{10-02}
39	1.016134_{10+00}	-1.6_{10-01}
40	7.185329_{10-01}	-2.4_{10-01}
41	4.951625_{10-01}	-3.3_{10-01}
42	3.281059_{10-01}	-4.2_{10-01}
43	2.040498_{10-01}	-5.3_{10-01}
44	1.131983_{10-01}	-6.5_{10-01}
45	4.847278_{10-02}	-7.7_{10-01}
46	4.934338_{10-03}	-9.0_{10-01}
47	-2.061928_{10-02}	-1.0_{10+00}
48	-2.991500_{10-02}	-1.2_{10+00}
49	-2.334529_{10-02}	-1.3_{10+00}
50	0.000000_{10+00}	---

method, but we have not yet obtained such results. What we have obtained so far attests, however, to the potential superiority of this method. This problem might be the touchstone of the efficiency of the numerical methods, and was used as such by Mehra and Davis [1] and by Fuji, Fujimoto, and Ono [1].

3. THE PROBLEM OF VERTICAL ASCENT OF A ROCKET

Without exaggeration, one may call this problem the classical test problem of optimal control. The statement and solution are given in Okhotsimskij [1], Ehneev [1], and Fedorenko [1]. The control process is described by the following differential equations:

$$\frac{dx^1}{dt} = -u, \quad \frac{dx^2}{dt} = x^3, \quad \frac{dx^3}{dt} = [Vu - Q(x)]\frac{1}{x^1} - g,$$
$$0 \leqslant t \leqslant T,$$

where $x^1(t)$ is the mass of the rocket; $x^2(t)$ is the altitude above the earth's surface (km); $x^3(t)$ is the rocket velocity (km/sec); $u(t)$ is the mass flow rate (sec^{-1}); $V = 2$ km/sec is the gas nozzle velocity; $q = 0.01$ km/sec^2 is the acceleration due to gravity (assumed constant); $Q(x)$ is the aerodynamic drag defined by the formula $Q(x) = 0.05 \exp(0.01x^2)(x^3)^2$.

The initial state of the rocket is: $x^1(0) = 1$, $x^2(0) = 0$, $x^3(0) = 0$; at the terminal time T = 100 sec., the final value of the mass should be 20% of the initial mass, i.e., $x^1(T) = 0.2$. We have thus the box constraint $0 \leqslant u(t) \leqslant 0.04$. It it required to find the mass flow rate at the rocket's peak altitude. According to our formulation of the general optimal control problem, the functional R is $R = -k_1 x^2(T)$ and the terminal equality-type constraint is $\Gamma = k_2[x^1(T) - 0.2] = 0$. The scale factors are $k_1 = 0.01$ and $k_2 = 10$.

To solve this problem numerically, the segment $[0,T]$ is divided into 100 equal parts ($q = 101$) and the system of differential equations is integrated by the Euler scheme with recalculation. The initial control is given by the function $u_0(t) \equiv 0.008$. The DISO system for solving optimal control problems calls for no special maneuvers to eliminate the box constraints. This helped avoid the harmful "sticking" of the control to the boundary (see Ehneev [1]). The problem was solved in a dialogue mode using a series of popular methods. The first iteration by the penalty function method OPT413 yielded: $x^1(T) - 0.147411$, $x^2(T) =$ 133.642 km; the next three iterations by OPT553 yielded: $x^1(T) = 0.200000614$, $x^2(T) = 132.133$ km. The optimal control law is graphed in Figure 9.

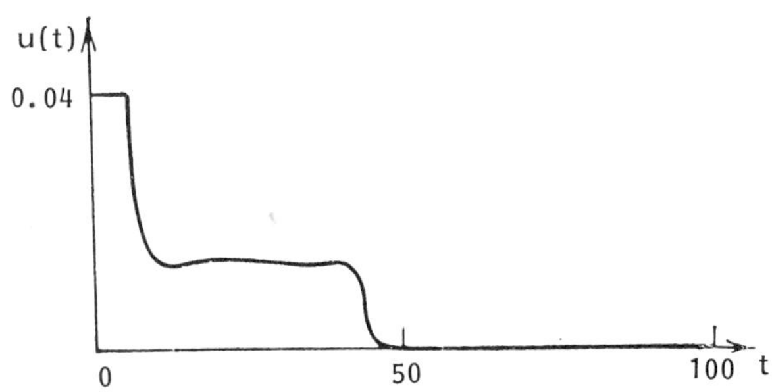

Figure 9

The comparison of these results with those obtained by Ehneev [1] demonstrates a qualitative and quantitative consistency of optimal control laws. During the first six seconds of the flight, the mass flow rate is maximal, i.e., 0.04 $\sec^{-1}$, then it drops drastically to 0.013/0.014 $\sec^{-1}$, remaining unchanged up to the 42nd second or near. The flow rate then drops to zero and from the 45th second to the end of the flight it does not change any more. According to Ehneev [1], this control law "possesses all the major well-known features of the optimal mass flow."

Table 5 contains the values of the functional and of the terminal mass $x^1(T)$ of the computations using the DISO-system methods, Ehneev's methods (I) [1], and Fedorenko's methods (II) [1].

Table 5

Method	OPT413		OPT553		I	II
Number of Iterations	0	1	1	1	50	12
$x^2(T)$	54.687	133.642	131.174	132.133	132.346	132.180
$x^1(T)$	0.20000	0.17471	---	0.2000006	0.19967	0.20000

The results obtained by the method OPT413 only are given in Table 6 and in Figure 10, with the curves of the initial control $u_0(t) \equiv 0.008$ and the control $u(t)$ after the first, second and fifth iterations.

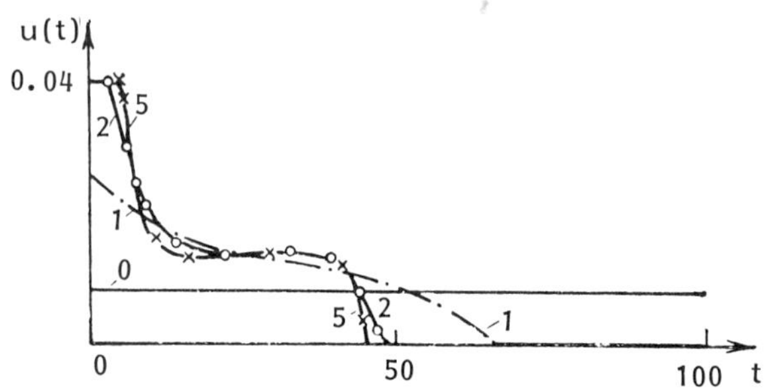

Figure 10

Table 7 contains the results of computations for discretization points $q = 201$, i.e., the integration step is half the size. The total time of computation has increased to 420 seconds (by a factor of 1.8).

Table 6

Number of Iteration	0	1	2	3	4	5
$x^2(T)$	54.687	133.642	139.270	135.403	133.761	132.898
$x^1(T)$	0.20000	0.17471	0.18617	0.19333	0.19673	0.19838

Table 7

Method	OPT413		OPT553		
Number of Iteration	0	1	1	2	3
$x^2(T)$	54.697	146.519	131.684	132.167	132.167
$x^1(T)$	0.20000	0.170345	--	--	0.199996

4. AN OPTIMAL TURNING PROBLEM FOR FLIGHT VEHICLES

The dynamic model used is based on the following assumptions: the flight vehicle is a material point moving in the three-dimensional right-handed coordinate system attached to the earth; the sidereal motion of the planet is ignored; the earth is assumed to be flat, the acceleration due to gravity is constant with respect to the altitude. The motion equations have the form

$$\dot{x}^1 = x^4 \cos(x^5)\cos(x^6), \qquad \dot{x}^2 = x^4 \sin(x^5),$$

$$\dot{x}^3 = -x^4 \cos(x^5)\sin(x^6), \qquad \dot{x}^4 = g\left[\frac{u^1 P\cos\alpha - C_x\bar{q}S}{x^7} - \sin(x^5)\right],$$

$$\dot{x}^5 = g\,\frac{u^2 N\cos(u^4) - \cos(x^5)}{x^4}, \qquad \dot{x}^6 = -g\,\frac{u^2 N\sin(u^4)}{x^4\cos(x^5)},$$

$$\dot{x}^7 = -C_s,$$

where x^1, x^2, x^3 are the Cartesian coordinates of the vehicle, x^2 is the altitude, x^4 is the velocity vector, x^5 is the flight path angle, x^6 is the course angle, x^7 is the vehicle mass, u^1 is the thrust related to maximum thrust P, u^2 is the overload relative to the maximum overload N, u^3 is the brake force relative to its maximum, u^4 is the bank angle, α is the angle of attack, $g = 9.81\ m/s^2$, S is the frontal area, $\bar{q}$ is the dynamic pressure, C_x is the aerodynamic drag coefficient, C_s is fuel consumption per second.

We use the following relations:

$$\bar{q} = \frac{\rho(x^2)(x^4)^2}{2},$$

$$\rho(x^2) = 3.3\cdot 10^{-10}(x^2)^2 - 1.155\cdot 10^{-5}x^2 + 0.125,$$

$$P = \left[10 - \frac{(x^4)^2}{a^2(x^2)}\right]\cdot\frac{25000 - x^2}{12.5},$$

$$a(x^2) = 340.3 - 4.08\cdot 10^{-3}x^2,$$

$$\alpha = \frac{u^2 N x^7}{u^1 P + 4.6\bar{q}S},$$

$$C_s = \frac{[0.7 + 2(u^1 - 0.3)^2]u^1 P}{3600},$$

$$S = 55\text{m}^2,$$

$$N = \min\left(\frac{\bar{q}S}{x^7}, \frac{150000}{x^7}, 8\right),$$

$$C_x = 0.02 + 3.174\alpha^2 + 0.03u^3.$$

Taking the approach suggested by Isaev and Sonin in [3], the function $N(x)$ has been "smoothed out."

In addition to the "box" constraints

$$0.05 \le u^1(t) \le 1, \quad 0.01 \le u^2(t) \le 1, \quad 0 \le u^3(t) \le 1, \tag{7.2}$$

the following "continuity" constraints are imposed:

$$\left|\frac{du^1}{dt}\right| \le 0.2 , \qquad \left|\frac{du^2}{dt}\right| \le 0.25 ,$$
$$\left|\frac{du^3}{dt}\right| \le 1 , \qquad \left|\frac{du^4}{dt}\right| \le 1.57 \text{ rad/sec} . \tag{7.3}$$

The initial state is:

$$x^1(0) = x^3(0) = 0 ,$$
$$x^2(0) = 5000\text{m} , \qquad x^4(0) = 300 \text{ m/s} ,$$
$$x^5(0) = x^6(0) = 0 ,$$
$$x^7(0) = 20000 \text{ kg} .$$

The initial values of the controls $u^2(0) = \frac{1}{N(x(0))}$, $u^4(0) = 0$, guaranteeing horizontal flight of the vehicle at an instant of time $t = 0$, are not changed during the computations. It is required that the terminal values of these controls satisfy the horizontal flight conditions, i.e.,

$$u^2(T) = \frac{1}{N(x(T))} , \qquad u^4(T) = 0 . \tag{7.4}$$

The state coordinates of the vehicle at the end of the flight need to be

$$x^2(T) = 7000\text{m} , \qquad x^5(T) = 0 , \qquad x^6(T) = -\pi . \tag{7.5}$$

We may formulate the following minimal-time problem: find the controls $u(t)$ satisfying the constraints (7.2), (7.3) and bringing the vehicle in the shortest time from horizontal flight at the altitude of 5000m to horizontal flight at the altitude of 7000m with the reversal of the velocity vector. Thus, time is taken as the functional $R = T$ and the conditions (7.4), (7.5) form a system of terminal equality-type constraints. Starting numerical solution of the problem, the variable t has been changed (see Section 1.8) and the differential equations have been integrated on $[0,1]$ by the Euler scheme with constant step $h = 0.02$ $(q = 51)$. The initial approximation was: $T_0 = 21$ sec and the functions $u^1(t) \equiv 1$, $u^2(t) \equiv 0.3$, $u^3(t) \equiv 0$, $u^4(t) = 1.5 \sin\left(\frac{\pi t}{T_0}\right)$.

Six different methods were used to solve the problem. In Figure 11 the graphs represent the functional T (in seconds) as

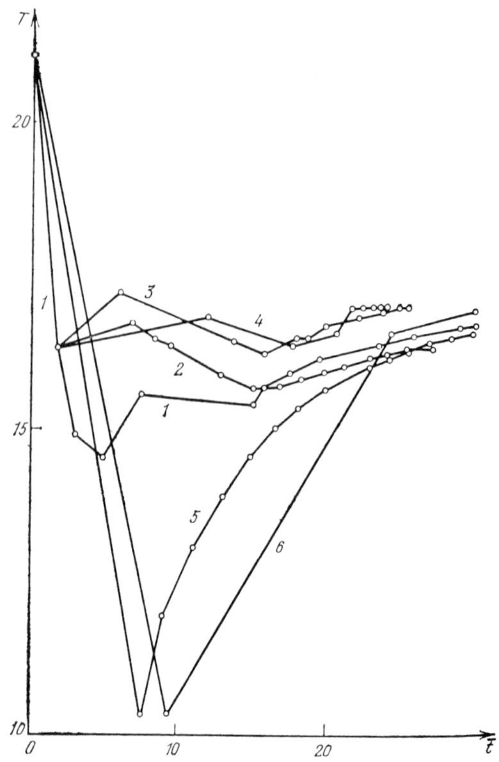

Figure 11

a function of the computational time $\bar{t}$ (in minutes) on a BESM-6 computer. The dots represent the values of T at the end of each kth iteration, interpolated between each two iterations. The numbers on the curves designate the following algorithms: 1 is the penalty method OPT41, 2 is the Lagrange multiplier method OPT7, 3 is the modified Lagrangian method OPT53, 4 is the simple iteration method OPT55, 5 and 6 are OPT45 and OPT46. The methods OPT7, OPT53 and OPT55 required the knowledge of the dual variables. Hence these three methods were applied only after one step made by the penalty method. Solution of the problem by each particular method was stopped if either two successive values of T differed from each other by a quantity smaller than 10^{-4}, or the allocated computer time had run out. In Figure 11 we can see that the methods OPT53 and OPT55 have the highest convergence rate. The feasible accuracy of solution was reached in 25 minutes. Using the penalty method OPT41, it took almost an hour to obtain similar results. For OPT46 the accuracy of solving the auxiliary problem had to be raised versus OPT45, otherwise the value T after the second step exceeded the optimal value $T_* \approx 16.96$ sec.

Solving this problem requires considerably longer computer time than the preceding cases. The reason is that the discretization yields a nonlinear programming problem of finding the minimum in 201 variables satisfying five equality-type constraints and 702 inequality-type constraints, with 302 box constraints taken into account int solving the auxiliary problem. The vector of dual variables had dimension 405.

In conclusion, let us examine the optimal programs $u(t)$ and the overload $n_y(t) = u^2(t)N(x(t))$ by OPT55: they are shown in Figures 12, 13, 14, 15. The corresponding values of the controls and state variables are such that $x^2(T_*) = 7001\text{m}$, $x^5(T) = 0$,

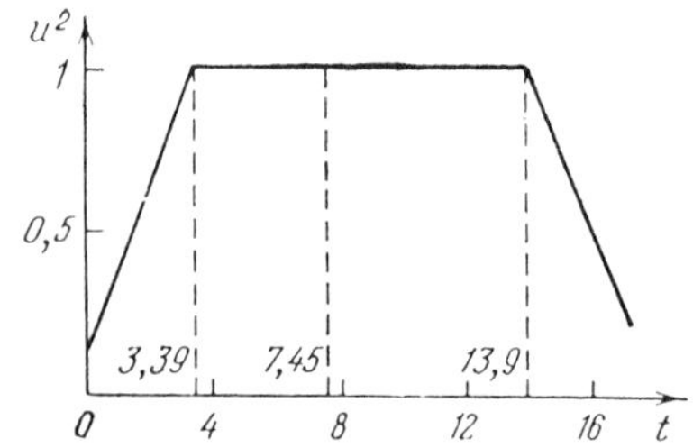

Figure 12

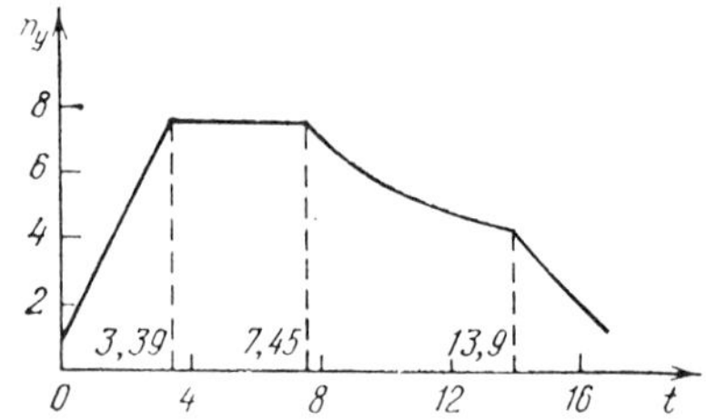

Figure 13

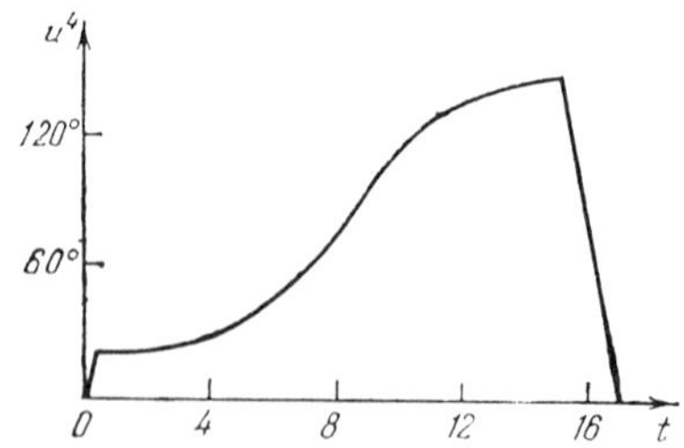

Figure 14

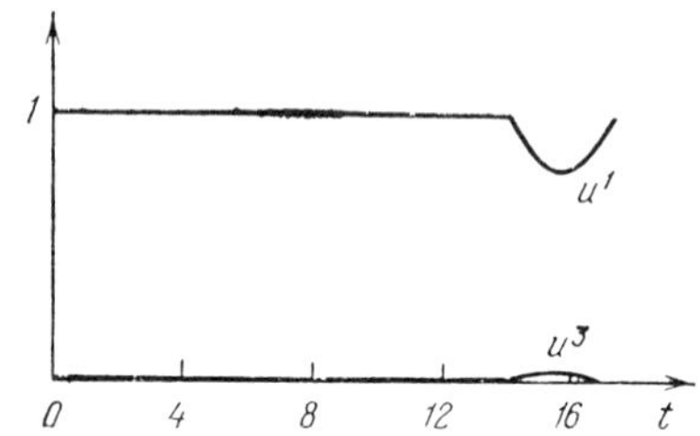

Figure 15

$x^6(T_*) = -180.2°$, $u^4(T_*) = 0$, $u^2(T_*)\cdot N(x(T_*)) - 1 = 10^{-4}$.

The control $u^2(t)$ is of a simple form and "boundary" in the sense of the constraints. At the start of the flight $(0 \le t \le 3.39 \text{ sec})$ $u^2(t)$ grows from 0.133 to 1 with the maximum velocity $(u^2 \approx 0.25)$, at the end of the flight $(13.90 \text{ sec} \le t \le T_*)$ $u^2(t)$ falls from 1 to 0.226 also with the maximum velocity $(\dot{u}^2 \approx -0.25)$, in the middle of the flight $(3.39 \text{ sec} \le t \le 13.90 \text{ sec})$ $u^2(t) \equiv 1$. The graphic representation of the corresponding overloads $n_t(t)$ is more difficult. The reason is that before $t \approx 7.45$ sec the function $N(x) = 150000/x^7$ and is practically constant $(\Delta N = 7.52 - 7.50 = 0.02)$ because of an insignificant change in weight $(\Delta x^7 = 19940\text{kg} - 20000\text{kg} = -60\text{kg})$. For $t \ge 7.45$ sec the function $N(x)$ is defined by the formula $N(x) = \frac{\bar{q}S}{x^7}$ for the first time and becomes essentially nonlinear.

This occurs since both the velocity and the atmosphere density decrease as the altitude increases. The corresponding fall in the dynamic pressure $\bar{q}$ results in $\bar{q}S < 150000$, and the maximum overload has to be computed by a different formula.

5. THE PROBLEM OF OPTIMAL DISTRIBUTION OF A STRUCTURE

Historically, most of the traditional problems of optimal control involved motion of moving objects, such as airplane, rockets, etc. As the methods of optimal control theory developed and improved, they were applied in other areas as well. As an example we consider the problem of designing a structure of minimum weight withstanding the design safe stress. Under some assumptions, the behavior of such structures is described by a system of differential equations instrumental to solution of optimization problems. These equations are due to G.I. Pshenichnov, who cooperated with this author in stating the solving the problem of optimal load distribution. The variables determining the problem are examined in detail in Grachev and Evtushenko [5] and Kashin, Pshenichnov and Flerov [1]. In the latter numerical results have been obtained by the methods described in this subsection. Hence optimization will be the focus of our discussion.

The state of the structure is described by the following system of equations:

$$\left.\begin{aligned} \frac{dx^1}{dt} &= \left(K - \frac{x^3}{EI}\right)x^2 - q(t), & \frac{dx^4}{dt} &= -Kx^5 + x^6, \\ \frac{dx^2}{dt} &= \left(\frac{x^3}{EI} - K\right)x^1 - p(t), & \frac{dx^5}{dt} &= \frac{x^1}{EF} + Kx^4 - \tfrac{1}{2}(x^6)^2, \\ \frac{dx^3}{dt} &= x^2 - m(t), & \frac{dx^6}{dt} &= -\frac{x^3}{EI}, \quad 0 \le t \le 1, \end{aligned}\right\} \tag{7.6}$$

where all the quantities are dimensionless. The functions $q(t)$, $p(t)$, $m(t)$, $K(t)$, $E(t)$ are specified. The controls $u(t)$ are parameters of the structural cross-section, and in the system (7.6) they appear as the area $F(u)$ and the moment of inertia $I(u)$ of the section.

The boundary conditions are also specified for the system (7.6):

$$\begin{aligned} x^4(0) &= x^5(0) = x^6(0) = 0 \, , \\ x^4(1) &= x^5(1) = x^6(1) = 0 \, . \end{aligned} \tag{7.7}$$

As control parameters we take the coordinates $x^1(0)$, $x^2(0)$, $x^3(0)$, not specified for $t = 0$. We pertain the conditions (7.7) to terminal equality-type constraints. Thus, the solution of the boundary value problem enabling us to override the static indefinability of the structure has been fitted into the general solution of optimal control problems.

In Figure 16 the exterior loads are balanced forces for a circular structure of radius R_1. The dimensionless parameters of the load are: $N = 0.5$, $q(t) = 2\sin(2\pi t)$, $p(t) \equiv 0$, $m(t) \equiv 0$. At the points 1, 2, where $t_1 = 0.12$, $t_2 = 0.88$, the massed loads N are projected onto the tangent and the normal to the contour:

$$x^1(t_j+0) = x^1(t_j-0) + N\sin(2\pi t_j) ,$$
$$x^2(t_j+0) = x^2(t_j-0) + N\cos(2\pi t_j) ,$$
$$j = 1,2 \, .$$

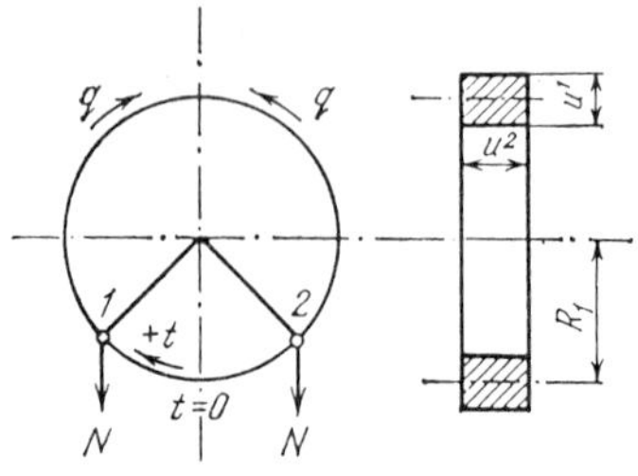

Figure 16

The rectangular cross-section has dimension $u^1 \times u^2$, where u^2 is the sectional width, u^1 is the vertical interval (in R_1-direction).

The sectional area and moment of inertia are computed by the formulas

$$f(u) = u^1u^2 , \qquad I(u) = \frac{u^2(u^1)^3}{12} .$$

It is required to find the functions $u^1(t)$, $u^2(t)$ determining the contour of the structure so as to minimize the stress:

$$R = \int_0^1 F(u(t))\, dt ,$$

satisfying the conditions

$$10^{-2} \le u^1(t) \le 10^{-1} , \qquad 10^{-2} \le u^2 \le 1 .$$

Furthermore, we impose the constraints on the safe stresses in each cross section:

$$|x^1(t)|u^1(t) + 6|x^3(t)| \le \frac{u^2(t)[u^1(t)]^2\sigma}{E_0} , \tag{7.8}$$

σ being the safe stress, $\frac{\sigma}{E_0} = 10^{-2}$. For the functions defining the constraints to be differentiable, we replace the condition (7.8) by an equivalent system of four inequalities:

$$\Gamma^i(x,u,t) = x^1(t)u^1(t)\delta^i - 6(-1)^i x^3(t) - \frac{u^2(t)[u^1(t)]^2\sigma}{E_0} \le 0 , \tag{7.9}$$

where $i \in [1:4]$, $\delta^1 = \delta^2 = 1$, $\delta^3 = \delta^4 = -1$.

The integration interval of the system (7.6) was partitioned into 50 parts, which led to a nonlinear programming problem with 103 variables, 200 inequality-type constraints and three equality-type constraints. As an initial approximation we took: $u_0^1(t) \equiv 10^{-2}$, $u_0^2 \equiv 10^{-2}$. In the first series of computations a linear version of the system (7.6) was used:

$$\dot{x}^1 = Kx^2 - q , \qquad \dot{x}^2 = -Kx^1 - p , \qquad \dot{x}^3 = x^2 - m ,$$
$$\dot{x}^4 = -Kx^5 + x^6 , \qquad \dot{x}^5 = \frac{x^1}{EF} + Kx^4 , \qquad \dot{x}^6 = -\frac{x^3}{EI} . \tag{7.10}$$

The integration of the systems (7.6), (7.10) by the Euler scheme has yielded large errors and unsatisfactory results. The Euler scheme with recalculation yielding an error of order $O(h_i^3)$ made it possible to complete solving the problem. The computations started via the penalty function method OPT413, then the dual method OPT553 was used. Solving the nonlinear problem (7.6) and the linear problem (7.10) led, in fact, to the same answer. In particular, the corresponding values of the functional are equal to $1.371 \cdot 10^{-4}$ and $1.378 \cdot 10^{-4}$, implying in turn that the design using the linear model does not lose stability under the specified stress.

As follows from physical considerations, the sectional width does not increase and remains equal to the initial values. The dependence of the optimal vertical interval u^1 on t is shown in Figure 17. Each constraint Γ^i is essential for some values of t. In Figure 18, one can see how Γ^1 and Γ^4 behave along the optimal

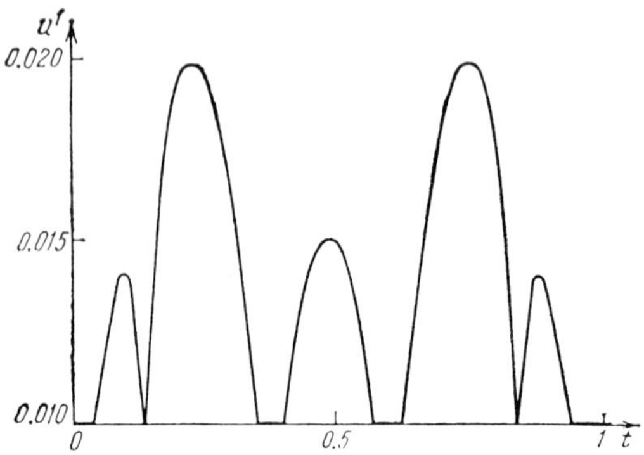

Figure 17

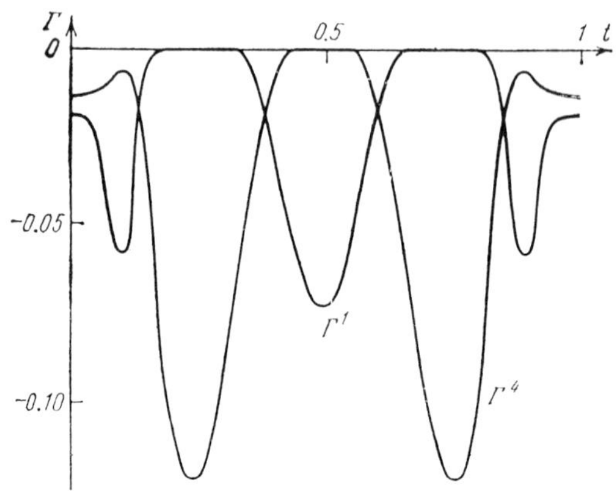

Figure 18

trajectory. From the diagram in Figure 17 we infer that the initial design was too lightweight for the specified stress, and the optimization raised the weight. At the same time, there are intervals in which $u^1(t) = 10^{-2}$ and the vertical intervals can be even smaller, which would lighten the structure. However, the lower box constraint is essential in this case. In actual design, these constraints are, as a rule, prescribed by engineering considerations.

The comparison of Figures 17 and 18 indicates that the increased vertical interval u^1 is due to reliability requirements. In terms of nonlinear programming the optimal solution is a boundary point of the feasible set given by both the box constraints and the constraints (7.9) along the trajectory. If there were no box constraints, then the maximum safe stress would be possible in each cross-section. This assertion corresponds to the "equirigidity" hypothesis, which indeed holds in the case considered. This approach extends to designing structures whose contours can be nonclosed or multiply connected, or can have cross-sections of a more complex configuration, for instance T- or I-shaped.

8. AN APPLICATION TO DIFFERENTIAL GAMES

1. THE STATEMENT OF THE PROBLEM

In recent years, many articles and monographs treating differential games have been published. Extensive reference lists are given in Isaacs [1] and Krassovsky and Subbotin [1]. Not much work is, however, available on numerical methods of solving game problems, because of arduous computations involved in seeking global extrema of multivariable functions.

Suppose the game is described by a system of differential equations:

$$\begin{aligned} \frac{dx}{dt} &= f(x(t),\, u(t),\, t)\,, \\ \frac{dy}{dt} &= g(y(t),\, v(t),\, t)\,, \qquad 0 \le t \le T\,. \end{aligned} \tag{8.1}$$

For simplicity, we assume that the interval T is fixed and then proceed to discrete approximation of this system by the Euler scheme:

$$x_{i+1} = x_i + h_i f(x_i, u_i, t_i) = F(z_i^x), \qquad z_i^x = [x_i, u_i, t_i], \tag{8.2}$$

$$y_{i+1} = y_i + h_i g(y_i, v_i, t_i) = P(z_i^y), \qquad z_i^y = [y_i, v_i, t_i]. \tag{8.3}$$

In the system, the constraints are mixed:

$$\Gamma^1(x_i, u_i) \le 0, \qquad \Gamma^2(y_i, v_i) \le 0, \qquad 1 \le i \le q. \tag{8.4}$$

Here, for simplicity's sake, the equality-type constraints are omitted and no terminal constraints are specified.

The game is estimated by the function $R(x_q, y_q)$ depending only on the terminal state coordinates. Suppose that the functions f, g, Γ^1, Γ^2, R determining the problem are everywhere continuously differentiable in x_i, y_i, u_i, v_i. We introduce the complete state vectors x, y and the complete control vectors u and v:

$$x = [x_1, x_2, \dots, x_q], \qquad u = [u_1, u_2, \dots, u_q],$$

$$y = [y_1, y_2, \dots, y_q], \qquad v = [v_1, v_2, \dots, v_q].$$

The vectors x, y, u, v need to be distinguished from the functions $x(t)$, $y(t)$, $u(t)$, $v(t)$ in the system (8.1).

Having defined the vectors u and v, we can define $x_q = x_q(u)$ and $y_q = y_q(v)$; using the formulas of Section 6.1, we obtain

$$\left.\begin{aligned}
\frac{dR}{du_i} &= H_u^1(z_i^x, p_{i+1}^x), & \frac{dR}{dv_i} &= H_v^2(z_i^y, p_{i+1}^y), \\
p_i^x &= H_x^1(z_i^x, p_{i+1}^x), & p_i^y &= H_y^2(x_i^y, p_{i+1}^y), \\
p_q^x &= R_x(x_q, y_q), & p_q^y &= R_y(x_q, y_q),
\end{aligned}\right\} \tag{8.5}$$

$$H^1(z_i^x, p_{i+1}^x) = \langle F(z_i^x), p_{i+1}^x \rangle, \qquad H^2(z_i^y, p_{i+1}^y) = \langle P(z_i^y), p_{i+1}^y \rangle.$$

Here the vectors p_i^x and p_i^y have the same dimensions as x_i and y_i, respectively.

Problem I consists in finding the quantity V_1 -- the best estimate of the composite function $R(x_q, y_q)$ for a player whose behavior is described by the system (8.3):

$$V_1 = \max_v \min_u R(x_q(u), y_q(v)) \quad . \tag{8.6}$$

Problem II consists in finding the best estimate for the opponent:

$$V_1 = \min_u \min_v R(x_q(u), y_q(v)) \; . \tag{8.7}$$

In both problems the extrema are sought under the constraints (8.4).

These are referred to as problems of finding the maximin and minimax estimates in the class of programmed strategies. The strict meaning of (8.6) and (8.7) was explained in Section 1.5. For the problem (8.7), in particular, the solution of the "interior" problem determines the point-set mapping

$$v(u) = \arg\max_v R(x_q(u), y_q(v)) \quad ; \tag{8.8}$$

the solution is next sought for the "exterior" problem of finding

$$u_* \in \arg\min_u \phi(u) \; , \qquad \phi(u) = R(x_q(u), y_q(v(u))) \; , \tag{8.9}$$

$$v_2 = \phi(u_*) \; .$$

According to the results of Section 1.5, among the solutions of the problems (8.6) and (8.7) we have the estimate $V_1 \le V_2$.

2. GLOBAL METHODS FOR SOLVING THE PROBLEMS

The simplest method of finding the minimax V_2 is based on minimization methods in a space of reduced dimension (see Section 2.6).

For fixed u one has an interior problem and needs to define $v \in v(u)$, that is a standard discrete optimal control problem. We assume that for each u the problem has a solution. Then we arrive at the problem of minimizing the composite function $\phi(u)$ reducible to finding the sequence u converging to u_*. If global methods are used to solve both interior and exterior problems, the problem (8.7) has a global solution. However, numerical implementation requires enormous amounts of computations and, naturally, only the simplest game problems can be solved on currently available computers. It is expedient to make simplifying assumptions, and this can help solve high-dimension problems, shrinking, however, the class of the problems.

3. NUMERICAL METHODS FOR FINDING A LOCAL MINIMAX

We are to solve the interior and exterior problems (8.8) and (8.9) using local methods, to obtain the local minimax. This approach requires a thorough analysis of the solutions obtained; one can do it only for an adequate initial approximation and a unique, continuous dependence $v(u)$. The interior problem (8.8) is solved for each complete control vector u_k. We assume that its solution $v_k \in v(u_k)$ exists, as well as the dual vector $\lambda_k = \lambda(u_k)$, and the necessary conditions of the maximum hold at the Kuhn-Tucker point $[v_k, \lambda_k]$:

$$\frac{d}{dv}\left[R(x_q(u_k),\ y_q(v_k)) - \langle \Gamma^2(y(v_k),\ v_k),\ \lambda_k \rangle\right] = 0 ,$$

$$\Gamma^2(y(v_k), v_k) \le 0 , \qquad \Gamma^{2s}(y(v_k), v_k)\lambda_k^s = 0 , \qquad \lambda_k \ge 0 ,$$

where Γ^{2s} is the sth coordinate of the vector Γ^2.

To solve the interior problem (8.8) one can apply any of the methods described in Section 6.3. Usually, a combination of different methods is needed: the first several steps by the penalty function method are followed by the dual methods. Thus, the values of the function $\phi(u)$ become known. The constrained minimization of $\phi(u)$ is carried out by the same methods. When the function $\phi(u)$ is not differentiable, a version of the penalty method involving nondifferentiable penalty functions is, for example, employed.

To implement methods using derivatives, an assumption is in order concerning the differentiability of the functions $v(u)$, $\lambda(u)$, which will hold, in particular, when the conditions of Theorem 1.7.7 are satisfied. The differentiability of $u(v)$, $\lambda(u)$ implies that $y_q(v(u))$ is differentiable. The systems (8.2) and (8.3) are integrated independently and are related only through the terminal functional R. Hence, instead of the functions $v(u)$, $\lambda(u)$, $y_q(u)$ we can use the dependence $\tilde{v}(x_q)$, $\tilde{\lambda}(x_q)$, $\tilde{y}_q(x_q)$. In order to take advantage of the formulas (8.5), we need to compute

$$p_q^x = \frac{dR(x_q,y_q)}{dx_q} = \frac{\partial R(x_q,y_q)}{\partial x_q} + \frac{\partial \tilde{y}_q}{\partial x_q}\,\frac{\partial R(x_q,y_q)}{\partial y_q} .$$

We showed in Section 1.7.6 that the latter summand is zero. Therefore

$$p_q^x = \frac{\partial R(x_q,y_q)}{\partial x_q} .$$

Note that if Γ^2 depended explicitly on u, additional terms would then appear in this formula.

This procedure is implementable for solving the problem (8.7), but it may involve elaborate computations. Hence it is more appropriate to use the method (2.6.2) requiring no solution of the interior problem. In that case, a transformation of the controls u and v is made:

$$u_{k+1} = u_k - \varepsilon\alpha \frac{dR(x_q(u_k), y_q(v_k))}{du_k},$$
$$v_{k+1} = v_k + \alpha \frac{dR(x_q(u_k), y_q(v_k))}{dv_k}. \tag{8.10}$$

Here, for simplicity, the constraints (8.4), $0 < \varepsilon << 1$, have been omitted.

4. NECESSARY MINIMAX CONDITIONS

We have described local and global procedures for solving the problem (8.7). An intermediate procedure is also possible, when the interior problem is solved through a global method and the exterior problem through local procedures. In that case, one can use properties similar to the discrete minimum principle of Section 6.4. We discuss this problem briefly, ignoring the constraints (8.4).

Let the control process be described by the difference relations (8.2) and (8.3). We stipulate the constraints $v_j \in V$, $j \in [1:q]$, on controls: $v \in V$. Let us fix all the components except u_i of the feasible complete control vector u. Furthermore, we consider the problem of finding

$$V_2 = \min_{u_i \in U} \max_{v \in V} R(x_q(u), y_q(v)). \tag{8.11}$$

In the interior problem this determines

$$v(u) = \arg\max_{v \in V} R(x_q(u), y_q(v)). \tag{8.12}$$

In the exterior problem we have the set of solutions

$$W_* = \arg\min_{u_i \in U} \phi(u), \qquad \phi(u) = R(x_q(u), y_q(v(u))).$$

We assume that the following conditions are satisfied:

•1. the vector function $F(x_i, u_i, t_i)$ and the function $R(x_q, y_q)$ are continuously differentiable with respect to the components of the vector x;

•2. for each $u_i \in U$ the interior problem has a solution; the condition (8.12) defines the point-set mapping $v(u_i)$ associated with the set of terminal (u_i) points $y_q(u_i)$.

Given the feasible vector $u_i \in U$, we can uniquely determine x_q from (8.2); by solving the interior problem we determine $y_q(u_i)$. The problem (8.11) becomes equivalent to the following problem:

$$V_2 = \min_{u_i \in U} \max_{y_q \in y_q(u_i)} R(x_q(u_i), y_q) . \tag{8.13}$$

We fix the point $u_i \in U$ and the point $y_q \in y_q(u_i)$. Then, using the formulas (8.5), we successively determine the vectors $p_q^x, p_{q-1}^x, \ldots, p_{i+1}^x$ depending on $x_{i+1} = F(x_i, u_i, t_i)$ and on y_q parametrically. In particular, we obtain

$$p_{i+1}^x(x_{i+1}, y_q) = \frac{dR(x_q, y_q)}{dx_{i+1}} .$$

Just as in Section 6.4, in solving the exterior problem we go over to the state space. Then the problem (8.13) is equivalent to the following problem:

$$V_2 = \min_{x_{i+1} \in \Omega} \max_{y_q \in y_q(u_i)} R(x_q, y_q) .$$

Here the set $\Omega = F(x_i, U, t_i)$ is convex and compact. By Theorem 1.5.4, to have $x_{i+1}^* = F(x_i, u_i^*, t_i)$, $u_i^* \in W_*$, it is necessary that the condition

$$\max_{y_q \in y_q(u_i^*)} \left\langle p_{i+1}^x(x_{i+1}^*, y_q), \; x_{i+1} - x_{i+1}^* \right\rangle \geq 0$$

be satisfied for any $x_{i+1} \in \Omega$, or passing to the control space, we obtain that for all $u_i \in U$ we have the inequality

$$\max_{y_q \in y_q(u_i^*)} \langle p_{i+1}^x(F(x_i,u_i^*,t_i), y_q), \ F(x_i,u_i,t_i) - F(x_i,u_i^*,t_i)\rangle \geq 0 . \tag{8.14}$$

We have thus arrived at the following theorem, known as the discrete minimax principle.

THEOREM 6.8.1. For the systems (8.2) and (8.3) let conditions 1 and 2 hold and let the set $\Omega = F(x_i,U,t_i)$ be convex and compact. Then, in order that $u_i^* \in W_*$ it is necessary and, if $\phi(u_i)$ is convex, also sufficient that the condition (8.14) holds for any $u_i \in U$.

The following assertion is an analog of the linearized minimum principle: for any $u_i \in U$ we have the inequality

$$\max_{y_q \in y_q(u_i^*)} \langle F_u(x_i,u_i^*,t_i)p_{i+1}^x(F(x_i,u^*_i,t_i), y_q), \ u_i - u_i^*\rangle \geq 0 .$$

These results make it possible to use the versions of the conditional gradient method and of the gradient projection method, as indicated in Section 6.5. The present problem is, however, more complex. In the preceding case, p_{i+1} was determined by the relations (1.5) and (1.7), whereas now, for the interior problem to have a nonunique solution, one needs to consider the set of vectors p_{i+1}^x defined by (8.5) for different values of y_q.

Taking this approach, it is easy to take the state space constraints into account and obtain the necessary minimax conditions.

5. AN EXAMPLE OF NUMERICAL COMPUTATIONS

These methods are extendable to the case where the equations (8.1) are not separate and there are no state space constraints. For illustration, we have taken from Isaacs [1] the dolichobrachistochrone problem. The game is described by the system

$$\frac{dx}{dt} = \sqrt{y}\cos u + \frac{1+v}{2},$$

$$\frac{dy}{dt} = \sqrt{y}\sin u + \frac{v-1}{2},$$

where u and v are controls satisfying the constraints: $0 \leq u \leq 2\pi$, $-1 \leq v \leq 1$.

A player controlling the function u tries to bring the state vector on the set

$$M = \{x,y:\ x=0,\ y \geq 0\}$$

in minimal time. His opponent handling the function v tries to prevent the entry on M or at least to delay it. Numerical solution of this problem begins by the method (8.10) and continues by the conditional gradient method, as indicated in the preceding subsection.

The results are diagrammed in Figure 19 as state trajectories on the plane $[x,y]$. The breakpoints of the trajectory lie on the so-called switching parabola, at which the second player -- "opponent" -- switches his control from -1 to $+1$. It is worthwhile to compare the diagram in Figure 19 with the similar diagram in Figure 5.2.2 in Isaacs [1], where the author seeks the solution of the game in closed-loop strategies. The state trajectories in the area above the switching parabola exactly correspond to the analytic solution obtained by Isaacs. However, the state

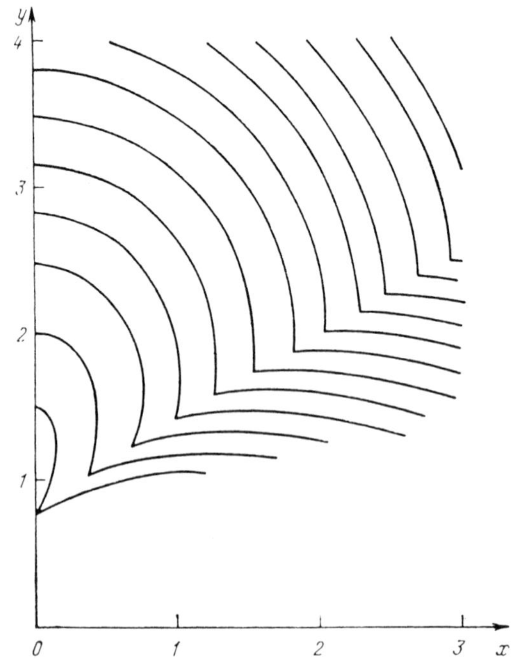

Figure 19

trajectories below the parabola have different form. Chigir' [1] points out the errors in Isaacs' qualitative description of the trajectory field. However, Chigir' has drawn incorrectly the state trajectories close to the set M, which should rather be perpendicular to the y-axis. One can easily see this from the analytic formulas and numerical results graphed in Figure 19.

Chapter 7

SEARCH FOR GLOBAL SOLUTIONS

In the majority of practical optimization problems, it is required to determine global solutions, and only in rare cases do local solutions suffice. Minimization of a convex function, for example, on a convex set can be solved using local methods since in this case the local minimum coincides with the global one. Numerical methods for seeking global solutions of multivariable problems, in spite of their practical importance, have been rather poorly developed. This is no doubt due to their exceedingly great complexity. We shall not detail all the available approaches to this problem. Instead, we shall concentrate on one most promising direction -- which is based on the idea of a non-uniform covering of a feasible set. This approach has turned out to be quite universal and, as we shall show, can be used not only for seeking global extrema of multivariable functions but also for nonlinear programming problems, for solving systems of equations and, most importantly, for multicriteria optimization. Problems which are solvable in reasonable computer time must be of limited dimension (of order 10 to 20); however, the use of multiprocessors, parallel computing and distributed processing substantially increase the possibilities of this approach.

The development of global methods calls for a new view on the numerical methods. Not only should these techniques yield the global solution; but it must also be verified that the solution is actually global. This requirement is crucial in problems of operations research and game theory, where the so-called guaranteed maximin and minimax estimates are required.

These demands on numerical methods compel us to reevaluate the available approaches to optimization problems. The local numerical methods described in the preceding chapters, such as the penalty function method, or parametrization method, are hardly suitable for finding the global minimum directly (for more detail, see Sections 3.4, 4.3, 6.2). However, utilization of local optimization methods as auxiliary aids to the basic method of global search substantially improves its efficiency.

1. THE GENERAL NOTION OF COVERINGS

1. *THE STATEMENT OF THE PROBLEM*

We examine first the problem of finding the global minimum of a multivariable function $f(x)$ on a feasible compact set $X \in R^n$:

$$f_* = \min_{x \in X} f(x) \quad . \tag{1.1}$$

In numerical computations this problem is usually simplified: one introduces the so-called set of ε-approximate solutions:

$$X_\varepsilon = \{x \in X: \; f(x) - \varepsilon \le f_*\} \quad . \tag{1.2}$$

Here $\varepsilon > 0$ is the specified accuracy of the computation. It is required to find some point $x_* \in X_\varepsilon$. In other words, it is necessary to find with the given accuracy the global minimum value of a function in n-variables and at least one point x_* at which the approximate value is reached. Let us delineate a few numerical methods for solving this problem.

2. RANDOM SEARCH

We randomly pick k feasible points $x_1, x_2, \ldots, x_k$. As a solution we take the point at which the minimal value $f(x_i)$ is attained, $i \in [1:k]$. The number of test points k is determined in such a way that the probability that at least one of them belongs to X_ε is sufficiently large. Obviously, the method is most effective when $f(x)$ is sufficiently "flat" and if so the ratio of the measure of the set X_ε to that of X is not small.

3. RANDOM SEARCH USING LOCAL PROCEDURES

As in the case above, we again define feasible points. Then, using local methods for finding the minimum of a multivariable function, we look for the local minima of $f(x)$ on X, taking $x_1, x_2, \ldots, x_k$ as initial points. As a result, we get the points $\bar{x}_i, \bar{x}_2, \ldots, \bar{x}_k$; as the solution x_* we take the point at which $f(x_i)$ has the minimal value. The number of points k necessary for the implementation of the method is determined by the condition of guaranteeing a sufficiently large probability that among the initial points $x_1, x_2, \ldots, x_k$ there is at least one point hitting in the region of attraction (the region of local convergence) of at least one point of the set of global solutions of the problem (1.1). It is appropriate to use this method when the number of local minima in the problem is not great. The method has two bad features: first, it does not guarantee finding a global extremum, rather it only gives a probability of the event, and achieving a given probability usually requires a lot of calculations. Second, in using the method a situation can occur when the local minimization methods will repeatedly seek the same local minima, which unreasonably increases the number of calculations. To devise more effective algorithms, more stringent conditions need to be imposed on the function to be minimized.

4. THE METHOD OF COVERING

We present the method in stages. First we give the general idea of the method and a little later give a more specific description of it.

Let the values of $f(x)$ be computed at a sequence of feasible points $\{x_k\} = [x_1, x_2, \ldots, x_k]$. Define the quantity

$$R_k = \min \, [f(x_1), \; f(x_2), \; \ldots, \; f(x_k)] \tag{1.3}$$

and call it a record; we call any points $x_i \in \{x_k\}$ satisfying $R_k = f(x_i)$ a record point. Define the set

$$Z_k = \{x \in R^n : \; R_k - \varepsilon \leq f(x)\} \quad . \tag{1.4}$$

Obviously,

$$R_k - \varepsilon \leq \min_{x \in Z_k} f(x) \quad .$$

The points of Z_k are of no interest for finding the global minimum since the exact global minimum of f on Z_k can improve the value of the record R_k by no more than ε. Hence the set Z_k can be omitted from consideration and it is sensible to continue the search of the minimum only on the set $X \setminus Z_k$. In particular, if

$$X \subset Z_k \tag{1.5}$$

then the initial problem is solved and the record point x_k is taken as an approximate solution; it is guaranteed that $x_k \in X$.

Thus the problem of finding the global extremum has been reduced to constructing a sequence of the points $\{x_k\}$ satisfying (1.5). The sequence R_k is monotonic decreasing, the sequence of sets Z_k is monotonic increasing, i.e.,

$$R_{k+1} \leq R_k \ , \qquad Z_k \subset Z_{k+1} \ . \tag{1.6}$$

5. LOCAL METHODS

The set Z_i essentially depends on the value of the record R_i and it is greatest for $R_i = f_*$, the value f_* being usually not known. Hence, to extend the set Z_k it is desirable that the record be as close as possible to f_*. The sequence $\{x_k\}$ will be chosen as to guarantee the condition (1.5), while to extend the set Z_i we use the auxiliary procedures of finding the local minimum in the problem (1.1). For $x_i \in X$ and

$$f(x_i) \leq R_{i-1} - \varepsilon \tag{1.7}$$

one goes to the program of local search of the minimum; if one thereby obtains a point $\bar{x}_i$ at which $f(\bar{x}_i) < f(x_i)$, then as

R_i one takes the quantity $f(\bar{x}_i)$. This technique substantially expedites the computations, and will be used in the sequel. The conditions (1.6) will be conserved in this case.

6. FUNCTIONS SATISFYING A LIPSCHITZ CONDITION

In numerical computations, finding Z_k from (1.4) is hard. It becomes, however, easier under the assumption that $f(x)$ satisfies a Lipschitz condition on X with constant ℓ, i.e., for any $x, y \in X$ one has

$$|f(x) - f(y)| \le \ell\|x - y\| . \tag{1.8}$$

Then

$$f(y) - \ell\|x - y\| \le f(x) , \tag{1.9}$$

yielding that the inequality

$$R_k - \varepsilon \le f(x) \tag{1.10}$$

will be satisfied for all x satisfying

$$R_k - \varepsilon \le f(y) - \ell\|x - y\| . \tag{1.11}$$

Let x_j belong to the sequence $\{x_k\}$, $j \le k$. Introduce the ball B_{jk} and the enveloping sphere S_{jk}:

$$\begin{aligned} B_{jk} &= \{x \in R^n : \|x - x_j\| \le r_{jk}\} \\ S_{jk} &= \{x \in R^n : \|x - x_j\| = r_{jk}\} . \end{aligned} \tag{1.12}$$

The center of B_{jk} is at the point x_j, the radius of S_{jk} is

$$r_{jk} = \frac{f(x_j) - R_k + \varepsilon}{\ell} . \tag{1.13}$$

From (1.12), (1.13), (1.19), it follows that (1.10) holds for all $x \in B_{jk}$. The set Z_k contains the union of all such balls:

$$\bigcup_{j=1}^{k} B_{jk} \subset Z_k .$$

Therefore the condition (1.5) is satisfied automatically if

$$X \subset \bigcup_{j=1}^{k} B_{jk} . \tag{1.14}$$

This representation suggests a constructive method for solving the problem. In broad terms, the method consists in the following. Suppose that for some sequence $\{x_k\}$ the record R_k is determined from (1.3). Let the sequence of points x_j and the radii r_{jk} of the sphere be stored in computer memory. If at the new point x_{k+1} the quantity $f(x_{k+1}) < R_k$, we set $R_{k+1} = f(x_{k+1})$ and replace the entire sequence r_{jk} by the sequence $r_{j(k+1)}$. If the sum of the balls $B_{j(k+1)}$ covers the set X, the computations end; otherwise, we take a new point x_{k+2} and go on. The set X is covered by balls of different radii. The shortest radius is at the points x_j at which $R_j = f(x_j)$, with $r_{jk} = \frac{\varepsilon}{\ell}$. If X is bounded, such a covering is accomplished in a finite number of steps.

Without changing the formulas above, one can replace the covering by balls by a covering of n-dimensional cubes whose faces are parallel to the coordinate planes. To this end, it suffices to assume that the Chebyshev norm is used in the Lipschitz condition (1.8) (see Appendix II). Then the "ball" B_{jk}

is a collection of points x satisfying

$$\max_{s \in [1:n]} |x^s - x_j^s| < r_{jk} , \tag{1.15}$$

i.e., B_{jk} turns into an n-dimensional cube. If the norm in (1.8) is not Chebyshev, from the equivalence of norms (see Appendix II) one can redefine the Lipschitz constant ℓ and use the Chebyshev norm. For example, if the Euclidean norm is in (1.8) then in (1.15) we need to take $\ell\sqrt{n}$ instead of ℓ.

This approach to solving the problem is not complete since it is not clear how to obtain a covering of X satisfying (1.14). We shall discuss this question in Section 7.3 and show that under very general assumptions the problem can be reduced to that of covering a parallelepiped containing the set X.

7. FUNCTIONS WHOSE GRADIENT SATISFIES A LIPSCHITZ CONDITION

Suppose the function f is differentiable in X and for any x and y of the convex compact set X we have

$$\| f_x(x) - f_x(y) \| \leq M \| x - y \| , \tag{1.16}$$

where M is constant. We will find estimates for functions of this class. It is easy to show that (1.16) implies that $f(x)$ and $\| f_x(x) \|$ are continuous and bounded on X. We use the Newton-Leibniz formula (see Appendix I):

$$f(x) = f(x_j) + \int_0^1 \langle f_x(x_j + \tau(x - x_j)), x - x_j \rangle \, d\tau .$$

Adding and subtracting from the right side the scalar product

$\langle f_x(x_j),\ x-x_j\rangle$ and using (1.16), we obtain

$$f(x_j) + \langle f_x(x_j),\ x-x_j\rangle - \frac{M}{2}\|x - x_j\|^2 \leq f(x) .$$

By the Cauchy formula we have

$$f(x_j) - [\|f_x(x_j)\| + \frac{M}{2}\|x - x_j\|] \cdot \|x - x_j\| \leq f(x) .$$

Using these inequalities, we obtain that (1.10) is satisfied for all x satisfying either the inequality

$$\frac{M}{2}\|x - x_j\|^2 - \langle f_x(x_j),\ x-x_j\rangle \leq f(x_j) + \varepsilon - R_k \tag{1.17}$$

or the inequality

$$\frac{M}{2}\|x - x_j\|^2 - \|f_x(x_j)\| \cdot \|x - x_j\| \leq f(x_j) + \varepsilon - R_k . \tag{1.18}$$

The boundary of the set (1.17) is the n-dimensional sphere S_{jk} centered at $\bar{x}_j$ with radius r_j given by

$$\bar{x}_j = x_j + \frac{1}{M} f_x(x_j) ,$$

$$r_{jk} = \frac{1}{M} [\|f_x(x_j)\|^2 + 2M(f(x_j) + \varepsilon - R_k)]^{\frac{1}{2}} .$$

The boundary of the set (1.18) is the n-dimensional sphere $\bar{S}_{jk}$ centered at x_j with radius ρ_{jk} equal to

$$\rho_{jk} = r_{jk} - \frac{1}{M}\|f_x(x_j)\| .$$

The shortest distance from x_j to the sphere $\bar{S}_{jk}$ is ρ_{jk}. Hence the sphere $\bar{S}_{jk}$ is inscribed in the sphere S_{jk}, and the points in the ball (1.18) are in the ball (1.17).

Let B_{jk} and $\bar{B}_{jk}$ denote the set of points in the balls

(1.17) and (1.18), respectively. The union of the balls B_{jk} lies in the set Z_k defined by (1.4). A numerical method of solving the problem consists in determining the sequence $\{x_k\}$ satisfying the condition (1.14).

2. COVERING A PARALLELEPIPED

1. THE STATEMENT OF THE PROBLEM

We consider a particular case of the problem (1.1) of finding

$$f_* = \min_{x \in P} f(x) \quad , \tag{2.1}$$

where P is an n-dimensional parallelepiped

$$P = \{x \in R^n: \quad a \le x \le b\} \quad . \tag{2.2}$$

The set of approximate solutions in the problem (2.1) is given by the formula (1.2), where P is taken as X. The function $f(x)$ satisfies the Lipschitz condition (1.8) with Euclidean norm. Then the cube inscribed in the sphere S_{jk} is defined as follows:

$$V_{jk} = \{x \in R^n: x_j - e\bar{r}_{jk} \le x \le x_j + e\bar{r}_{jk}\} \quad , \tag{2.3}$$

where

$$\bar{r}_{jk} = p + \frac{f(x_j) - R_k}{\bar{\ell}} \quad , \qquad \bar{\ell} = \ell\sqrt{n} \quad , \quad p = \frac{\varepsilon}{\bar{\ell}} \quad , \tag{2.4}$$

$e \in R^n$ is the vector with all coordinates equal to 1.

The center of the cube V_{jk} is at x_j, the length of the edge is $2\bar{r}_{jk}$, the lateral faces are parallel to the coordinate axes. According to the arguments in the preceding section, if

$f(x_j)$ and R_j are known, then one can exclude the cube V_{jj} from the parallelepiped P. If it happens that $R_k < R_j$ for some $k > j$, then one can omit the bigger cube V_{jk} containing the cube V_{jj}. From (2.4) it is not hard to derive a formula for recomputing the half-length of the face of the new enlarged cube.

$$\bar{r}_{jk} = \bar{r}_{jk} + \frac{R_j - R_k}{\bar{\ell}} , \qquad k \ge j . \tag{2.5}$$

The problem (2.1) will have a solution if the sequence of cubes V_{jk} completely covers the parallelepiped P, i.e.,

$$P \subset \bigcup_{j=1}^{k} V_{jk} . \tag{2.6}$$

A great variety of such coverings are possible. We shall describe one which had to satisfy two conditions: the smallest possible computer memory and the simplest possible computer program should be used. The first condition was due to the fact that the computations were run on a computer with a relatively small core memory. The second condition could be met because ALGOL-60 in which the program was written allows recursive procedures. Relaxing these conditions would undoubtedly permit developing more sophisticated ways of covering.

2. THE METHOD OF COVERING

This author developed several variants of the method of covering; he was also fully acquainted with other authors' methods. The simplest and free from ambiguity method was chosen in preference to other methods. We shall now describe it, following the program's implementation as close as possible.

The method is implemented through a recursive procedure (we used ALGOL-60 terminology), i.e., with a recursive call of the basic program. The procedure is referred to as $N(i,a,b)$, where $i \in [1:n]$, i being the index of the coordinate of the vector x, which changes in the process of the covering P, the vectors a and b in the primal version of the method coinciding with the corresponding vectors determining P in (2.2). We introduce now the auxiliary variable vector $v_s \in R^n$. First we take all the components of v_1 equal to u, where u is any sufficiently large number $\left(u > \max_{i \in [1:n]} [b^i - a^i]\right)$. Let some initial point $x_1 \in P$ be known that is the initial approximation of solving the problem (2.1). Let

$$x_2 = a + pe \quad ,$$

$$R_2 = \min [f(x_1), f(x_2)] \quad .$$

We construct a sequence of points $x_3, x_4, \ldots, x_k$ from P satisfying the covering condition (2.6) using a recursive procedure, which we now describe.

<u>THE PROCEDURE N(i,a,b)</u>. At input time the vectors x_s, v_s, the record R_{s-1} and the value $f(x_{s-1})$ are known.

Suppose

$$x_s^i < b^i + p \quad . \tag{2.7}$$

Then we make the following operations. If $x_s^i > b^i$, we set $x_s^i = b^i$. First we consider the case $i = 1$.

We compute $f(x_s)$, R_s. If it turns out that

$$R_s < R_{s-1} , \tag{2.8}$$

we take R_s as the record.

If, in addition,

$$R_s + \delta < R_{s-1} , \tag{2.9}$$

where δ is some given number, we turn to the problem of finding the local minimum in (2.1). As a result, we get a point $\bar{x} \in P$, where $f(\bar{x}) < R_s$. We take $\bar{x}$ as the record point, setting $R_s = f(\bar{x})$ and keeping the point x_s. The quantity δ measures the accuracy of the local minimum.

If at least one of the conditions (2.8) or (2.9) holds, we redefine v_s. In accordance with (2.5) we take

$$v_s^j + \frac{R_{s-1} - R_s}{\bar{\ell}} \tag{2.10}$$

as the jth coordinate of v_s.

We define

$$h_s = 2p + \frac{f(x_s) - R_s}{\bar{\ell}} . \tag{2.11}$$

Set

$$x_{s+1}^1 = x_s^1 + h_s ,$$

$$v_{s+1}^1 = \min [v_s^1, h_s] . \tag{2.12}$$

All components of x_{s+1} and v_{s+1}, except the first ones that have been redefined here, are the same as the corresponding components of x_s and v_s.

When $i = 1$ the control in this place is transferred to the beginning of the procedure, where the condition (2.7) is checked. The process is continued until (2.7) is violated, and then the procedure $N(1,a,b)$ stops computing. We take x_s to be the one for which (2.7) was satisfied for the last time, in spite of the fact that x_{s+1} was defined by (2.12). But at x_{s+1} the value of f was not calculated, so it may be viewed only as a test point.

Now consider the case $i > 1$.

We call the procedure $N(i-1, a, b)$. Upon completion, we obtain the vectors x_j, v_j and the set

$$\begin{aligned} v^i_{j+1} &= \min [v^{i-1}_j, v^i_j] \ . \\ x^i_{j+1} &= x^i_j + v^{i-1}_j \ , \end{aligned} \tag{2.13}$$

The remaining components of the vectors x_{j+1} and v_{j+1} are the same as the corresponding components of x_j and v_j. After this the control is transferred to the beginning of the procedure $N(i,a,b)$, where (2.7) is checked. The process is continued until (2.7) is violated. Then $N(i,a,b)$ stops computing.

It has been assumed throughout that the current points x and v, the record point, and the record are global variables, i.e., any changes in them in the process of running $N(i,a,b)$ are reflected in all procedures $N(s,a,b)$ $(i,s \in [1:n])$.

The computations are initiated by calling $N(n,a,b)$. As a result, a sequence of points satisfying the condition (2.6) is produced.

This concludes the formal description of the procedure N.

We indicate next an illustration of how it works first for a one-variable function.

3. THE ONE-DIMENSIONAL CASE

Let $n = 1$ in problem (2.1). The cube (2.3) in this case is the segment

$$V_{jk} = \{x \in R^1: x_j - r_{jk} \le x \le x_j + r_{jk}\} ,$$

where $r_{jk} = p + \frac{f(x_j) - R_k}{\ell}$. For simplicity, we omit the step of using local auxiliary methods, and from (2.10) - (2.12) we obtain the following formulas:

$$x_2 = a + p_1 , \qquad R_s = \min [R_{s-1}, f(x_s)] ,$$

$$h_s = r_{ss} + p . \tag{2.14}$$

If $R_s < R_{s-1}$, we take $v_s + \frac{R_{s-1} - R_s}{\ell}$ as r_s; otherwise v_s does not change. Next we have

$$\begin{aligned} x_{s+1} &= x_s + h_s , \\ v_{s+1} &= \min [v_s, h_s] . \end{aligned} \tag{2.15}$$

The method stops computing as soon as k is found such that

$$x_k \le b + p < x_{k+1} . \tag{2.16}$$

Thus, this method produces a monotonic increasing sequence of points $\{x_k\}$ with variable step h_s, which is distinct from r_{ss}. This is because after computing the value of f at x_s the next calculation of f will be made at x_{s+1}, where automatically $r_{(s+1)(s+1)} \geq p$. Hence we can omit at least the segments

$$|x_s - x| \leq r_{ss} , \qquad |x_{s+1} - x| \leq p .$$

We combine both segments and arrive at (2.15). Hence noting the subsequent calculations, the step h_s can be made p units bigger than r_{ss}. Here, however, we need to specially check the condition for ending the process. From the sequence resulting from the formulas (2.14) - (2.16) one can form a sequence of segments $[x_i-p, x_{i+1}-p]$, $i \in [2:k]$ completely covering the segment $[a,b]$. Hence for each point $x \in [a,b]$ we can find an $x_i \in \{x_k\}$ such that

$$x_i - p \leq x \leq x_i + h_i - p .$$

We use the inequality (1.9) and determine (1.3), and we obtain

$$f(x) \geq f(x_i) - \ell|x - x_i| \geq f(x_i) - \varepsilon \geq R_k - \varepsilon . \qquad (2.17)$$

Since x is arbitrary in the segment $[a,b]$, we reach the conclusion that the inequality (2.17) is valid everywhere on $[a,b]$. Taking the minimum in x on the left side of the inequality, we get that $f_* \geq R_k - \varepsilon$. Hence the method guarantees that on $[a,b]$ in a finite number of steps the global minimum of the arbitrary function $f(x)$ satisfying the Lipschitz condition (1.8) will be found within an error ε.

The step h_s turns out to be minimal and equal to $2p$ at points where $f(x_s) = R_s$ and maximal where $f(x_s) >> R_s$. This has a simple intuitive interpretation: at points where the values of $f(x)$ are considerably larger than the current value of the record, we can vary x with a big step, without being apprehensive of missing the required points. Conversely, we have to vary x with caution near the points where $f(x_s)$ and R_s are close, so that we do not miss the points at which (2.17) is violated. This method is frequently called an exhaustive search on a non-uniform mesh. One of the most adverse cases for this method is the case of a constant function $f(x)$; here the step of varying x is constant and equal to $2p$. The method turns into a full-scale search with constant step; the number of steps required for the computations is roughly:

$$K_1 \approx \frac{(b-a)\ell}{2\varepsilon} .$$

In the most favorable case the function being minimized has $f(x) = c_1 + \ell(x - c_2)$, where c_1, c_2 are arbitrary scalars. The number of necessary steps to solve the problem is

$$K_2 \approx \log_2 K_1 .$$

For an arbitrary function satisfying (1.8) the necessary number of steps is contained in the interval $[K_2, K_1]$.

In getting the estimate K_1, no assumptions were made about whether methods of local search were used in the counting process, and no additional information about the properties of the function being minimized was used. Moreover, it should be kept in mind that an almost constant function has a Lipschitz constant close to zero so that the number of computations for functions close to constant is not significant.

The quantity v_s in the given one-dimensional case has no influence on the performance of the algorithm. It is required, however, in the multidimensional case. At the end of the computations, we obtain some value v_k equal to the minimal radius of the spheres $S_{2k}, S_{3k}, \ldots, S_{kk}$.

4. THE MULTIDIMENSIONAL CASE

Let $n = 2$. Now the set P is a rectangle and the set V_{jk} is a square. First we set

$$x_2 = a + pe ,$$

$$x_2 = [x_2^1, x_2^2] ,$$

$$v_2 = [u,u] .$$

We call $N(2,a,b)$. If $x_2^2 < b^2 + p$, then $N(1,a,b)$ is called. In the process of running this procedure, the second components x^2, v^2 of the vectors x, v do not change, whereas for the first components we obtain from (2.11) and (2.12) two sequences $x_3^1, \ldots, x_k^1$, $v_3^1, \ldots, v_k^1$ analogous to those considered in the preceding subsection. If condition (2.8) is satisfied for some s, we redefine the jth component of v by (2.10), $j = 1,2$. The last point k is found from the condition (2.16). Using (2.13), we let

$$x_{k+1}^2 = x_k^2 + v_k^2 ,$$

$$v_{k+1}^2 = \min\,[v_k^1, v_k^2] .$$

If $x_{k+1}^2 < b^2 + p$, then a new call is made to $N(1,a,b)$, and so on. The process stops when m is found such that

$$x_m^2 \le b^2 + p < x_{m+1}^2 .$$

Let us comment on this stage. Consider the sequence $x_3^1, \dots, x_k^1$ obtained as a result of the first pass using $N(1,a,b)$. Introduce the rectangle

$$W = \{x \in R^2 : a^1 \le x^1 \le b^1, \quad a^2 \le x^2 \le a^2 + v_k^1 - p\} .$$

Let $x \in W$. Then we can find an x_i such that

$$\begin{aligned} x_i^1 - p \le x^1 \le x_i^1 + h_i - p = x_i^1 + \bar{r}_{ii} , \\ a^2 \le x^2 \le v_k^1 - p = \bar{r}_{ii} . \end{aligned} \tag{2.18}$$

The set of points defined by these inequalities is contained in the cube V_{ii}, so the inequality $f(x) \ge R_k - \varepsilon$ holds for all the points satisfying (2.18). Since x is arbitrary in W, we reach the conclusion that $\min_{x \in W} f(x) \ge R_k - \varepsilon$. Therefore the set W can be omitted from further consideration. This can be done by changing the parallelepiped P in (2.1). Let

$$P_1 = \{x \in R^2 : a^1 \le x^1 \le b^1, \quad a^2 + v_k^1 - p \le x^2 \le b^2\}$$

and repeat the above computations on P_1. Arguing just as in the preceding subsection one can show that in determining W the inequality $x^2 \le a^2 + v_k^1 - p$ may be relaxed by setting $x^2 \le a^2 + v_k^1$.

The two cases $n = 1$, $n = 2$ clarify the idea of covering the parallelepiped P. Hence we will not describe the covering process in more detail since all the details merely follow from the procedure N.

The numerical computations noticeably speed up if a good

initial value x_1 is known and the value $f(x_1)$ is close to f_*. Hence, before calling N it is useful to find at least a rough estimate of f_*. To do this, one can either use the random search method or call the procedure N, taking a reduced Lipschitz constant ℓ, or increase ε.

If the value of the Lipschitz constant ℓ is a priori unknown, one can start with some value ℓ_0 to solve the problem with Lipschitz constants $2\ell_0$, $4\ell_0$ and so on, till the result (the value $f(x_k)$) is not different from the preceding one by more than ε. This criterion is a necessary but not sufficient condition for the value of the Lipschitz constant obtained to be greater than or equal to the true value of the Lipschitz constant.

Experience with numerical computations shows that the use of local methods speeds up the computations essentially. This approach carries over to the case of functions satisfying (1.16), examined in the preceding section.

Here we end the description of the method for solving (2.1), which we shall refer to as the primal method in the sequel. This method can be refined by removing the constraints introduced above, with respect to the computer memory and the simplicity of the program used. Various modifications of this method have been devised, and we shall mention some of them next.

Let us first point out the simplest modification connected with the possibility of covering P by parallelepipeds instead of by cubes. We illustrate the idea using a two-dimensional case as an example. Let the procedure N(1,a,b) run and let some value $v_{s-1}^2 > 2p$ be known. Let $f(x_s)$ and r_{ss} be determined at x_s, and suppose that r_{ss} is larger than the difference $v_{s-1}^2 - p$. Then we shall inscribe a rectangle in the circle S_{ss} instead of a square (see (1.12)), such that one side is equal to $2v_{s-1}^2$ and the other is equal to

$$2\rho_s = 2\sqrt{r_{ss}^2 - (v_{s-1}^2 - p)^2} .$$

From the fact that $r_{ss} > (v_{s-1}^2 - p)\sqrt{2}$ it follows that $\rho_2 > \bar{r}_{ss}$. Hence one can omit all the points of the rectangle

$$x_s^1 - \rho_s \leq x^1 \leq x_s^1 + \rho_s ,$$

$$x_s^2 - v_{s-1}^2 + p \leq x^2 \leq x_s^2 + v_{s-1}^2 - p$$

from further analysis. In (2.12) one can take $h_s = p + \rho_s$, which is bigger than the step given by (2.11).

This modification makes it possible to increase the search interval for the first coordinate, without decreasing the size of the current interval for the second coordinate. This idea carries over to the multidimensional case. Let us now turn to a more radical modification of the method.

5. A FIRST MODIFICATION

Numerical computations show that the effectiveness of the method is decreased appreciably as the dimension of the vector x is increased. Especially laborious were problems in which the global minimum is attained on some "plateau." In this method, the components of v decrease as the number of coordinates increases, and for multidimensional problems the search in the last coordinates often has to be done with minimal step equal to $2p$. To remove this defect at least partially, the following modification is offered.

We modify the method so that the minimal step-size for the second, third and last coordinates is always not less than $2pq$, i.e., $v^i \geq 2pq$, $i \in [2:n]$, where q is a given number. For this we proceed as follows. At some point x_s suppose we find that $h_s < 2pq$. We define the coordinates of the vectors $\bar{a}, \bar{b} \in R^n$ by the formulas

$$\bar{a}^i = x^i - p \quad ,$$

$$\bar{b}^i = \min [b^i, x^i + 2pq] \quad , \qquad i \in [1:n] \; .$$

We call $N(n,\bar{a},\bar{b})$. As a result, the global minimum of $f(x)$ on the parallelepiped

$$\bar{P} = \{x \in R^n : \; \bar{a} \leq x \leq \bar{b}\}$$

will be determined with error ε. In the process of solving this auxiliary problem, a new record point might be found, which will then be taken as R_s. After solving this problem, we set

$$x_{s+1}^1 = x_s^1 + 2pq + p \quad ,$$

$$v_s^1 = 2pq + p \quad ,$$

and the computations are carried further by the initial method. This recursive call of the basic program is simple to implement in ALGOL-60, and it is also easy to assure that in solving the auxiliary problem, the local methods seek the minimum of $f(x)$ on P.

This modification substantially improves the operation of the method since it makes it possible to refine the step size for inspecting the set P locally, just near the points of the minimum of $f(x)$.

It is hard to make general recommendations for selecting q. Obviously, if $q \le 1$ then no changes in the operation of the method are introduced, since the minimal step is always greater than or equal to $2p$. On the other hand, if one takes q so large that $2pq \sim b^i - a^i$, then there will still be no reduction in the volume of calculations. For moderate values of q we have been able to cut the volume of calculations by a factor of 2 to 3, in a number of problems.

6. A SECOND MODIFICATION

In analyzing the set (2.18), it was shown to be contained in the cube V_{ii}. Here is the one essential drawback of the method. The requirement for dense sequential packing of rectangles leads to the situation that the segment $[x_i^1 - p,\ x_i^1 + h_i - p]$ is dropped, although a larger segment $[x_i^1 - h_i + p,\ x_i^1 + h_i - p]$ could be omitted instead. Hence it is desirable to place the points x_i inside the segment $[a^1, b^1]$ not too closely at once. Of course, one needs to remember all the segments among which the search has to be continued and the completeness of the covering to be checked. We illustrate this with a one-dimensional example. Let us denote by $[d,g]$ the set of points $d \le x \le g$. In case $g \le d$ this segment is omitted from the sequences of segments defined below.

First we take $x_2 = \frac{a+b}{2}$ and calculate $f(x_2)$. Search for the minimum is continued on the two segments

$$[a, x_2 - r_{22}] , \qquad [x_2 + r_{22}, b] .$$

Let $x_3 = \frac{x_2 - r_{22} + a}{2}$ and calculate $f(x_3)$.

We continue the search on the three segments

$$[a, x_3 - r_{33}], \qquad [x_3 + r_{33}, x_2 - r_{23}], \qquad [x_2 + r_{23}, b] .$$

Upon computing the value of f at x_4, the midpoint of the last segment, we obtain a set of four segments

$$[a, x_3 - r_{34}] , \qquad [x_3 + r_{34}, x_2 - r_{24}] ,$$

$$[x_2 + r_{24}, x_4 - r_{44}] , \qquad [x_4 + r_{44}, b] .$$

We choose the largest segment and take its midpoint as x_5. If in the process of these computations the quantity R changes, then r_{ij} changes and in accordance with (2.5) the lengths of the segments on which the minimum has to be sought decrease, with some segments vanishing. This process is continued until all the segments have disappeared from the variable set of segments in which the minimum is to be sought. Possibly this occurs right away after computing $f(x_2)$ if

$$f(x_2) - R_2 + \varepsilon \geq \frac{\ell(b-a)}{2} .$$

If f was computed the last time at x_k, then we set

$$v_k^1 = \min_{i \in [2:k]} [r_{ik}] .$$

The procedure carries over to the general n-dimensional

case and becomes a recursive procedure analogous to N. The collections of segments on which the minimum is to be sought on each coordinate should be stored in the computer memory. Only for the nth coordinate will these segments be stored, split and reduced from throughout the computation. For the ith coordinate and each new call of N(i,a,b) this system of segments is determined anew for the coordinates with indices 1,2,...,i. This modification is seen to be quite effective, in spite of the fact that in the most unfavorable case when f is constant, it leads to roughly the same result as the basic method does. The segment [a,b] will be covered by a uniform mesh, the number of calculations of f(x) will be close to K_1.

Implementation of this modification does not require very large computer memory, since the number of stored segments does not exceed K_1. By a recursive procedure analogous to N, the above-described covering carries over to a multidimensional case.

7. PARALLEL COMPUTATIONS

The operating efficiency of this method and its modifications depends on the sequence of coordinatewise coverings of the parallelepiped. Before the computations, it is hard to determine the best sequence. Multiprocessor computers are most helpful in such cases. We show how the procedure N is modified in this case. Let us introduce variable n-dimensional vectors a_i, $b_i \in R^n$. At the beginning of the computations we put $a_1 = a$, $b_1 = b$. Suppose that according to the basic method, the computations are carried on with the procedure N(i,a,b) which we write here as

$N(i,a,b,w_1)$, where the vector $w_1 \in R^n$ designates the sequence in which the covering occurs. In the basic method $w_1 = [1,2,\ldots,n]$. After the first call of $N(n-1,a,b,w_1)$, the value of the record R_k is determined and the value of v_k^n is computed.
Hence in the further computations we need to find the minimum of f on the parallelepiped $P_k = \{x \in R^n: a_k \le x \le b_k\}$, where all the components a_k and b_k coincide with the corresponding components a and b except $a_k^n = a^n + v_k^n$.

Suppose that simultaneously with the first processor in which the procedure $N(n,a,b,w_1)$ was processed, a second processor operated in which the procedure $N(n,a,b,w_2)$ was used, with the sequence of covering coordinates given as $w_2 = [2,3,4,\ldots,n,1]$. After the first call of $N(n-1,a,b,w_2)$, the record R_s and the quantity v_s^1 are obtained. As a result of operating both processors, further search can be continued on the parallelepiped

$$P_s = \{x \in R^n: a^i \le x^i \le b^i, \quad i = [2,3,\ldots,n-1],\ a^1+v_s^1 \le x^1 \le b^1, \\ a^n+v_k^n \le x^n \le b^n\} \quad .$$

For the simultaneous operation of two processors an exchange of the refined records and reductions of the parallelepiped is needed.

If the computer has n processors, then on the ith processor one can call the procedure $N(n,a,b,w_i)$, where

$$w_i = [i,\ i+1,\ \ldots,\ n,\ 1,\ \ldots,\ i-1] \quad .$$

The refined record and reductions of the parallelepiped are exchanged between all the processors.

Other variants are also possible for computations using multiprocessor computers. For example, the parallelepiped P can be partitioned into s parts, s being the number of processors.

8. NUMERICAL RESULTS

The problem of finding

$$f_* = \max \left[\frac{1}{6} \sum_{j=1}^{6} \sin 2\pi\left(x^j + \frac{j}{5}\right)\right]^2 ,$$

where $0 \leq x^j \leq 1$ for all j, provided a test problem. Within the feasible set this function has 30 isolated local maxima and two global maxima, $f_* = 1$, with $\varepsilon = 0.01$, $\ell = 0.7$. If for solving this problem one uses a complete search on a uniform mesh, it is necessary to evaluate the objective function $4 \cdot 10^{11}$ times. The computations using a nonuniform covering were carried out in two stages: first the problem with $\ell = 0.2$ was solved, and next with $\ell = 0.7$. The computations by the primary scheme needed $3 \cdot 10^4$ evaluations of the function; better modifications can reduce it to a few thousands. Local methods could find the global solution with high accuracy much quicker. Nevertheless, the method can find a guaranteed result also for more complex problems, but, of course, does not finish computing before the feasible set has been covered completely.

3. SOLUTION OF NONLINEAR PROGRAMMING PROBLEMS

1. THE STATEMENT OF THE PROBLEM

The approach described in the preceding sections carries over to solving nonlinear programming problems. Suppose the global minimum is sought:

$$f_* = \min_{x \in P \cap X} f(x) , \qquad (3.1)$$

where

$$X = \{x \in R^n: h(x) \leq 0\} ,$$

P being an n-dimensional parallelepiped defined by (2.2) and $h: R^n \to R^c$.

Analogously to (1.2) we define the set of approximate global solutions of the problem (3.1):

$$X_\varepsilon = \{x \in P \cap X: f(x) - \varepsilon \leq f_*\} . \qquad (3.2)$$

The set X can have a very general form; it can be nonconvex and non-simply connected. It is assumed that the functions f and $h^i(x)$ satisfy a Lipschitz condition on P with the same constant ℓ, i.e., for any x, y P one has (1.8) and

$$|h^i(x) - h^i(y)| \leq \ell \|x - y\| .$$

In this case (3.1) has a solution. As before, we denote by X_* the set of all global solutions. Introduce the function

$$\psi(x) = \max_{i \in [1:c]} h^i(x) .$$

Let Γ denote the set of boundary points of X. Using the function ψ one can write $\Gamma = \{x \in R^n: \psi(x) = 0\}$.

For each $x_j \in \Gamma$ we define the ball

$$B(x_j) = \left\{ x \in R^n: \|x - x_j\| < \frac{\varepsilon}{2\ell} \right\} .$$

The union of all such balls when x_j takes on all possible values from Γ will be denoted by Y. This set is thus some open cover of the set Γ.

Using Theorem 1.5.2, it is easy to show that $\psi(x)$ satisfies a Lipschitz condition with constant ℓ, i.e.,

$$|\psi(x) - \psi(y)| \leqslant \ell \|x-y\| \quad . \tag{3.3}$$

Hence for each $x \in B(x_j)$ we have the estimate $|\psi(x)| < \frac{\varepsilon}{2}$ and for each point $x \in Y \cap X$ the inequalities $-\frac{\varepsilon}{2} \leqslant \psi(x) \leqslant 0$ are satisfied.

Let $Z = P \cap (X \backslash Y)$. We replace problem (3.1) by the following problem:

$$\bar{f} = \min_{x \in Z} f(x) \ . \tag{3.4}$$

In somewhat nonstandard fashion, we define the set of approximate solutions of this problem:

$$Z_\varepsilon = \left\{ x \in Z\colon \ |f(x) - \bar{f}| \leqslant \frac{\varepsilon}{2} \right\} \ . \tag{3.5}$$

Suppose that $Z \neq \emptyset$. Then we have the following lemma.

<u>LEMMA 1</u>. Let the set X have interior points and let the closure of the interior coincide with X. Then $Z_\varepsilon \subset X$.

<u>Proof</u>. Let $x_1 \in Z_\varepsilon$. We show that

$$0 \leqslant f(x_1) - f_* \leqslant \varepsilon \ . \tag{3.6}$$

We now compare (3.1) and (3.4). From the obvious inclusion $Z \subset P \cap X$ it follows that $f_* \leqslant \bar{f}$. Hence for every $x_1 \in Z$ (and a posteriori for a point in Z_ε) the left inequality in (3.6) is satisfied. From the definition (3.5) it follows that

$$f(x_1) - \bar{f} \leqslant \frac{\varepsilon}{2} \ . \tag{3.7}$$

Consider the case where there is at least one point $x_* \in X_*$ such that $x_* \in (X \backslash Y) \cap P$. Then $f_* = \bar{f}$ and from (3.7) one has

$$f(x_1) - f_* \leqslant \frac{\varepsilon}{2} \ ,$$

which is stronger than (3.6).

We consider next a second case: $X_* \subset Y$. The assumption of the Lemma that the closure of the interior of X coincides with X excludes sets with "tentacles." Hence for any point $x_* \in X_*$ we can find a point $x_2 \in Z$ such that $\|x_2 - x_*\| \leqslant \frac{\varepsilon}{2\ell}$. Using a Lipschitz condition, we obtain

$$f(x_2) - \ell\|x_2 - x_*\| \leqslant f(x_*) \ . \tag{3.8}$$

Since $x_2 \in Z$, it follows that $\bar{f} \leqslant f(x_2)$. Hence from (3.8) we find

$$\bar{f} - f_* \leqslant \frac{\varepsilon}{2} \ .$$

Combining this inequality with (3.7), we arrive at the right side of (3.6).

Comparing (3.2) with the definition (3.5) of the set X_ε, we reach the conclusion that $x_1 \in X_\varepsilon$; and by the arbitrariness of x_1 we arrive at the desired inclusion $Z_\varepsilon \subset X_\varepsilon$. The Lemma enables us to replace the problem of finding the global minimum of (3.1) by an approximate solution of the problem (3.4). The error of determining the minimal value of f is ε, and if the global minimum in (3.1) is attained at least at one point of $X\backslash Y$, the error does not exceed $\frac{\varepsilon}{2}$. The set Y was deliberately included in Z_ε, in spite of the fact that all points in Y are not feasible in the problem (3.4) since $Z\backslash Y = \emptyset$. We did it because in solving (3.4) we may obtain points not belonging to Z but belonging to X instead. For example, this can happen in using auxiliary procedures for finding a local minimum. It seems natural to take into account the best points obtained in determining the record.

2. A METHOD FOR SOLVING PROBLEMS

Suppose the values of $f(x)$ in the sequence of points $\{x_k\}$ belonging to the parallelepiped P are calculated. We use (1.3), taking the minimum just over the values $f(x_i)$ at points $x_i \in X$. Thus, as before, the R_k can be called a record since these are the minimal values of $f(x)$ at feasible points of the $\{x_k\}$.

All points satisfying the inequality

$$R_k - \frac{\varepsilon}{2} \leqslant f(x) \tag{3.9}$$

are of no interest in solving (3.4) since the minimum of $f(x)$ on this set cannot improve the value of the record by more than $\frac{\varepsilon}{2}$. If $x_j \in X$ and $f(x_j) \geqslant R_k$, the condition (3.9) is satisfied at all points which belong to the ball centered at x_j and are enclosed by the sphere of radius

$$r_{jk} = \frac{f(x_j) - R_k + \frac{\varepsilon}{2}}{\ell} . \tag{3.10}$$

If the union of such balls covers the set $(X \setminus Y) \cap P$, the problem (3.4) will be solved. We can do this by covering the entire parallelepiped P, but we still need to derive formulas for covering unfeasible points.

Let $x_j \notin X$. We determine a neighborhood of x_j which can be omitted from consideration. Here the following is of interest:

- 1. one can omit the points for which (3.9) is satisfied;
- 2. one can omit the points for which $\psi(x) \geq 0$ since they are unfeasible in (3.4);
- 3. one can omit the points which lie inside the sphere centered at x_j with radius less than the shortest distance from

x_j to $X \setminus Y$ since these points are also unfeasible in (3.4).

Using Lipschitz conditions (1.8) and (3.3), we can show that conditions 1 and 2 are satisfied, respectively, by the points lying inside the balls:

$$B_{1k} = \left\{ x \in R^n : \|x - x_j\| \le \frac{f(x_j) - R_k + \frac{\varepsilon}{2}}{\ell} \right\} ,$$

$$B_2 = \left\{ x \in R^n : \|x - x_j\| \le \frac{\psi(x_j)}{\ell} \right\} .$$

Let us show now that condition 3 is satisfied by points in

$$B_3 = \left\{ x \in R^n : \|x - x_j\| \le \frac{\psi(x_j) + \frac{\varepsilon}{2}}{\ell} \right\} .$$

Obviously, $B_2 \subset B_3$. $\psi(x) \ge 0$ everywhere on B_2 and vanishes only on the boundary. Hence the interior of B_2 contains no points of X, and boundary points of X can lie on the boundary of B_2. Hence the point x_* from $X \setminus Y$ that is closest to x_j is either inside the layer $H = B_3 \setminus B_2$ or on the boundary of B_2, or outside B_3.

We show that there is no point of $X \setminus Y$ inside H. Assume that such an x_* exists. Then

$$\|x_* - x_j\| < \frac{\psi(x_j)}{\ell} + \frac{\varepsilon}{2\ell} . \tag{3.11}$$

Since $x_* \in X$ and $x_* \notin Y$, it follows that inside the sphere

$$B_4 = \{x : \|x_* - x\| = \frac{\varepsilon}{2\ell} \}$$

there are no points which do not belong to X_*, in particular there are no points of B_3. Points of X and possibly of B_2

lie on B_4. We construct the straight line through x_j and x_*. From the two points of intersection of the line with the sphere B_4 we take the one closest to x_j, which we denote $\bar{x}$. Since $\bar{x} \in B_4$, it follows that $\bar{x} \in X$. We use the representation

$$||x_j - x_*|| \leq ||x_j - \bar{x}|| + ||\bar{x} - x_*|| .$$

Noting (3.11) and the fact that $\bar{x} \in B_4$, we obtain

$$||x_j - \bar{x}|| < \frac{\psi_1(x_j)}{\ell} ,$$

i.e., $\bar{x} \in \text{int } B_2$, which contradicts the fact that $\bar{x} \in X$. Hence it follows that the interior of B_3 has no points from $X \backslash Y$.

From the formulas obtained for the balls B_1, B_2 and B_3 we find the formula for the largest ball. Let

$$B_{jk} = \left\{ x \in R^n : ||x - x_j|| \leq \frac{\frac{\varepsilon}{2} + d_{jk}}{\ell} \right\} , \tag{3.12}$$

where

$$d_{jk} = \max [\psi(x_j), f(x_j) - R_k] .$$

If (3.12) is used for feasible points with $x_j \in X$, we then obtain that $\psi(x_j) \leq 0$ and

$$d_{jk} = f(x_j) - R_k , \qquad f(x_j) \geq R_k ,$$

the radius of the sphere enclosing S_{kj} coinciding with (3.10). Thus, (3.12) is a general formula for determining balls with feasible or unfeasible centers, which can be omitted. We arrive at the following lemma.

<u>LEMMA 2</u>. Let the sequence of points $\{x_k\}$ in P be such that the union of the balls B_{jk} $(j \in [1:k])$ completely covers the

parallelepiped P. Then every point $x_j \in X$ such that $f(x_j) = R_k$ belongs to the set of approximate solutions X_ε of the problem (3.1.)

This lemma enables one to solve the problem (3.1) by constructing a covering of P. For this one can use the results of the preceding section. The parallelepiped will be covered by balls enclosed by spheres of distinct radii. The smallest radii will be near the points at which $f(x)$ takes on values close to the current record R_k. Far away from the boundary Γ (more precisely, where $\psi(x) >> 0$) and at points where $f(x) >> R_k$, the radii of the spheres will be larger.

To reduce the calculations it is essential to know as accurately as possible the value of the current record. Hence at the $x_i \in X$ at which is satisfied (1.7) one should go to the local methods of solving (3.1), sharpen the value of R_i, and enlarge the balls B_{jk} $(j = 1,\dots,i)$ if they are saved during the computation.

3. TAKING INTO ACCOUNT EQUALITY CONSTRAINTS

It may happen that $X \backslash Y = \emptyset$. This occurs, for example, if among the constraints on the feasible set X there are equality-type constraints. Such constraints give sets with empty interior.

Consider first a particular case. Let $\psi(x) > 0$ everywhere on P with the exception of a finite number of points with $\psi(x) = 0$. Instead of this problem we consider the problem of finding the global minimum on the intersection of P and

$$X_\delta = \{x: \psi(x) < \delta\} ,$$

where $\delta > 0$ specifies the accuracy.

On the one hand, it is desirable to take the value δ charactizing the accuracy of fulfilling the constraints as small as possible, and on the other hand, in order to use our approach, δ should be such that $X_\delta \setminus X_\varepsilon \neq \emptyset$. Here Y_ε denotes the union of all the balls

$$\|x - x_i\| \le \frac{\varepsilon}{2\delta} ,$$

as x_i ranges over all possible values of the boundary of X_δ. If $\psi(x)$ satisfies the Lipschitz condition (3.3), then $X_\delta \setminus Y_\varepsilon \neq \emptyset$ for $\delta > \varepsilon$.

In the general case of seeking a global minimum under equality constraints, one can use an analogous technique for passing to inequality constraints. The feasible set X given by (1.6.2) can then be replaced by the set

$$X_\varepsilon = \{x \in R^n : \|g(x)\| \le \delta ,\ h(x) \le 0\} .$$

Let us solve the problem (3.1) in which X_δ is taken as X, using this method. The approximate solution will satisfy the inequality constraints exactly; the equality constraints $g(x) = 0$ will be satisfied approximately with error not larger than δ.

4. CONCLUSIONS

It is worthwhile to compare problem (2.1) of finding the minimum of $f(x)$ on P with the problem (3.1) under the additional constraints $x \in X$. At first glance, it seems paradoxical, although it is true, that finding the global solution in (3.1) is simpler than in (2.1). The constraint $x \in X$ provides an additional possibility to increase the radii of the covering balls on

P\X. Hence, the additional constraints merely simplify the problem of finding global solutions. The auxiliary procedures of local search in the problem (3.1) are not much more complex than in the problem (2.1), if we are concerned with the problem of finding the global minimum. In that case, the employment of the penalty function method becomes absolutely irrational for finding the global solution of (3.1). We exclude a rather unusual case where the upper estimate of the Lagrange multipliers is known sufficiently precisely at the global minimum point (see Section 3.2) and the problem reduces to a one-step minimization of the exact penalty function. In the general case where the penalty function method is used, passing from (3.1) to the multiple solution of the problem (2.1) substantially complicates the computation.

Thus, the penalty function method, though a very effective tool for finding local solutions, is not advantageous for finding global solutions. The same can be said about other local methods which require multiple minimization in x of auxiliary functions (the cost-function parametrization method, the method of modified Lagrangians, among others). We point out once more that all these methods play an important but only auxiliary role in finding global solutions.

Let us consider another essential property of this method. To reduce the number of computations, the exploitation of necessary conditions for a minimum as additional constraints gives a basis for expecting good results. We illustrate our statement, using a simple version of problem (2.1) as an example. Suppose that $f(x)$ is differentiable on P and attains a global minimum at an interior point of P. We introduce the necessary condition for a minimum $f_x(x) = 0$ as an additional constraint in relaxed form $\|f_x(x)\| \leqslant \delta$. In this case, instead of (1.12) one can use (3.12), where

$$d_{jk} = \max \left[\|f_x(x_j)\| - \delta, \quad f(x_j) - R_k \right] .$$

This makes it possible to enlarge the volume of the balls B_{jk}. If one drops the requirement that the global minimum be attained inside P, then the necessary conditions of the minimum are written in the form

$$\frac{\partial}{\partial x^i} f(x_i) \cdot (x^i - a^i)(b^i - x^i) = 0 , \qquad i \in [1:n] .$$

We determine the feasible set in (3.1) as follows:

$$X = \{x \in R^n : \| f_x^T(x) D(x-a) D(b-a) \| \leq \delta\} .$$

This idea carries over not only to (3.1) but also to other, more complex problems in which some properties of the solutions are known.

We shall dwell briefly on yet another approach to solving (3.1). Let the feasible set be defined by the scalar function $\phi(x)\colon R^n \to R^1$:

$$X = \{x \in P\colon \phi(x) = 0\} ,$$

where the function ϕ satisfies on P a Lipschitz condition

$$|\phi(x) - \phi(z)| \leqslant \ell \, \| x - z \| .$$

In contrast to (3.3), the set of approximate solutions is:

$$X_\varepsilon = \{x \in P\colon \phi(x) \leqslant \varepsilon\} .$$

To each point $x_i \in P$ we associate the ball

$$B_{jk} = \{x \in R^n: \|x - x_j\| \leqslant r_{jk}\} ;$$

here the radius is given by

$$r_{jk} = \frac{1}{\ell} \max[\varepsilon + f(x_j) - R_k, \quad \phi(x_j) - \delta] ,$$

where $0 < \delta < \varepsilon$.

<u>THEOREM</u>. Let the set $X \neq \emptyset$ and let the sequence of points $\{x_i\}$ of P be such that

$$P \subset \bigcup_{j=1}^{k} B_{jk} .$$

Then it is guaranteed that each record point $x_i \in X_\varepsilon$.

The proof follows the same arguments above. As the function $\phi(x)$ one may take, for instance, the Hölder norm of the vector $F(x)$ which was defined in Section 3.2. In this case the function $\phi(x)$ satisfies a Lipschitz condition if all components of the vector functions $g(x)$, $h(x)$ satisfy a Lipschitz condition.

4. SOLUTION OF SYSTEMS OF ALGEBRAIC EQUATIONS

1. *GENERAL SCHEME OF COMPUTATIONS*

On the n-dimensional parallelepiped P let a mapping $F: P \rightarrow R^m$, $m \leq n$, be defined and P be given by (2.2). The problem consists in finding solutions of the system

$$F(x) = 0 , \qquad x \in P . \tag{4.1}$$

Approximate solutions of the problem will be given by the points of the set $X_\varepsilon = \{x \in R^n: ||F(x)|| \leq \varepsilon,\ x \in P\}$. To solve the problem, it suffices to find at least one point x_* in X_ε. We assume that further computations for sharpening the solution of (4.1) will involve local methods (see Section 2.5). When X_ε is empty, the algorithm should guarantee that the assertion concerning the absence of approximate solutions be true.

We suppose that the mapping $F(x)$ satisfies a Lipschitz condition on P with constant ℓ, i.e., for any $x, y \in P$ one has

$$\|F(x) - F(y)\| \leq \ell\|x - y\| . \tag{4.2}$$

The problem (4.1) is equivalent to the minimization of the norm of $F(x)$ on P,

$$f_* = \min_{x \in P} f(x) , \tag{4.3}$$

where $f(x) = \|F(x)\|$. If (4.1) has a solution, then $f_* = 0$; otherwise, $f_* > 0$. To solve (4.3) one can use the results of the preceding sections. For the sequence of points $\{x_k\}$ in P we use (1.3) and (1.4) to determine the record and the set Z_k. If $R_k \leqslant \varepsilon$, then (4.1) is solved; otherwise, the set Z_k can be excluded from the search. If x_j is an interior point of Z_k, then along with x_j one can omit from consideration some ball B_{jk} centered at x_j. For all points of this ball, the condition (1.10) has to be satisfied. We use the known inequality

$$| \, \|a\| - \|b\| \, | \leq \|a - b\| .$$

Then (4.2) yields

$$|f(x_j) - f(x)| \leq \|F(x_j) - F(x)\| \leq \ell\|x_j - x\| .$$

Hence (1.10) a priori holds if (1.11) holds, whence we conclude that the ball B_{jk} and the sphere enclosing it are determined from (1.12) and the radius of the sphere from (1.13). Further constructions are made following the schemes described above. The covering of P yields either at least one point from X_ε or guarantees that X_ε is empty; also we find a point x_* furnishing the norm f with the smallest values R_k on P (with error not larger than ε).

2. A SIMPLIFIED VERSION

When it is known that X_ε is not empty, the computations can be made really simple. Take $\varepsilon_1 < \varepsilon$, $\varepsilon_1 > 0$. Let $f(x) = ||F(x)||$ be calculated at the points $\{x_k\}$. If the R_k determined from (1.3) is less than ε, then the problem is solved, otherwise we omit from consideration the points x which satisfy $\varepsilon_1 \leq f(x)$. Using (4.2), we obtain that this condition is satisfied by the points lying inside the sphere with centers at points x_j and radii

$$r_j = \frac{f(x_j) - \varepsilon}{\ell} . \tag{4.4}$$

From the inequality $f(x_j) > \varepsilon$ it follows that all the radii r_j are not less than $(\varepsilon - \varepsilon_1)/\ell$. It $R_j - \varepsilon \geq \varepsilon_1$, then the comparison of (1.11) and (4.4) implies that $r_j \geq r_{jk}$, which makes it possible in the given case to cover P by balls of larger radius. We complete the covering and obtain $R_k < \varepsilon$; the problem is solved. If the initial hypothesis that $X_\varepsilon \neq \emptyset$ does not hold, we can see it for sure by the results; however, R_k need not coincide in this case with the minimum of $||F(x)||$ on P.

3. PARAMETRIZATION

Solving the problem (2.1) of finding the global minimum on P can be reduced to finding solutions of the equation

$$\phi(f(x) - \eta) = 0 , \qquad x \in P , \tag{4.5}$$

where $\phi(z)$ is a strictly monotonic increasing function of z, $\phi(0) = 0$. If the auxiliary parameter $\eta = f_*$, the solution of (4.5) yields points $x \in X_*$; if $\eta < f_*$, then (4.5) has no real solutions. One can reduce the problem of finding the global minimum on P to finding the smallest value η for which (4.5) has a solution. This technique is analogous to the cost-function parametrization method (see Section 3.3). However, the wisdom of such an approach in the present case is questionable. The point is that in finding the global solution to (2.1) it was required to perform only one covering of P, and we obtained an approximate solution. But for an appropriate choice of the parameter η it is necessary to make a covering of P at least for several values of η, which is much more difficult to do than immediately solving the initial problem. Therefore, it is not sensible to reduce the problem (2.1) to finding a solution of the equation (4.5).

5. SOLUTION OF MINIMAX PROBLEMS

We examine here the problem (1.5.6) of finding the minimax of $F(x,y)$ that was studied in Chapter 1:

$$V_1 = \min_{x \in X} \max_{y \in Y} F(x,y) . \tag{5.1}$$

By (1.5.8) we define the maximum function $\phi(x)$, the point-set mapping $B(x)$ and the set X_*.

By Theorem 1.5.2, if $F(z) = F(x,y)$ on $\Omega = X \times Y$ satisfies a Lipschitz condition in z with constant ℓ, the maximum function $\phi(x) = F(x, B(x))$, where $B(x) = \operatorname{Arg}\max_{y \in Y} F(x,y)$, defined by (1.58), also satisfies a Lipschitz condition with the same constant ℓ. This property opens broad possibilities to use the method of finding global extrema for sequential minimax problems. The same method can be used sequentially for solving interior as well as exterior problems. The subprograms of local search usually are chosen differently, since the functions $F(x,y)$ are often differentiable and their local maximization is carried out using properties of smoothness of F in y. The function ϕ is only directionally differentiable, and it has to be locally minimized by other methods.

Comparing the problem (5.1) with the problem of finding the global extremum of F in z on Ω, we can conclude that (5.1) has an important advantage. Indeed, let the value of $\phi(x_1)$ be known at some point x_1. If for some other point x_2 one needs to find the value of $\phi(x_2)$, the process of maximization of ϕ in x can be stopped as soon as at least one point $y_2 \in Y$ has been found such that $F(x_2,y_2) \geqslant \phi(x_1)$ since in this case a priori $\phi(x_2) \geq \phi(x_1)$ and the value $\phi(x_2)$ does not improve the current approximate value of the minimax of $\phi(x_1)$. This property makes it possible in a number of cases to terminate the process of solving the interior problem.

We use a very simple scalar case where

$$X = \{x \in R^1 : a \le x \le b\} \quad ,$$

$$Y = \{y \in R^1 : c \le y \le d\} \quad .$$

For simplicity, we consider the case where both the interior and exterior problems are solved by the basic method described in Section 6.2. The accuracy of solving the interior problem (maximization in y) is given by ε_1 and that of the exterior problem (minimization in x) is given by ε_2.

In the process of solving the exterior problem, suppose the approximate values $\bar{\phi}(x_1)$, $\bar{\phi}(x_2), \ldots,$ $\bar{\phi}(x_s)$ are determined, for which we have the estimate

$$\phi(x_i) - \bar{\phi}(x_i) \le \varepsilon_1 \quad .$$

We determine the record R_s and the approximate record $\bar{R}_s$:

$$R_s = \min\,[\phi(x_1), \ \ldots, \ \phi(x_s)] \quad ,$$

$$\bar{R}_s = \min\,[\bar{\phi}(x_1), \ \ldots, \ \bar{\phi}(x_s)] \quad .$$

This implies that $R_s - \bar{R}_s \le \varepsilon_1$. Passing from the sth to the $s{+}1^{st}$ point is made by the formulas (2.12):

$$x_{s+1} = x_s + \frac{2\varepsilon_2 + \bar{\phi}(x_3) - \bar{R}_s}{\ell} \quad . \tag{5.2}$$

The solution of the exterior problem continues until (2.7) is satisfied.

In the process of the last solution of the interior problem, suppose the values $f(x_s,y_1), \ \ldots, \ f(x_s,y_k)$ are calculated. We determine the record

$$H_k(x_s) = \max [f(x_s, y_1), \ldots, f(x_s, y_k)] .$$

It is obvious that

$$H_k(x_s) \le \max_{c \le y \le d} f(x_s, t) = \phi(x_s) .$$

If $\bar{R}_s \le H_k(x_s)$, then further maximization can be stopped and one can proceed to the new point x_{s+1} using (5.2), taking there the value $H_k(x_s)$ as $\bar{\phi}(x_s)$. Another possibility is to continue the maximization process in y. Then we set

$$y_{k+1} = y_k + \frac{2\varepsilon_1 + H_k(x_s) - f(x_s, y_k)}{\ell} .$$

The process can be continued till $y_i \le d + \frac{\varepsilon_1}{\ell}$. If the last such point was y_q, we change x by (5.2), letting $\bar{\phi}(x_s) = H_q(x_s)$. If $H_q(x_x) > H_k(x_s)$, these additional computations are justified since the set of variation of x by (5.2) has grown by the quantity $\frac{H_q(x_s) - H_k(x_s)}{\ell}$. The applicability of any version of the method depends in general on the dimension of the vectors x and y, as well as the behavior of $f(x,y)$.

Theoretically, this approach makes it possible to solve sequential minimax problems and opens the door to solving the problems of discrete approximation of differential games. The exploitation of local methods can, as usual, enhance the effectiveness of the basic method.

6. SOLUTION OF MULTICRITERIA PROBLEMS

1. THE STATEMENT OF THE PROBLEM

Practical construction of complex multipurpose program packages leads one to solving so-called multicriteria problems. Let $x \in R^n$ be the vector of parameters of the product to be manufactured and the set X be given to which the vectors x have to belong. The product quality is estimated by the set of characteristics $F^i(x)$, $i \in [1:m]$. We denote this set by $F(x) := (F^1(x), \ldots, F^m(x))$.

The designer prefers to choose a feasible point $x \in X$ such that all the components of the vector $F(x)$ take on the smallest possible values. However, this condition is usually unfeasible: when one component is minimized the other components increase. Hence the term "solution of the multicriteria optimization problem" requires special clarification. By

$$\min_{x \in X} F(x) \tag{6.1}$$

we denote the problem of multicriteria minimization of $F(x)$ on X. By solving this problem we mean finding points from the so-called Pareto set X_*. We say that the point x_* belongs to the Pareto set X_* if $x \in X$, there are no other points $x \in X$ such that $F^i(x) \le F^i(x_*)$ for all $i \in [1:m]$ and for at least one $j \in [1:m]$ the strict inequality $F^j(x) < F^j(x_*)$ is satisfied. The set of all points having this property is referred to as the Pareto set and is denoted by X_*.

Next we introduce the images of the sets X and X_* under the mapping $F(x)$:

$$Y = F(X) , \qquad Y_* = F(X_*) .$$

The set Y_* is the Pareto set for the following very simple multicriteria problem:

$$\min_{y \in Y} y . \qquad (6.2)$$

We say that X_* is the Pareto set in the space of constructive parameters and its image Y_* is the Pareto set in the space of criteria.

If it happens that for two points $x_1, x_2 \in X$ the inequalities $y_1 = F(x_1) \le y_2 = F(x_2)$, $y_1 \ne y_2$ are satisfied, we say that the point y_1 is more efficient than y_2, or that y_2 is less efficient than y_1.

We assume that each component of F satisfies a Lipschitz condition with the same constant ℓ, i.e., for any x_1 and x_2 we have

$$|F^i(x_1) - F^i(x_2)| \le \ell\|x_1 - x_2\| , \qquad i \in [1:m]$$

yielding the vector inequality

$$F(x_1) - e\ell\|x_1 - x_2\| \le F(x_2) , \qquad (6.3)$$

where $e \in R^n$ is the vector of all ℓ's.

2. THE CONSTRUCTION OF THE NET

The structure of the Pareto set even for the simplest problems turns out, as a rule, to be very complex. This set is frequently nonconvex and non-simply connected. Hence it is hard to approximate. We will attempt to construct a finite set A_k reminiscent of the usual notion of an ε-net of the set Y_*. Take a set of points $A_k = \{y_1,\dots,y_k\}$, $y_i = F(x_i)$, $x_i \in X$, for all $i \in [1:k]$. We assume that along with A_k the corresponding set of points x_i from the feasible set X is stored in the computer memory, or is easily calculated.

Besides feasibility, we impose two other conditions on the set of points A_k:

• 1. for any $y_* \in Y_*$ there exists a vector $y_i \in A_k$ such that

$$y_i \le y_* + \varepsilon e \quad ; \tag{6.4}$$

• 2. for any $y_j \in A_k$ there does not exist a vector $y_i \in A_k$ such that $y_j \le y_i$.

We say that the set of points A_k satisfying these conditions is an ε-net of the Pareto set, and refer to the conditions as the first and second net conditions, respectively.

For $y_i \in Y$ we define the set

$$M_i = \{y \in R^m\colon\ y_i \le y + \varepsilon e\} \quad ,$$

which contains all the points less efficient than $y_i - \varepsilon e$.

Let $y_i \in A_k$. Then the set M_i is of no interest to us from the viewpoint of constructing the ε-net. Indeed, each y_* in Y_* that belongs to M_i satisfies the first net condition (6.4). Hence the set M_i can be omitted from consideration.

We take the union of the sets M_i over all i of A_k and denote it by $Z_k = \bigcup_{i=1}^{k} M_i$. This set can be expressed in terms of

$$Z_k = \left\{y \in R^m\colon\ \max_{i\in[1:k]}\ \min_{j\in[1:m]}\ [\varepsilon + y^j - y_i^j] \ge 0\right\} \quad .$$

The set A_k varies in the computational process. If we could find a point $\bar{y} \in Y$ such that $\bar{y} \leq y_i$, where $y_i \in A_k$, we take y_i out of A_k and replace it by $\bar{y}$. Several points can be removed simultaneously. Thanks to this, the second net condition of the Pareto set holds automatically. On the other hand, if it happens that $\bar{y}$ does not belong to Z_k, then it is included in A_k, which is now written A_{k+1}. If, as a result of the construction of the set A_k it happens that

$$Y \subset Z_k \quad , \tag{6.5}$$

then A_k forms the ε-net of the Pareto set. Indeed, for each point $y_* \in Y_* \subset Y$ we can find a point $y_i \in A_k$ such that (6.4) holds. The problem of constructing the ε-net of the Pareto set has thus been reduced to constructing the set of points A_k satisfying (6.5). As in the preceding sections, to do this we use the corollary (6.3) of the Lipschitz condition.

At $x_* \in X$ let the value $y_* = F(x_*)$ be calculated, and suppose it turned out that $y_* \in M_i$. From (6.3) it follows that

$$F(x_*) - e\ell \|x - x_*\| \leq F(x) \quad .$$

If x is such that

$$F(x_i) - \varepsilon e \leq F(x_*) - e\ell \|x - x_*\| \quad ,$$

then $y = F(x) \in M_i$. Hence all points in X satisfying

$$e\ell \|x_* - x\| \leq F(x_*) - F(x_i) + \varepsilon e \tag{6.6}$$

can be omitted in covering the set X. This set contains a ball with center at x_*:

$$B_i = \left\{ x \in R^n : \ell \| x - x_* \| \le \varepsilon + \min_{s \in [1:m]} [F^s(x_*) - F^s(x_i)] \right\} .$$

If $y_i = y_*$, then the radius is smallest, equal to $\frac{\varepsilon}{\ell}$. In the case when A_k contains several points more efficient than y_*, we introduce the index set

$$I(y_*) = \{ i \in [1:k] : y_i \le y_*, \; y_* \in A_k \} .$$

This set contains the whole collection of indices of vectors in A_k which are more efficient than y_*. If $I(y_*)$ is nonempty, then after determining $y_* = F(x_*)$ one can omit all the points x for which (6.6) holds for at least one $i \in I(y_*)$. Hence it is optimal to choose an i such that the corresponding ball B_i has the largest radius. This radius is computed by the formula

$$\rho = \frac{1}{\ell} \left| \varepsilon + \max_{i \in I(y_*)} \min_{s \in [1:m]} (F^s(x_*) - F^s(x_i)) \right| . \qquad (6.7)$$

Construction of the ε-net of the Pareto set has thus been reduced to covering the set X by balls of the form (6.7). To implement this process one can employ the techniques described in Sections 7.2 and 7.3. If X is bounded, then its covering is accomplished in a finite number of steps, and the ε-net will also be finite. Here, and in all the cases considered earlier in this chapter, in order to speed up the computations it is useful to use the methods of local search. Such methods for determining the points in the Pareto set are now being successfully developed.

The set A_k thus found is provided to the customer who, according to his own considerations, chooses the most preferable set of design parameters. If A_k turns out to be too large, it can be reduced by discarding close points. The distance between points can be defined in the space of criteria and in the space of design parameters. The user gives a number N determining the smallest distance between the points, and a special program "sifts" through the set A_k, leaving only those points which are separated by the distance greater than N.

The basic result of this section is embodied in the following assertion: approximate solution of the multicriteria problem (6.1), in terms of computational labor, is equivalent to the problem of finding the global minimum (1.1). There is, of course, some complication connected with the fact that it is necessary to calculate m values of $F^i(x)$ instead of calculating the value of $f(x)$, as well as to store the set of points A_k. However, the basic computations with respect to the covering of X are roughly the same.

In the literature on multicriteria optimization, the following three ideas for solving the problem are most popular.

♦1. Forming the convolution of criteria: For example, under the assumption that all criteria are nonnegative, the convex hull is:

$$P(x,\alpha) = \sum_{i=1}^{m} \alpha^i F^i(x) , \qquad \alpha^i \geq 0 .$$

Then for a fixed set α^i one solves the problem

$$\min_{x \in X} P(x,\alpha) = P(x_*(\alpha), \alpha) \tag{6.8}$$

and then has $x_*(\alpha) \in X_*$. For another vector $\bar{\alpha}$ a new problem of seeking the minimum of $P(x,\bar{\alpha})$ is solved, etc. In the case when the required number of points from the Pareto set is greater than one, this approach is not efficient since its implementation requires a multiple search for the global minimum.

♦2. One can pose the problem of nonlinear programming of finding $\min_{x \in W} F^1(x)$, where the feasible set $W = \{x \in X: F^i(x) \geq \alpha^i, i \in [2:m]\}$ is specified by specifying the vector α. Again, to obtain only one point from X_* one has to solve a global minimum problem.

♦3. The parametrized function is:

$$P(x,\alpha) = \sum_{i=1}^{m} (F^i(x) - \alpha^i)^2$$

and its minimum is sought on the set X for equivalent sets of α. This approach also requires a correctly specified vector α and leads to a large volume of calculations.

All these approaches can be useful if they are employed as local procedures for determining the initial approximate ε-net of the Pareto set. Then the succeeding solutions of the auxiliary problems do not require a large volume of calculations.

This concludes the description of a new trend in global optimization. The results of this chapter should convincingly demonstrate the universality as well as the potential of the method of nonuniform coverings.

Appendix I

DIFFERENTIABILITY

1. DIFFERENTIABLE FUNCTIONS

Let the real function $f(x)$ of the n-dimensional vector $x \in R^n$ be defined on an open set X. We say that $f(x)$ is differentiable at $x \in X$ if there exists a bounded vector $p \in R^n$ such that for all $h \in R^n$ satisfying the condition $x + h \in X$ we have

$$f(x+h) = f(x) + \langle p,h \rangle + \|h\| \alpha(x,h) , \qquad (A1.1)$$

where the function α has the property

$$\lim_{\|h\| \to 0} \alpha(x,h) = 0 .$$

Suppose that f is differentiable at x. Then f is continuous at x, its partial derivatives with respect to all coordinates exist and the n-dimensional vector

$$f_x(x) = \left| \frac{\partial f(x)}{\partial x^1}, \frac{\partial f(x)}{\partial x^2}, \ldots, \frac{\partial f(x)}{\partial x^n} \right|^T ,$$

which is called the gradient of f at x, is defined. In (A1.1) one can put $p = f_x(x)$.

We say that f is differentiable on an open set X if it is differentiable at each $x \in X$.

By f being differentiable on the set X (perhaps closed)

we mean that f is differentiable on an open set containing X.

We say that f is differentiable (everywhere) if it is differentiable at each $x \in R^n$.

A function $f(x)$ that has continuous partial derivatives at a point x (or on an open set X) is said to be continuously differentiable at x (on X).

Let h denote an n-dimensional vector with norm 1. The limit

$$\lim_{t\to+0} \frac{f(x+th) - f(x)}{t} = \langle h, f_x(x) \rangle ,$$

if it exists, is said to be the derivative of $f(x)$ at x in the direction h and is denoted $\frac{\partial f(x)}{\partial h}$.

If f is differentiable at x, it has derivatives in any direction at x.

<u>THEOREM A.1</u>. If $f(x)$ is differentiable on an open convex set $X \subset R^n$ and $x_1, x_2 \in X$, then there exists a number $0 < \tau < 1$ such that the following equalities hold:

$$f(x_2) - f(x_1) = \langle f_x(x_1 + \tau(x_2 - x_1)), x_2 - x_1 \rangle , \tag{A1.2}$$

$$f(x_2) - f(x_1) = \int_0^1 \langle f_x(x_1 + \theta(x_2 - x_1)), x_2 - x_1 \rangle \, d\theta . \tag{A1.3}$$

The formula (A1.2) is usually known as the Lagrange formula, and (A1.3) as the Newton-Leibniz formula.

We say that a function $f(x)$ defined on X satisfies a Lipschitz condition with constant ℓ on X if for any $x_1, x_2 \in X$ we have the inequality

$$|f(x_1) - f(x_2)| \le \ell \|x_1 - x_2\| . \tag{A1.4}$$

It is obvious that the function f satisfying a Lipschitz condition on X is continuous on X.

THEOREM A.2. Let $f(x)$ be differentiable on an open convex set X, where its gradient $f_x(x)$ satisfies a Lipschitz condition with constant ℓ. Then for any $x_1, x_2 \in X$ and any $0 \le \tau \le 1$ we have

$$f(x_1 + \tau(x_2 - x_1)) \le f(x_1) + \tau\langle f_x(x_1), x_2 - x_1\rangle + \frac{\tau^2}{2}\ell \|x_2 - x_1\|^2 ,$$

$$f(x_2) - f(x_1) = \langle f_x(x_1), x_2 - x_1\rangle + \eta(x_2 - x_1) , \tag{A1.5}$$

where the function $\eta(z)$ is such that $\eta(0) = 0$ and $|\eta(z)| \le \ell\|z\|^2$.

2. TWICE DIFFERENTIABLE FUNCTIONS

Let $f(x)$ be defined on an open set $X \subset R^n$. We say that $f(x)$ is twice differentiable at $x \in X$ if for all $h \in R^n$ satisfying $x + h \in X$ we have

$$f(x+h) - f(x) = f_x^T(x)h + \tfrac{1}{2} h^T f_{xx}(x)h + \|h\|^2 \beta(x,h) ,$$

where the function β has the property

$$\lim_{\|h\| \to 0} \beta(x,h) = 0 .$$

Here $f_{xx}(x)$ is the $n \times n$ matrix of second derivatives of f at x (the Hessian) with the $(i,j)^{th}$ element

$$\frac{\partial^2 f(x)}{\partial x^i \partial x^j} .$$

THEOREM A.3 (Taylor's Formula). If f is twice differentiable on an open convex set $X \subset R^n$ and $x_1, x_2 \in X$, then there exists a number $0 < \tau < 1$ such that we have the equality

$$f(x_2) - f(x_1)$$
$$= f_x^T(x_1)(x_2-x_1) + \tfrac{1}{2}(x_2-x_1)^T f_{xx}(x_1 + \tau(x_2-x_1))(x_2-x_1) \quad .$$

3. DIFFERENTIABILITY OF MAPPINGS

Let a mapping $g\colon X \to R^m$ be given, where X is a set in R^n. We say that the mapping g is differentiable at an interior point x of X if there exists $n \times m$ matrix A such that for all $h \in R^n$ satisfying the condition $x + h \in X$ we have

$$g(x+h) - g(x) = A^T h + \|h\| \Phi(x,h) \quad ,$$

where the vector function Φ has the property

$$\lim_{\|h\| \to 0} \|\Phi(x,h)\| = 0 \quad .$$

The matrix A is called the derivative of the mapping $g(x)$ at x and denoted $g_x(x)$.

The differentiability defined in this way is often called a Fréchet differentiability. This definition implies that if the mapping $g\colon X \to R^m$ is differentiable at $x \in \text{int}\, X$, $X \subset R^n$, then the condition

$$\lim_{\|h\| \to 0} \frac{1}{\|h\|} \|g(x+h) - g(x) - g_x^T(x)h\| = 0$$

is satisfied.

The Newton-Leibniz formula (A1.3) extends to the case of mappings, while (A1.2) does not. Hence the following inequality is what we call the Lagrange formula for mappings:

$$||g(y) - g(x)|| \le \sup_{0 \le \tau \le 1} ||g_x(x + \tau(y-x))|| \; ||y-x|| \; . \qquad (A1.6)$$

THEOREM A.4. Let the mapping $g: X \to R^m$ be differentiable on an open convex set X. Then for any $x, y \in X$ the inequality (A1.6) holds.

If for any $x_1, x_2 \in R^n$ we have

$$||g(x_1) - g(x_2)|| \le \ell ||x_1 - x_2|| \quad ,$$

$g(x)$ is called a Lipschitz mapping with constant ℓ.

4. DIFFERENTIABILITY OF COMPOSITE FUNCTIONS

THEOREM A.5. Let the function $g: x \to R^m$ be defined on an open set $X \subset R^n$ and let the function ϕ be defined on R^m. Then, if g is differentiable at $\bar{x} \in X$ and ϕ is differentiable at $\bar{y} = g(\bar{x})$, the composite function $\gamma(x) = \phi(g(x))$ is differentiable at $\bar{x}$ and

$$\frac{d\gamma(\bar{x})}{dx} = g_x(\bar{x}) \; \phi_g(g(\bar{x})) \quad .$$

This result carries over to the case where $\gamma(x) = \phi(x, g(x))$, where ϕ is defined on the Cartesian product $R^n \times R^m$. The formula for calculating the derivative of the composite function $\gamma(x)$ has the form

$$\frac{d\gamma(x)}{dx} = \phi_x(x, g(x)) + g_x(x) \; \phi_g(x, g(x)) \quad .$$

Here ϕ_x is the partial derivative of ϕ with respect to the explicit vector x and ϕ_g is the partial derivative of $\phi(x,g)$ with respect to g.

Appendix II

SOME PROPERTIES OF MATRICES

1.

By a matrix A we mean a rectangular table of numbers

$$A = \begin{bmatrix} a_{11}, & a_{12}, & \dots, & a_{1n} \\ a_{21}, & a_{22}, & \dots, & a_{2n} \\ \cdot & \cdot & \cdot & \cdot \\ a_{m1}, & a_{m2}, & \dots, & a_{mn} \end{bmatrix} .$$

If $m = n$, the matrix is called square, the number $m = n$ is called its order. In the general case when m and n are not equal, the matrix is called a rectangular matrix of dimension $m \times n$. The numbers forming the matrix are called its elements (components). At the intersection of the ith row and the jth column there is the element a_{ij} which is called the $(i,j)^{th}$ element of A. The elements a_{ii}, $i \in [1:n]$ for the principal diagonal of a square matrix A.

Let A and B have elements a_{ij} and b_{ji}, respectively, where $i \in [1:m]$, $j \in [1:n]$. Then by the product of A by B we mean the matrix C whose $(i,q)^{th}$ element is

$$c_{iq} = \sum_{k=1}^{n} a_{ik} b_{kq} .$$

Let $D(z)$ denote the diagonal matrix, all off-diagonal elements of which are zero and the ith diagonal element is z^i (the ith component of the vector z). The order of $D(z)$ is determined by the dimension of the vector z.

If $C = AD(z)$, $B = D(z)A$, then $c_{ij} = a_{ij}z^j$, $b_{ij} = z^i a_{ij}$. Thus, when an $m \times n$ rectangular matrix A is multiplied on the right by a diagonal matrix $D(z)$ of order n, all columns of A are multiplied by the numbers $z^1, z^2, \ldots, z^n$. When the matrix A is multiplied on the left by a diagonal matrix $D(z)$, all rows of A are multiplied by the numbers $z^1, z^2, \ldots, z^n$.

An $m \times n$ matrix, all elements of which are zero is called the zero matrix and denoted 0_{mn}. A square matrix of order n, with 1 on the main diagonal and zero elsewhere, is called an identity matrix and denoted I_n.

A square matrix A of order n is called symmetric if $a_{ij} = a_{ji}$ for all $i,j \in [1:n]$. Such a matrix coincides with its transpose: $A = A^T$.

Let A be a square matrix of order n. The matrix A is called degenerate (or singular) if its determinant is equal to zero: $|A| = 0$. The roots of the equation

$$|A - \lambda I_n| = 0 \tag{A2.1}$$

are called the characteristic values or eigenvalues of A. The equation (A2.1) is called the characteristic equation of the matrix A. The determinant and the eigenvalues of A are continuous functions of the elements of A. If the equation

$$Ax = \lambda x \tag{A2.2}$$

has a nonzero solution $x \in E^n$, x is called an eigenvector of A corresponding to the eigenvalue λ. The matrices A and A^T have identical eigenvalues.

If A is a real symmetric matrix, then all its eigenvalues are real and we have the inequalities

$$\lambda_1 x^T x \leq x^T A x \leq \lambda_n x^T x \qquad x \in E^n ,$$

where $\lambda_1 \leq \lambda_2 \leq \cdots \leq \lambda_n$ are the eigenvalues of A. If $\lambda_1 > 0$, A is positive definite $(A > 0)$; if $\lambda_1 \geq 0$, A is said to be positive semidefinite or nonnegative definite $(A \geq 0)$.

LEMMA (R. Finsler). If $x^T Ax > 0$ for all nonzero x satisfying the condition $x^T Bx = 0$, where B is a nonnegative definite square matrix, there exists a number τ_* such that the quadratic form $x^T Ax + \tau x^T Bx$ is positive definite for $\tau > \tau_*$.

The matrix B is called a square root of A if $A = BB$. Each positive definite matrix has such a positive definite square root. If the square matrix A is positive and B is nonsingular, then $B^T AB$ is positive definite. In particular, for any nonsingular matrix B the matrix BB^* is positive definite.

If A and B are square matrices of the same order, the following formulas hold true:

$$|AB| = |A| \cdot |B| , \qquad (AB)^T = B^T A^T .$$

The matrix A^{-1} is called the inverse of A if $AA^{-1} = I_n$. Any nonsingular matrix has an inverse. If A and B are nonsingular matrices of the same order, then

$$|A^{-1}| = |A|^{-1} , \qquad (A^{-1})^T = (A^T)^{-1} , \qquad (AB)^{-1} = B^{-1}A^{-1} .$$

2.

A mapping $\|\cdot\|: R^n \to R^1$ is called a norm if for any $x \in R^n$ the following three conditions are satisfied:

- 1. $0 \le \|x\|$, and $\|x\| = 0$ iff $x = 0$;
- 2. $\|\alpha x\| = |\alpha| \cdot \|x\|$, $\alpha \in R^1$;
- 3. $\|x+y\| \le \|x\|+\|y\|$ for any $x,y \in R^n$.

For the scalar product in E^n the Cauchy inequality is satisfied:

$$|\langle x,y\rangle| \le \|x\|_2 \cdot \|y\|_2 .$$

For a vector norm $\|\cdot\|_p$ in R^n there exists a vector $\|\cdot\|_q$ called the dual or the conjugate norm to $\|\cdot\|_p$, which is defined by the condition

$$\|x\|_q = \sup_{\|y\|_p=1} \langle x,y\rangle .$$

If as $\|\cdot\|_p$ one takes the Hölder norm

$$\|x\|_p = \left[\sum_{i=1}^{n} |x^i|^p\right]^{\frac{1}{p}}, \qquad 1 \le p \le \infty ,$$

then the dual norm is given by

$$\|x\|_q = \left[\sum_{i=1}^{n} |x^i|^q\right]^{\frac{1}{q}}, \qquad \frac{1}{p} + \frac{1}{q} = 1 ,$$

and we have:

$$|\langle x,y\rangle| \le \|x\|_p \cdot \|y\|_q .$$

In particular, for the so-called sum norm

$$\|x\|_1 = \sum_{i=1}^{n} |x^i| ,$$

the dual is the norm

$$\|x\|_\infty = \max_{i\in[1,n]} |x^i|$$

usually called the Chebyshev norm or maximum norm. The initial and the dual norms coincide for $p = q = 2$:

$$\|x\|_2 = \left[\sum_{i=1}^{n} (x^i)^2\right]^{\frac{1}{2}} .$$

Such a norm is called Euclidean or spherical. It can be obtained starting from the scalar product

$$\|x\|_2^2 = \langle x,x\rangle = x^T x .$$

For the norms of n-dimensional vectors thus defined, we have the following estimates:

$$\|\cdot\|_\infty \le \|\cdot\|_1 \le n\|\cdot\|_\infty$$

$$\|\cdot\|_\infty \le \|\cdot\|_2 \le \sqrt{n}\,\|\cdot\|_\infty$$

$$\|\cdot\|_2 \le \|\cdot\|_1 \le \sqrt{n}\,\|\cdot\|_2$$

$$n^{-\frac{1}{2}}\|\cdot\|_1 \le \|\cdot\|_2 \le \|\cdot\|_1$$

Hence

$$\|\cdot\|_\infty \le \|\cdot\|_2 \le \|\cdot\|_1 .$$

3.

Each norm of any matrix A of dimension $m \times n$ must satisfy the following three conditions:

- 1. $0 \le ||A||$, and $||A|| = 0$ iff $a = 0_{mn}$,
- 2. $||\alpha A|| = |\alpha| \cdot ||A||$, $\alpha \in R^1$,
- 3. $||A+B|| \le ||A|| + ||B||$ for any $m \times n$ matrix B.

We say that the matrix norm $||\cdot||$ is associated with the given vector norm $||\cdot||$ if for any matrix A of dimension $m \times n$ and any n-dimensional vector x the inequality

$$||Ax|| \le ||A|| \cdot ||x||$$

is satisfied.

We present the rule for constructing the smallest matrix norm associated with a given vector norm. For norm of A we take the maximum of the vector norms Ax as x runs through the set of all vectors with unit norm:

$$||A|| = \max_{||x||=1} ||Ax|| .$$

Starting from this rule one can show that to the vector norms $||\cdot||_1$, $||\cdot||_\infty$, $||\cdot||_2$ there correspond the following associated matrix norms:

$$||A||_1 = \max_{j \in [1:n]} \sum_{i=1}^{m} |a_{ij}| ,$$

$$||A||_\infty = \max_{i \in [1:m]} \sum_{j=1}^{n} |a_{ij}| ,$$

$$||A||_2 = \sqrt{\lambda} ,$$

where λ is the maximum eigenvalue of A^TA. In particular, if A is an $n \times n$ symmetric matrix with eigenvalues $\lambda_1, \lambda_2, \ldots, \lambda_n$, then

$$\|A\|_2 = \max_{i \in [1:n]} |\lambda_i| . \qquad \text{(A2.3)}$$

By the spectral radius $S(A)$ of the square matrix A we mean the maximum modulus of the eigenvalues of A. From (A2.2) and the properties of the norms we obtain

$$|\lambda| \; \|x\| \leq \|A\| \cdot \|x\| .$$

No norm of a matrix is less than its spectral radius. From (A2.3) it follows that $S(A) = \|A\|_2$. At the same time, one can show that for any given $\varepsilon > 0$ there always exists a norm $\|A\|_\delta$ such that $S(A) + \varepsilon \geq \|A\|_\delta$.

LEMMA (Neumann). Let the spectral radius of the square matrix A be strictly less than one. Then the matrix $I_n - A$ is nonsingular and

$$(I_n - A)^{-1} = \lim_{s\to\infty} \sum_{i=0}^{s} A^i .$$

4.

Let the matrices A, B and C of dimensions $n \times n$, $n \times m$ and $m \times n$, respectively, be given; and let the matrix A be nonsingular and $m \leq n$. The matrix $A + BC$ is nonsingular iff $I_m + CA^{-1}B$ is nonsingular, and if so we have the Sherman-Morrison-Woodbury formula:

$$(A + BC)^{-1} = A^{-1} - A^{-1}B(I_m + CA^{-1}B)^{-1} CA^{-1} . \qquad \text{(A2.4)}$$

For $m = 1$, the vectors b and c are taken as B and C, respectively; then from (A2.4) we have the so-called Sherman-Morrison formula:

$$(A + bc^T)^{-1} = A^{-1} - (1 + c^T A^{-1} b)^{-1} A^{-1} bc^T A^{-1} .$$

5.

Let the nonsingular matrix M be decomposed into blocks

$$M = \left[\begin{array}{c|c} A & B \\ \hline C & P \end{array}\right],$$

where A, B, C, P have dimensions $n \times n$, $n \times q$, $q \times n$, $q \times q$, respectively. Let the square matrix A be nonsingular. The Frobenius formula for the inverse matrix holds:

$$M^{-1} = \left[\begin{array}{c|c} A^{-1}+A^{-1}BH^{-1}CA^{-1} & -A^{-1}BH^{-1} \\ \hline -H^{-1}CA^{-1} & H^{-1} \end{array}\right],$$

where $H = P - CA^{-1}B$.

Using this formula, inversion of a square matrix of order $n+q$ reduces to inversion of two matrices of orders n and q, respectively, and to the operations of addition and multiplication of matrices of dimensions $n \times n$, $q \times q$, $n \times q$, $q \times n$.

If instead of the condition $|A| \neq 0$ we introduce the assumption that $|P| \neq 0$, the Frobenius formula has a different form:

$$M^{-1} = \left[\begin{array}{c|c} K^{-1} & -K^{-1}BP^{-1} \\ \hline -P^{-1}CK^{-1} & P^{-1}+P^{-1}CK^{-1}BP^{-1} \end{array}\right],$$

where $K = A - BP^{-1}C$.

More detailed information concerning properties of matrices may be found, for example, in Gantmacher [1], Lancaster [1], Bellman [2], Voevodin [1], and Coddington and Levinson [1].

Appendix III

SOME PROPERTIES OF MAPPINGS

1. SINGLE-VALUED MAPPINGS

To each element x of a set $X \subset E^n$ let there correspond a unique element $T(x)$ of a set $Y \subset E^m$. In this case we say that we have a single-valued mapping T of the set X into the set Y and we write $T: X \to Y$. For any set $S \subset X$ we define its image:

$$T(s) = \{y \in Y : \exists \ x \in X \ \text{ such that } \ y = T(x)\} \quad .$$

There exist two equivalent formulations of the notion of continuity for mappings, one in terms of convergence and one in the sense of Cauchy.

●1. The mapping T is continuous at x if for any sequence $\{x_k\}$ in X converging to the limit x as $k \to \infty$ the limit exists:

$$\lim_{k\to\infty} T(x_k) = T(x) \quad .$$

●2. The mapping T is continuous at x if for any neighborhood V of $T(x)$ we can find a neighborhood G of x such that

$$T(G) \subset V \quad .$$

Of special interest are the fixed points of mappings, i.e., points belonging to the following set:

$$X_* = \{x \in E^n: x = T(x)\} ,$$

where $T: E^n \to E^n$. In constructing numerical methods of solving systems of equations, nonlinear programming problems and optimal control problems, it is often appropriate to make a reduction to a problem of finding fixed points of certain mappings. One of the first results on the existence of fixed points of mappings was obtained at the beginning of the century by the Dutch mathematician Brouwer.

BROUWER'S FIXED-POINT THEOREM: Let X be a nonempty compact convex set in E^n and the mapping $T: X \to X$ be continuous. Then T has a fixed point.

2. POINT-SET MAPPINGS

Suppose some rule W setting each point of the space E^n into correspondence with that of the space E^m is defined. Such a correspondence is called multivalued or a point-set mapping and is denoted $W: E^n \to 2^{E^m}$. The set $W(x)$ is called the image of the point x. Obviously, a single-valued mapping is a particular case of a point-set mapping under which the image of each point is a set consisting of a single point. Cases are possible when W is defined on a set $X \subset E^n$ and to each $x \in X$ there corresponds a subset of X, then we write $W: X \to 2^X$.

If the notations of continuity in terms of convergence and in the Cauchy sense are carried over to multivalued mappings, we arrive at the following two distinct notions.

A multivalued mapping $W: E^n \to 2^{E^m}$ is called closed at x if the conditions $\lim_{k\to\infty} x_k = x$, $\lim_{k\to\infty} y_k = y$ and $y_k \in W(x)$ imply $y \in W(x)$.

A multivalued mapping $W: E^n \to 2^{E^m}$ is called upper semi-continuous at x if for any neighborhood V of the image $W(x)$ there is a neighborhood G of x such that $W(G) \subset V$.

A multivalued mapping W is called closed on the set $X \subset E^n$ if it is closed at each $x \in X$. We say that the mapping W is closed if it is closed at all points at which it is defined. One extends analogously the notion of upper semi-continuity.

The point x_* is called a fixed point of the point-set mapping $T: E^n \to 2^{E^n}$ if

$$x_* \in W(x_*) \quad .$$

Brouwer's theorem has been generalized to multivalued mappings:

<u>KAKUTANI'S THEOREM</u>. Let X be a nonempty compact convex set in E and let $W: X \to 2^X$ be a multivalued mapping satisfying the conditions:

- a. for each $x \in X$ the set $W(x)$ is a nonempty convex subset of X;
- b. the mapping W is closed.

Then the mapping W has a fixed point.

Proofs of Brouwer's and Kakutani's Theorems may be found in Nikaido [1].

NOTES AND COMMENTS

CHAPTER 1

Sections 1-4. The theory of convex functions and sets, necessary and sufficient conditions for extrema of functions are described in detail in many published works. We refer the reader to Vasil'ev [1], Gol'shtein [1], Eremin and Astaf'ev [1], Karmanov [1], Mangasarian [1], Rockafellar [1], and Pshenichnyj [3].

Section 5. The first published investigation of general properties of minimax problems was apparently Chebotarev [1]. A detailed bibliography of later studies can be found in Dem'yanov and Malozemov [1], [2] and Danskin [1]. Theorems 1.5.7 - 1.5.8 have been borrowed from Evtushenko [3].

Section 6. Theorem 1.6.1 is an obvious generalization of the result obtained by Uzawa (see Chapter 3 in Arrow et al. [1]). Theorem 1.6.4 is due to Evtushenko [12]. Lemma 1.6.1 has been taken from Ky Fan et al. [1]. The assertion of Theorem 1.6.7 on the existence of a saddle point without equality-type constraints was first proved by Kuhn and Tucker [1]; and that with equality-type constraints is due to Uzawa (see Chapter 3 in Arrow et al. [1]).

Section 7. A more detailed description of conditions for constraint qualifications and their relatedness is given in Mangasarian [1]. A proof of Theorem 1.7.2 can be found in McCormick [1] and in Fiacco and McCormick [1]. Proofs of Theorem 1.7.3 and Lemma 1.7.5 have been taken from Han and Mangasarian [1]. The formulation of Theorem 1.7.5 is due to Evtushenko and Zhadan [3].

Section 8. A detailed presentation of the optimal control theory is contained in Pontryagin et al. [1], Moiseev [2], and Gabasov and Kirillova [2].

CHAPTER 2

Sections 1, 2. The fundamental theorems of stability theory were obtained at the end of the last century by Lyapunov [1]. A detailed discussion of this subject can be found in numerous works in stability theory; for instance, Barbashin [1], Demidovich [1], Krassovskij [1], and Malkin [1].

Theorem 2.2.6 was proved by Barbashin and Krassovskij. Theorem 2.2.7 is due to Evtushenko [9], a close result is obtained in Venets and Rybashov [1]. Several theorems on convergence of solutions to systems of ordinary differential equations are given in Evtushenko and Zhadan [2]. Rybashov [1], [2] examines the applications of stability theory methods in studying optimization methods.

Section 3. A very thorough investigation of the convergence of iterative processes is contained in Ortega and Rheinboldt [1], Faddeev and Faddeeva [1], Khalanaj and Veksler [1], Gaevskij, Greger, and Zakharias [1], among others. The assertion of Theorem 2.3.7 has been stated by many authors; for example, Skalkina [1].

Section 4. The theory of Fejér mappings has been developed by Eremin and his students. Their works are listed in Eremin and Mazurov [1]. An investigation of distinct variants of the generalized gradient method is contained, in particular, in Ermol'ev [1], [2] and Shorr [1].

Section 5. A very detailed discussion of methods for solving systems of nonlinear equations, plus an extensive bibliography are contained in Ortega and Rheinboldt [1] and Rheinboldt [1].

Studies of Newton's method are contained in Kantorovich and Akilov [1]. A discrete variant of Newton's method is examined in Shamanskij [2] and in more detail in Ortega and Rheinboldt [1]. Rules 1-5 for the choice of a step-length in Newton's method, with norms of special kind are formulated in Danilin and Panin [1], Kuptsov and Shurshkova [1], Panin [1], Pshenichnyj [2], Polak and Teodoru [1], and Stoer [1]. Rules 1-5 with norms of the arbitrary type are formulated and studied in Burdakov [2], [3]. A survey of quasi-Newton methods can be found in Dennis and Moré [2] and Spedicato [1], [2], Spedicato and Greenstadt [1], Shanno and K. Phua [1], among others, demonstrate that these methods are very efficient in solving systems of equations and minimizing functions.

Methods (5.27) and (5.28) have been developed by Broyden [1], (5.29) by Pearson [2], (5.30) by McCormick (see Pearson [2]), (5.31) by Davidon [1], Broyden [2], among others; (5.32) by Thomas [1], (5.36) by Davidon [1], Fletcher and Powell [1]; (5.37) by Broyden [3], Fletcher [1], Goldfarb [1], Shanno [1]; (5.38) by Powell [2]. Adachi [1] suggests a whole class of quasi-Newton methods which while minimizing quadratic functions generate conjugate directional vectors--some of those methods we give in Section 5. A proof of the n-step quadratic rate of convergence of methods with these properties towards the minimum of a nonquadratic function is given in McCormick and Ritter [1], Dixon [1], Danilin [1], Stoer [2], Baptist and Stoer [1], among others. Estimates of the convergence rate, as high as those in the (n+1)-point secant method (with the order of convergence rate being a root of the equation $t^{n+1} - t^n - 1 = 0$) have been obtained by Schuller [1]. The secant method for solving systems of nonlinear equations has been suggested for the first time by Bittner [1] and Wolfe [1]. Methods incorporating the secant method have been treated in Tornheim [1], Anderson [1], Barnes [1], Shamanskij [1], and Ul'm [1], [2]. The secant method has been studied also in Polak [2], Danilin and Pshenichnyj [1], [2], and Pshenichnyj and Danilin [1] for solving programming problems. A symmetric variant of the secant method is due to Burdakov [1].

Section 6. Many works deal with numerical methods for finding the minimax; for example, Dem'yanov and Malozemov [1], [2], Dem'yanov and Rubinov [1], Germeyer [1], and Shor [1]. A survey of methods for finding saddle points can be found in Dem'yanov and Pevnyj [1].

Theorem 2.6.1 has been borrowed from Grachev and Evtushenko [4], and the results of Subsections 2 and 3 from Evtushenko [3], [4]. Many of these methods have been generalized to the solution of game problems involving nonantagonistic players in Grachev and Evtushenko [1]. Method (6.27) has been developed by Volkonskij [1].

Methods for finding the global extremum are described in Strongin [1] and Dixon and Szegö [1]. Methods for seeking the global extremum on a nonuniform network are described in detail in Evtushenko [1] and [5].

CHAPTER 3

Section 1. The most detailed discussion of the penalty function method can be found in Fiacco and McCormick [1]. This method has also been treated in works on nonlinear programming numerical methods, for example, Vasil'ev [1], Zangwill [2], Karmanov [1], Polak [1], Pshenichnyj and Danilin [1]. In describing the penalty function method the present author has deviated from the traditional path; the presentation here follows Evtushenko [9], including Theorem 3.1.3. An analogous result is due to Shepilov [1].

Section 2. Eremin was the first who pointed out the possibility of using exact penalty functions in solving convex programming problems. Among later studies, there is Zangwill [1], Skarin [1], Pietrzykowski [1], Charalambous [1], Han and Mangasarian [1] from which Theorem 3.2.2 has been borrowed. In proving Theorem 3.2.1 we were able to do without the traditional conditions for convexity. Estimate (2.9) can be found in many works (see, for example, Skarin [1]. Studies of the estimation of an accuracy of the penalty function method through the method of asymptotic expansions appear for the first time. Polak [2] treats similar problems.

Section 3. References to works on this subject are cited throughout the text.

Section 4. The interior penalty function method is discussed in more detail in Fiacco and McCormick [1].

Section 5. References are cited throughout the text.

CHAPTER 4

In the introductory part of this chapter, a number of works treating modified Lagrangians are listed.

Section 1. The discussion of the simplest modification of the Lagrangian (1.1) is based on Evtushenko [6], [8], [9] and [10]. A study of Newton's method (1.11) and the method (1.21) for $\varepsilon = 1$ without equality-type constraints is contained in Polyak [1]. One of the first implementations of Newton's method (1.11) has been presented in Grachev and Evtushenko [2].

Sections 2-5. The discussion is based on Evtushenko [11], [12], Golikov and Evtushenko [1], and Evtushenko and Pavlovskij [1]. Results close to those described in Section 4 have been obtained by Kort and Bertsekas [1]. An approach close to that described in Section 5 has been developed by Mangasarian [2].
Section 6. The discussion follows Golikov and Zhadan [1]. Another approach to studying the primal methods is examined in Antipin [1].

CHAPTER 5

Section 1. The idea of the reduced gradient method has been enunciated by many authors; see, for example, the article of Arrow and Solow referred to in Arrow, Hurwicz, and Uzawa [1].
Sections 2, 3. The discussion follows Evtushenko [7] and Evtushenko and Zhadan [1], [3]. Some results of the numerical computations using the method (2.4) are given in Efimenko and Zagorujko [1]. A different generalization of the reduced gradient method to problems involving equality-type constraints can be found in Rosen [1]. A similar approach has been followed in Akim and Ehneev [1] and Ehneev [1].
Sections 4, 5. Properties of the gradient projection method and the conditional gradient method have been studied in detail in Karmanov [1], Dem'yanov and Rubinov [1], and Vasil'ev [1].

CHAPTER 6

Some trends in the development of numerical methods for solving optimal control problems are delineated in the introductory part of this chapter. These methods have been studied extensively and a complete bibliography is hard to list.
Sections 1, 2, 3. The discussion follows Grachev and Evtushenko [5], [6] and Evtushenko [12]. A different approach to deriving the formulas (1.7) is developed in Polyak [1].
Sections 4, 5. The discussion is based on the approach suggested by Evtushenko in [12]. The concept of the quasiminimum principle and its justification are presented in Gabasov and Kirillova [1]. The interpretation and proof of the quasiminimum principle has been

slightly modified. Still another approach to deriving the discrete maximum principle can be found in Boltyanskij [1] and Propoj [2].

Section 6. In addition to the works cited throughout the text, we refer the reader to Dem'yanov [1] and Dem'yanov and Rubinov [1].

Section 7. This section has been written by this author together with N.I. Grachev. The computational results for the problem (7.1) obtained via Newton's method have been a courtesy of V.A. Purtov.

Section 8. The results obtained in solving Isaacs' problem have been taken from Grachev and Evtushenko [5]. It is particularly simple to solve game problems in the case where the program strategies have a saddle point; in this situation many well-known methods are no longer applicable (see Dem'yanov and Pevnyj [1] and Grachev and Evtushenko [3]).

CHAPTER 7

Section 1. The described approach has been suggested for the first time in Evtushenko [1], [5].

Section 2. The program to implement the basic method in ALGOL-60 has been given in Evtushenko [1] and an improved program in Evtushenko [2]. Neither program, however, did account for the increasing components of the vector v, which has been accomplished by M.A. Potapov. Another discription of this method can be found in Vasil'ev [1], where the presentation is rather cumbersome due to the absence of recursive procedures.

Section 3. The discussion follows Evtushenko [5].

Section 5. The discussion follows Evtushenko [2], which also contains a program for implementing the method of finding the sequential minimax in ALGOL-60. This program will essentially improve if the modifications of the basic method described in Section 2 have been used.

Section 6. The idea of generalizing the method of nonuniform coverings to solve multicriteria problems was expressed in Evtushenko and Potapov [1]. Problems of finding the extremum of a function are treated in Strongin [1] and Dixon and Szegö [1], [2].

REFERENCES

Adachi, N.
[1] "On Variable Metric Algorithm." *J. Optimiz. Theory Applications*, vol.7, no.6 (1971): 391-409.

Adamenko, G.M.
[1] *O reshenii klassov zadach minimizatsii* (On a Class of Minimization Problems). Minsk: Institut Matematiki Akademii Nauk BSSR, vol.1 (1976). Preprint.

Akim, Eh.L., and Ehneev, T.M.
[1] "Opredelenie parametrov dvizheniya kosmicheskogo letatel'nogo apparata po dannym traektornykh izmerenij (Estimation of Motion Parameters of a Space Vehicle on Flight Data)." *Kosmicheskie issledovaniya*, vol.1, no.1 (1963): 5-28.

Anderson, Donald G.
[1] "Iterative Procedures for Nonlinear Integral Equations." *Journal of the Association for Computing Machinery*, vol.12, no.4 (1965): 547-560.

Anorov, A.S.
[1] "Printsip maksimuma dlya protsessov s ogranicheniyami obshchego vida (The Maximum Principle for Processes With General Constraints)." *Avtomatika i Telemekhanika*, 3 (1967): 5-15.

Antipin, A.S.
[1] *Metody nelinejnogo programmirovaniya, osnovannye na pryamoj i dvojstvennoj modifikatsii funktsii Lagranzha* (Nonlinear Programming Methods Based on Primal-Dual Modification of the Lagrangian). Moskva: VNIISI, 1979.

Arrow, K.J., Hurwicz, L., and Uzawa, H.
[1] *Studies in Linear and Non-linear Programming*. Stanford, California: Stanford University Press, 1958.

Bakhvalov, N.S.
[1] *Chislennye metody* (Numerical Methods). Moskva: Nauka, 1973.

Baptist, P., and Stoer, J.
[1] "On the Relation Between Quadratic Termination and Convergence Properties of Minimization Algorithms." *Numerical Math.*, vol.28, no.4 (1977): 367-391.

Barbashin, E.A.
[1] *Vvedenie v teoriyu ustojchivosti* Introduction to Stability Theory). Moskva: Nauka, 1967.

Barnes, J.G.P.
[1] "An Algorithm for Solving Non-linear Equations Based on the Secant Method." *The Computer Journal*, 8 (1965): 66-72.

Batishchev, D.I.
[1] *Poiskovye metody optimal'nogo proektirovaniya* (Search Methods in Optimal Design). Moskva: Sovetskoe radio, 1975.

Bellman, Richard E.
[1] *Dynamic Programming*. Princeton, New Jersey: Princeton University Press, 1957.
[2] *Introduction to Matrix Analysis*. New York: McGraw-Hill, 1960. (2nd edition, 1970).

Bertsekas, D.
[1] "Combined Primal-Dual and Penalty Methods for Constrained Minimization." *SIAM J. Control*, 13 (1975): 521-544.
[2] "Multiplier Methods: A Survey." *Automatika*, 12 (1976): 133-145.
[3] "On the Convergence Properties of Second-Order Multiplier Methods." *J. Optim. Theory Applications*, vol.25, no.3 (1978): 443-449.

Bittner, L.
[1] "Eine Verallgemeinerung des Sekantenverfahrens (redula falsi) zur näherungsweisen Berechnung der Nullstellen eines nichtlinearen Gleichungssystems." *Wissen. Zeit der Technischen Hochschule Dresden*, 9 (1959): 325-329.

Boltyanskii, Vladimir G.
[1] *Optimal Control of Discrete Systems*. (Transl. from the Russian) New York: Wiley; Jerusalem: Israel Program for Scientific Translations, 1978.

Bryson, A., and Denham, B.
[1] "The Application of Steepest Descent to Optimal Control Problems." *Raketnaya tekhnika i kosmonavtika*, no.2 (1964).

Broyden, C.G.
[1] "A Class of Methods for Solving Nonlinear Simultaneous Equations." *Mathematics of Computation*, vol.19, no.92 (1967): 577-593.
[2] "Quasi-Newton methods and Their Application to Function Minimization." *Mathematics of Computation*, vol.21, no.99 (1967): 368-381.
[3] "A New Double-rank minimization Algorithm." *Notices of the American Mathem. Society*, vol.16, no.4 (1969): 670.

Budak, B.M., Berkovich, E.M., and Solov'eva, E.N.
[1] "The Convergence of Difference Approximations for Optimal Control Problems." *U.S.S.R. Comput. Maths. Math. Phys.*, vol.9, no.3 (1969): 30-65.

Burdakov, O.P.
[1] "Ob odnom printsipe postroeniya kvazi-Newtonovskikh metodov resheniya sistem uravnenij (On Quasi-Newton Methods for Solving Systems of Equations." *Vsesoyuznyj Nauchno-tekhnicheskij seminar "Chislennye metody nelinejnogo programmirovaniya.* Tezisy dokladov. Chast' I, 38-40. Moskva, 1979.
[2] "O nekotorykh sposobakh vybora dliny shaga v metode Newtona (On some Techniques for Choosing Step Length in Newton's Method)." *Issledovanie operatsij*, vyp.7, 111-115. Moskva: VTs AN SSSR, 1979.
[3] "Some Globally Convergent Modifications of Newton's Method for Solving Systems of Nonlinear Equations." *Soviet Math., Doklady*, vol.22, no.2 (1980): 376-379.

Burova, N.K., Stanevichene, L.I., Stanevichus, A., and Shklyar, P.Eh.
[1] *Sistema linejnogo programmirovaniya LP-BECM-6* (A Linear Programming System LP-BECM-6). Moskva: VTs Akademii Nauk SSSR, 1981.

Charalambous, C.
[1] "A Lower Bound for the Controlling Parameters of the Exact Penalty Functions." *Math. Programming*, 15 (1978): 278-290.

Chebotarev, N.G.
[1] "Ob odnom kriterii minimaksa (On a Minimax Criterion)." *Doklady AN SSSR*, 39 (1943): 373-376. (Also, see *Sobranie sochinenij*, Tom 2. Moskva: Izd-vo Akademii Nauk SSSR, 1949.)

Chentsov, I.P.
[1] "O primenenii gradientnykh metodov k resheniyu nekotorykh zadach optimal' nogo upravleniya (On the Application of Gradient Methods to Discontinuous Optimal Control Problems)." *Kibernetika*, 1 (1976): 87-91.

Chernous'ko, F.L., and Banichuk, N.V.
[1] *Variatsionnye zadachi mekhaniki i upravleniya* (Variational Problems of Mechanics and Control). Moskva: Nauka, 1973.

Chigir', S.A.
[1] "Ob igrovoj zadache o dolikhobrakhistrone (On a Game Problem of Dolichobrachistrone)." *Prikladnaya matematika i mekhanika*, vol.40, no.6 (1976): 1003-1013.

Coddington, E.A., and Levinson, N.
[1] *Theory of Ordinary Differential Equations*. New York: McGraw-Hill, 1955.
Danilin, Yu.M.
[1] "Skorost' skhodimosti metodov sopryazhennykh napravlenij (The Rate of Convergence in Conjugate Gradient Methods)." *Kibernetika*, 6 (1977): 97-105.
Danilin, Yu.M., and Panin, V.M.
[1] "O nekotorykh metodakh poiska sedlovykh tochek (On Some Methods for Finding Saddle Points)." *Kibernetika*, 3 (1974): 119-124.
Danilin, Yu.M., and Pshenichnyj, B.N.
[1] "On Methods of Minimization With Accelerated Convergence." *U.S.S.R. Comput. Maths. Math. Phys.*, vol.10, no.6 (1970): 4-19.
[2] "Estimates of the Rate of Convergence of a Class of Minimization Algorithms." *Soviet Math., Doklady*, vol.213, no.2 (1973): 1681-1685.
Danskin, John M.
[1] *The Theory of Max-min and Its Application to Weapons Allocation Problems*. New York Berlin Heidelberg: Springer-Verlag, Inc. 1967.
Davidenko, D.F.
[1] "Ob odnom novom metode chislennogo resheniya sistem nelinejnykh uravnenij (On a New Method for Numerical Solution of Systems of Nonlinear Equations)." *Doklady Akademii Nauk SSSR*, 88 (1953): 601-602.
[2] "An Application of the Method of Variation of Parameters to the Construction of Iterative Formulas of Higher Accuracy for the Determination of the Elements of an Inverse Matrix." *Soviet Math., Doklady*, vol.162, no.4 (1965): 738-742.
Davidon, W.
[1] *Variable Metric Method for Minimization*. Argonne Nat. Lab. Rep. NANL-5990, 1959.
Davydov, Eh.G.
[1] "O raspredelenii resursov na grafakh (On the Display of Resources on Graphs)." *Sistemy raspredeleniya resursov na grafakh*. Moskva: VTs AN SSSR, 1970.
[2] *Metody i modeli teorii antagonisticheskikh igr* (Methods and Models of the Theory of Competitive Games).Moskva: Izd-vo MGU, 1978.
Demidovich, B.P.
[1] *Lektsii po matematicheskoj teorii ustojchivosti* (A Course in Stability Theory). Moskva: Nauka, 1967.
Dem'yanov, V.F.
[1] "On the Solution of Optimum Problems in Non-linear Automatic Control Systems." *U.S.S.R. Comput. Maths. Math. Phys.*, vol.6, no.2 (1966): 32-46.
[2] "Paketnyj printsip minimaksa (Packaging Minimax Algorithms)." *Vestnik Leningragskogo Universiteta*, 6 (1976): 35-39.
Dem'yanov, V.F., and Malozemov, V.N.
[1] *Vvedenie v minimaks* (Introduction to Minimax Theory). Moskva: Nauka, 1972. Transl. from the Russian: *Einführung in minimax-probleme*. Leipzig, Akad. Verlagsgesell., Geest & Portig, 1975.
[2] *Voprosy teorii i elementy programmnogo obespecheniya minimaksnykh zadach* (Software for Minimax Problems). Leningrad: Izd-vo LGU, 1977.
Dem'yanov, V.F., and Pevnyi, A.B.
[1] "Numerical Methods for Finding Saddle Points." *U.S.S.R. Comput. Maths. Math. Phys.*, vol.12, no.5 (1972): 11-52.
Dem'yanov, V.F., and Rubinov, A.M.
[1] *Approximate Methods in Optimization Problems*. New York: American Elsevier, 1970.

Dennis, J.
[1] "On Some Methods Based on Broyden's Secant Approximation to the Hessian." *In Numerical Methods for Nonlinear Optimization*. Edited by F. Lootsma. London: Academic Press, 1972.
Dennis, J., and Moré, J.
[1] "A Characterization of Superlinear Convergence and Its Application to Quasi-Newton Methods." *Mathematics of Computation*, vol.28, no.126 (1974): 549-560.
[2] "Quasi-Newton Methods: Motivation and Theory." *SIAM Review*, vol.19, no.1 (1977): 46-89.
Dixon, L.
[1] "On Quadratic Termination and Second Order Convergence: Two Properties of Unconstrained Optimization Algorithms." In *Towards Global Optimization*. Edited by L.Dixon and G. Szego, 211-228. Amsterdam: Noth-Holland, 1975.
Dixon, L., and Szego, G.
[1] *Towards Global Optimization*. Amsterdam: North-Holland, 1975.
Dorodnitsyn, A.A.
[1] "Asymptoticheskoe reshenie uravneniya Van-der-Polya (An Asymptotic Solution of Van-der-Pohl's Equation)." *Prikladnaya Matematika i mekhanika*, vol.11, no.3 (1947): 313-328.
Dubovitskii, A.Ya., and Milyutin, A.A.
[1] "Extremum Problems in the Presence of Restrictions." *U.S.S.R. Comput. Maths. Math. Phys.*, vol.5, no.3 (1965): 1-80.
Efimenko, V.V., and Zagorujko, A.S.
[1] "Programming of Direct Integration Methods for a Multi-dimensional Minimization Problem." *U.S.S.R. Comput. Maths. Math. Phys.*, vol.20, no.3 (1980): 74-81.
Ehneev, T.M.
[1] "O primenenii gradientnogo metoda v zadachakh teorii optimal'nogo upravleniya (On the Application of the Gradient Method in Problems of Optimal Control Theory)." *Kosmicheskie issledovaniya*, vol.4, no.5 (1966): 651-669.
Eremin, I.I.
[1] "O metode shtrafov v vypuklom programmirovanii (On the Penalty Function Method in Convex Programming)." *Tezisy kratkikh nauchnykh soobshchenij mezhdunarodnogo matematicheskogo kongressa*. Sektsiya 14, Vychislitel'naya matematika. Moskva, 1966.
[2] "The Penalty Method in Convex Programming." *Soviet Math., Doklady*, vol.173, no.4 (1966): 459-462.
Eremin, I.I., and Astaf'ev, N.N.
[1] *Vvedenie v teoriyu linejnogo i vypoklogo programmirovaniya* (Introduction to the Theory of Linear and Convex Programming). Moskva: Nauka, 1976.
Eremin, I.I., and Mazurov, V.D.
[1] *Nestatsionarnye protsessy matematicheskogo programmirovaniya* (Nonstationary Processes of Mathematical Programming). Moskva: Nauka, 1979.
Ermol'ev, Yu.M.
[1] "Metody reshcniya nelinejnykh ekstremal'nykh zadach (Methods for Solving Nonlinear Extremal Problems)." *Kibernetika*, 4 (1966): 1-17.
[2] *Metody stokhasticheskogo programmirovaniya* (Methods of Stochastic Programming). Moskva: Nauka, 1976.
Ermol'ev, Yu.M., and Gulenko, V.P.
[1] "Konechno-raznostnyj metod v zadachakh optimal'nogo upravleniya (The Finite-difference Method in Optimal Control Problems)." *Kibernetika*, 3 (1967): 1-20.
Ermol'ev, Yu.M., Gulenko, V.P., and Tsarenko, T.I.
[1] *Konechno-raznostnyj metod v zadachakh optimal'nogo upravleniya* (The Finite-difference Method in Optimal Control Problems). Kiev: Naukova Dumka, 1978.

Evtushenko, Yu.G.
[1] "Numerical Methods for Finding Global Extrema (Case of a Non-uniform Mesh)." *U.S.S.R. Comput. Maths. Math. Phys.*, vol.11, no.6 (1971): 38-54.
[2] "A Numerical Method for Finding Best Guaranteed Estimates." *U.S.S.R. Comput. Maths. Math. Phys.*, vol.12, no.1 (1972): 109-128.
[3] "Some Local Properties of Minimax Problems." *U.S.S.R. Comput. Maths. Math. Phys.*, vol.14, no.3 (1974): 129-138.
[4] "Iterative Methods for Solving Minimax Problems." *U.S.S.R. Comput. Maths. Math. Phys.*, vol.14, no.5 (1974): 52-63.
[5] "Metody poiska global'nogo ekstremuma (Methods for Finding the Global Extremum)." *Issledovanie operatsij*, vyp. 4, 39-68. Moskva: VTs SSSR, 1974.
[6] "Nekotorye chislennye metody nelinejnogo programmirovaniya (Some Numerical Methods of Nonlinear Programming)." *Doklady konferentsii IFIP po metodam optimizatsii*. Novosibirsk, 1974.
[7] "Two Numerical Methods of Solving Nonlinear Programming Problems." *Soviet Math., Doklady*, vol. 215, no.2 (1974): 420-423.
[8] "Numerical Methods in Nonlinear Programming." *Soviet Math., Doklady*, vol. 221, no.5 (1975): 443-447.
[9] "Numerical Methods for Solving Problems of Non-linear Programming." *U.S.S.R. Comput. Maths. Math. Phys.*, vol.16, no.2 (1976): 24-42.
[10] "Generalized Lagrange Multiplier Technique for Nonlinear Programming." *J. Optimiz. Theory Applications*, vol.21, no.2 (1977): 121-135.
[11] "Primenenie obobshchennykh funktsij Lagranzha dlya resheniya zadach nelinejnogo programmirovaniya (The Application of Generalized Lagrangians to Nonlinear Programming Problems)." *Issledovanie operatsij*, vyp.7, 3-23. Moskva: VTs AN SSSR, 1979.
[12] *Chislennye metody resheniya ekstremal'nykh zadach i ikh primenenie v sistemakh optimizatsii* (Numerical Methods for Solving Extremal Problems and Their Application in System Optimization). Avtoreferat Doktorskoj Dissertatsii. Moskva: VTs AN SSSR, 1980.

Evtushenko, Yu.G., and Potapov, M.
[1] "Space Covering Technique for Multicriteria Optimization." *Lecture Notes in Control and Information Sciences*, 59. System Modelling and Optimization. Proceedings of the 11th IFIP Conference. New York: Springer-Verlag, Inc. 1984.

Evtushenko, Yu.G., and Pavlovsky, Yu.N.
[1] "Integrated Optimization-Simulation System for Industry and Regional Planning." IFIP, 651-658. Amsterdam: North-Holland, 1980.

Evtushenko, Yu.G., and Zhadan, V.G.
[1] "Numerical Methods of Solving Some Operations Research Problems." *U.S.S.R. Comput. Maths. Math. Phys.*, vol.13, no.3 (1973): 56-77.
[2] "Application of the Method of Lyapunov Functions to the Study of the Convergence of Numerical Methods." *U.S.S.R. Comput. Maths. Math. Phys.*, vol. 15, no.1 (1975): 96-108.
[3] "A Relaxation Method for Solving Problems of Non-linear Programming." *U.S.S.R. Comput. Maths. Math. Phys.*, vol.17, no.4 (1977): 73-87.

Faddeev, D.K., and Faddeeva, V.N.
[1] *Computational Methods of Linear Algebra*. Transl. from the Russian. San-Francisco: Freeman, 1963.

Fan, K., Glicksberg, I., and Hoffman, A.
[1] "Systems of Inequalities Involving Convex Functions." *Proc. Amer. Math. Soc.*, 8 (1957): 617-622.

Fedorenko, R.P.
[1] *Priblizhennoe reshenie zadach optimal'nogo upravleniya* (Approximate Solution of Optimal Control Problems). Moskva: Nauka, 1978.
Fiacco, A.V., and McCormick, G.P.
[1] *Nonlinear Programming: Sequential Unconstrained Minimization Techniques.* New York: Wiley, 1968.
Fikhtengol'ts, G.M.
[1] *Infinite Series: Rudiments.* Rev. English ed. New York: Gordon and Breach, 1970.
Fletcher, R.
[1] "A New Approach to Variable Metric Algorithms." *Comp. J.*, 13 (1970): 317-322.
Fletcher, R., and Powell, M.
[1] "A Rapidly Convergent Descent Method for Minimization." *Comp. J.*, vol.6, no.2 (1963): 163-168.
Frank, M., and Wolfe, P.
[1] "An Algorithm for Quadratic Programming." *Nav. Res. Log. Quart.*, 3 (1956): 95-110.
Fujii, S., Fujimoto, H., and Ono, M.
[1] " Shifting Method for Constrained Optimal Control Problems." *Bull. JSME*, vol.20, no.140 (1977): 176-182.
Gabasov, R.
[1] "The Theory of Optimum Processes in Discrete Systems." *U.S.S.R. Comput. Maths. Math. Phys.*, vol.8, no.4 (1968): 99-123.
Gabasov, R., and Kirillova, F.M.
[1] "K voprosy o rasprostranenii printsipa L.S. Pontryagina na diskretnuyu sistemy (On the Generalization of Pontryagin's Maximum Principle to Discrete Systems)." *Avtomatika i telemekhanika*, 11 (1966): 46-51.
[2] *Printsip maksimuma v teorii optimal'nogo upravleniya* (The Maximum Principle in Optimal Control Theory)." Minsk: Nauka, Tekhnika, 1974.
Gaevskij, Kh., Greger, K., and Zakharias, K.
[1] *Nelinejnye operatornye uravneniya i operatornye differentsial'nye uravneniya* (Nonlinear Operator Equations and Operator Differential Equations). Moskva: Mir, 1978.
Gantmacher, F.R.
[1] *Teoriya matrits* (The Theory of Matrices). Moskva: Nauka, 1967. Transl. from the Russian: New York, Chelsea Publ. Co., 1959.
Garcia Palomares, U.M., and Mangasarian, O.
[1] "Superlinearly Convergent Quasi-Newton Algorithms for Nonlinearly Constrain ed Optimization Problems." *Math. Programming*, vol.11, no.1 (1976): 1-13.
Gay, D.
[1] "Some Convergence Properties of Broyden's Method." *SIAM J. Numer. Anal.*, vol.16, no.4 (1969): 623-630.
Gay, D., and Schnabel, R.
[1] "Solving Systems of Nonlinear Equations by Broyden's Method with Projected Updates." In *Nonlinear Programming*. Edited by O. Mangasarian, S. Robinson, and R. Meyer, 254-281. New York: Academic Press, 1978.
Germeyer, Yu.B.
[1] *Vvedenie v teoriyu issledovaniya operatsij* (An Introduction to Operations Research Theory). Moskva: Nauka, 1971.
Gleyzal, A.
[1] "Solution of Nonlinear Equations." *Quart. J. Appl. Math.*, 17 (1959): 95-96.

Glushkov, V.M., Kaspshitskaya, M.F., and Sergienko, I.V.
[1] "Formalization and Solution of a Class of Discrete Optimization Problems." *U.S.S.R. Comput. Maths. Math. Phys.*, vol.20, no.6 (1980): 23-36.

Glushkov, V.M., and Oleyarsh, G.B.
[1] "Dialogovaya sistema planirovaniya DISPLAN (An Interactive Planning System)." *Upravlyayushchie sistemy i mashiny*, 4 (1976): 123-124.

Goldfarb, D.
[1] "A Family of Variable-Metric Methods Derived by Variational Means." *Mathematics of Computation*, vol.24, no.109 (1970): 23-26.

Golikov, A.I., and Evtushenko, Yu.G.
[1] "On a Class of Methods for Solving Nonlinear Programming Problems." *Soviet Math., Doklady*, vol.19, no.2 (1978): 352-356.

Golikov, A.I., and Zhadan, V.G.
[1] "Iterative Methods for Solving Nonlinear Programming Problems Using Modified Lagrangians." *U.S.S.R. Comput. Maths. Math. Phys.*, vol.20, no.4 (1980): 62-78.
[2] "Two Modifications of the Linearization Method in Nonlinear Programming." *U.S.S.R. Comput. Maths. Math. Phys.*, vol.23, no.2 (1983): 36-44.

Gol'shtein, E.G.
[1] *Teoriya dvojstvennosti v matematicheskom programmirovanii i ee prilozheniya* (Duality Theory in Mathematical Programming and Applications). Moskva: Nauka, 1971.

Gol'shtein, E.G., and Tret'yakov, N.V.
[1] "Modifitsirovannye funktsii Lagranzha (Modified Lagrangians)." *Ekonomika i matematicheskie metody*, vol.10, no.3 (1974): 568-591.

Grachev, N.I., and Evtushenko, Yu.G.
[1] "Nekotorye chislennye metody resheniya igr s neprotivopolozhnymi interesami (Some Numerical Methods for Solving Cooperative Games)." *Issledovanie Operatsij*, vyp.4, 219-235. Moskva: VTs AN SSSR, 1974.
[2] "Variant metoda Newtona dly resheniya obshchej zadachi nelinejnogo programmirovaniya (A Version of Newton's Method for Solving a General Problem of Nonlinear Programming)." *Issledovanie operatsij*, vyp.5, 54-58. Moskva, VTs AN SSSR, 1976.
[3] "Chislennye metody otyskaniya sedlovykh tochek (Numerical Methods for Finding Saddle Points)." *Tezisy II Vsesoyuznogo Seminara*, 138-142. Khar'kov, 1976.
[4] "Application of the Singular Perturbation Method for Solving Minimax Problems." *Soviet Math., Doklady*, vol.18, no.2 (1977): 370-373.
[5] *Paket programm dlya resheniya zadach optimal'nogo upravleniya* (Program Package for Solving Optimal Control Problems). Moskva: VTs AN SSSR, 1978.
[6] "A Library of Programs for Solving Optimal Control Problems." *U.S.S.R. Comput. Maths. Math. Phys.*, vol.19, no.2 (1979): 99-119.

Gragg, B., and Steward, G.
[1] "A Stable Variant of the Secant Method for Solving Nonlinear Equations." *SIAM J. Numer. Anal.*, vol.13, no.6 (1976): 889-903.

Haarhoff, P., and Buyes, J.
[1] "A New Method for the Optimization of a Nonlinear Function Subject to Nonlinear Constraints." *Comp. J.*, vol.13, no.2 (1970): 171-177.

Hadamard, J.S.
[1] "Sur les Transformations Ponctuelles." *Bulletin de la Société mathématique de France*. T. XXXIV (1906): 71-84.

Halkin, H.
[1] "A Maximum Principle of the Pontryagin Type for Systems Described by Nonlinear Difference Equations." *SIAM J. Control*, vol.4, no.1 (1966).
Hamala, M.
[1] "Quasibarrier Method for Convex Programming." *Abstracts of the IX Intern. Symposium Math. Programming*, pp. 110-111. Budapest: Bolaya Math. Society, 1976.
Han, S.
[1] "Superlinearly Convergent Variable Metric Algorithms for General Nonlinear Programming Problems." *Math. Programming*, vol.11, no.3 (1976): 263-281.
[2] "A Globally Convergent Method for Nonlinear Programming." *J. Optim. Theory Applic.*, vol.22, no.3 (1977): 297-309.
Han, S., and Mangasarian, O.
[1] "Exact Penalty Functions in Nonlinear Programming." *Math. Programming*, vol.17 (1979): 251-269.
Haselgrove, C.
[1] "The Solution of Nonlinear Equations with Two-point Boundary Conditions." *Comp. J.*, vol.4, no.3 (1961): 255-259.
Hestenes, M.
[1] "Multiplier and Gradient Methods." *J. Optimiz. Theory Appl.*, vol.4, no.5 (1969): 303-320.
Isaacs, Rufus
[1] *Differential Games; a Mathematical Theory with Applications to Warfare and Pursuit, Control and Optimization.* New York: Wiley, 1965.
Isaev, V.K., and Sonin, V.V.
[1] "On a Modification of Newton's Methods for the Numerical Solution of Boundary Problems." *U.S.S.R. Comput. Maths. Math. Phys.*, vol.3, no. 6 (1963): 1525-1528.
[2] "Numerical Aspects of the Problem of Optimum Flight as a Boundary Value Problem." *U.S.S.R. Maths. Math. Phys.*, vol.5, no.2 (1965): 117-129.
[3] "Novyj podkhod k probleme approksimatsii i ego prilozheniya k variatsionnym i minimaksnym zadacham (A New Approach to Approximation Problems and Applications to Variational and Minimax Problems)." *Trudy TsAGI*, pp. 3-23. Moskva, vyp. 1646, 1975.
Ivanilov, Yu.P.
[1] "Dva algoritma resheniya zadachi vypuklogo programmirovaniya (Two Algorithms for Solving a Convex Programming Problem)." In *Teoriya optimal'nykh reshenij.* Kiev: Izd-vo IK AN Ukrainskoj SSR, vyp. 4, 1968.
Ivanov, V.V., and Lyudvichenko, V.A.
[1] "Ob odnom metode posledovatel'noj bezuslovnoj minimizatsii resheniya zadach matematicheskogo programmirovaniya (On a Method of Sequential Unconstrained Minimization for Solving Mathematical Programming Problems)."*Kibernetika*, 2 (1977): 1-8.
John, F.
[1] "Extremum Problems with Inequalities as Side Conditions," In *Studies and Essays Presented to R. Courant on his 60th Birthday, January 8, 1948.* New York: Interscience Publishers, 1948.
Kantorovich, L.V., and Akilov, G.P.
[1] *Functional Analysis.* New York: The Macmillan Company, 1964.
Karmanov, V.G.
[1] *Matematicheskoe programmirovanie* (Mathematical Programming). Moskva: Fizmatgiz, 1975.

Karpov, V.Ya., Koryagin, D.A., and Samarskij, A.A.
[1] "Principles of the Development of Applied Program Packages for Problems of Mathematical Physics." *U.S.S.R. Comput. Maths. Math. Phys.*, vol.18, no.2 (1978): 149-155.

Kashin, G.M., Pshenichnov, G.I., and Flerov, Yu.A.
[1] *Metody avtomatizirovannogo proektirovaniya samoleta* (Methods of Automated Design of an Aircraft). Moskva: Mashinostroenie, 1979.

Khalanaj, A., and Veksler, D.
[1] *Kachestvennaya teoriya impul'snykh sistem* (Qualitative Theory of Impulse Systems). Moskva: Mir, 1971.

Kornilova, G.F., et al.
[1] "Paket "Optima-II" resheniya zadach optimal'nogo planirovaniya dlya mashiny BESM-6 (The 'Optima-II' Package for Solving Optimal Design Problems on a BESM-6 Computer)." In *Programmy Optimizatsii*. Sverdlovsk: Izd-vo IMM UNTs AN SSSR, vyp.5, 1974.

Kort, B., and Bertsekas, D.
[1] "Combined Primal-Dual and Penalty Methods for Convex Programming." *SIAM J. Control*, vol.14, no.2 (1976): 268-294.

Kostina, M.A.
[1] "Nekotorye sposoby upravleniya iteratsionnymi protsedurami v metodakh tipa shtrafnykh funktsij (Some Techniques for Iterative Procedure Using Penalty Function Methods)." In *Metody vypuklogo programmirovaniya i prilozheniya*, pp. 32-41. Sverdlovsk: Izd-vo IMM UNTs AN SSSR, 1973.

Kowalik, J., Osborne, M.R., and Ryan, D.M.
[1] "A New Method for Constrained Optimization Problems." *Operations Research*, vol.17, no.4 (1969): 973-983.

Krassovskij, N.N.
[1] *Nekotorye zadachi teorii ustojchivosti dvizheniya* (Some Problems of Stability Theory). Moskva: Fizmatgiz, 1959.

Krassovskij, N.N., and Subbotin, A.I.
[1] *Pozitsionnye differentsial'nye igry* (Closed-Loop Differential Games). Moskva: Nauka, 1974.

Krylov, I.A., and Chernous'ko, F.L.
[1] "On a Method of Successive Approximations for the Solution of Problems of Optimal Control." *U.S.S.R. Comput. Maths. Math. Phys.*, vol.2, no.6 (1963): 1371-1382.

Kuhn, H., and Tucker, A.
[1] "Nonlinear Programming." *Proc. Berkeley Symp. Math. Statistics, Probability*, pp. 481-492. Berkeley and Los Angeles: University of California Press, 1951.
[2] *Linear Inequalities and Related Systems.* (Edited by Kuhn, H., and Tucker, A.) Princeton, NJ: Princeton University Press, 1956.

Kuptsov, V.I., and Shurshkova, E.G.
[1] "O skhodimosti modifitsirovannogo metoda Newtona (On the Convergence of a Modified Newton Method)." *Vychislitel'nye metody i programmirovanie*. Moskva: Izd-vo MGU, vyp. 14, 1970.

Lancaster, P.
[1] *Theory of Matrices*. New York: Academic Press, 1969.

Lebedev, V.N.
[1] *Paschet dvizheniya kosmicheskogo apparata s maloj tyagoj* (Calculation of the Trajectory of a Low-Thrust Space Vehicle). Moskva: VTs AN SSSR, 1963.

Lootsma, F.
[1] "Convergence Rates of Quadratic Exterior Penalty-Function Method for Solving Constrained-Minimization Problems." *Philips Research Reports*, vol.29, no.1 (1974): 2-12.

Lyapunov, A.M.
[1] *Obshchaya zadacha ob ustojchivosti dvizheniya* (A General Problem of Stability of Motion). Moskva: Gostekhizdat, 1950.
McCormick, G.
[1] "Second-Order Conditions for Constrained Minima." *SIAM J. Appl. Math.*, vol.15, no.3 (1967): 641-652.
McCormick, G., and Ritter, K.
[1] "Methods of Conjugate Directions Versus Quasi-Newton Methods." *Math. Programming*, vol.3, no.1 (1972): 101-116.
Malkin, I.G.
[1] *Teoriya ustojchivosti dvizheniya* (Stability Theory). Moskva: Nauka, 1966.
Mangasarian, O.
[1] *Nonlinear Programming.* New York: McGraw-Hill, 1969.
[2] "Unconstrained Lagrangians in Nonlinear Programming." *SIAM J. Control*, vol.13, no.4 (1975): 772-791.
Martinez, J.
[1] "Three New Algorithms Based on the Sequential Secant Method." *BIT*, vol.19, no.2 (1979): 236-243.
Mehra, R., and Davis, R.
[1] "A Generalized Gradient Method for Optimal Control Problems with Inequality Constraints and Singular Arcs." *IEEE Trans. Automatic Control*, no.1 (1972): 69-79.
Meleshko, V.I., and Pesina, R.I.
[1] "Razrabotka paketov programm bezuslovnoj optimizatsii na modul'nom printsipe (Development of Packages for Unconstrained Optimization Modular Program)." *Upravlyayushchie sistemy i mashiny*, no.2 (1977): 35-40.
Mifflin, R.
[1] "Convergence Bounds for Nonlinear Programming Algorithms." *Math. Programming*, vol.8, no.3 (1975): 251-271.
Mishchenko, E.F., and Pontryagin, L.S.
[1] "Dokazatel'stvo nekotorykh asimptoticheskikh formul dlya resheniya differentsial'nykh uravnenij s malym parametrom (Proof of Some Asymptotic Formulas for Solution of Differential Equations Containing a Small Parameter)." *Doklady Akademii Nauk SSSR*, vol.120, no.5 (1958): 643-660.
Mishchenko, E.F., and Rozov, N.Kh.
[1] *Differentsial'nye uravneniya s malym parametrom i relaksatsionnye kolebaniya* (Differential Equations With Small Parameter and Relaxation Oscillations). Moskva: Nauka, 1975.
Moiseev, N.N.
[1] *Metody optimizatsii* (Optimization Methods). Moskva: VTs AN SSSR, 1969.
[2] *Chislennye metody v teorii optimal'nykh sistem* (Numerical Methods in System Optimization). Moskva: Nauka, 1971.
Moiseev, N.N., Ivanilov, Yu.P., and Stolyarova, E.M.
[1] *Metody optimizatsii* (Optimization Methods). Moskva: Nauka, 1978.
Morrison, D.
[1] "Optimization by Least Squares." *SIAM J. Numerical Analysis*, vol.5, no.1 (1968): 83-88.
von Neumann, J., and Morgenstern, O.
[1] *Theory of Games and Economic Behavior.* New York: John Wiley, 1964.
Nikaido, H.
[1] *Vypuklye struktury i matematicheskaya ekonomika* (Convex Structures and Mathematical Economics). Moskva: Mir, 1972.

Okhotsimskij, D.E.
[1] "K teorii dvizheniya raket (On Rocket Motion Dynamics)." *PMM*, vol.10, no.2 (1946): 251-272.
Okhotsimskij, D.E., and Ehneev, T.M.
[1] "Nekotorye variatsionnye zadachi, svyazannye s zapuskom iskusstvennogo sputnika Zemli (Some Variational Problems Related to the Launching of a Satellite of the Earth)." *UFN*, vol.63, no.1a (1957): 36-51.
Ortega, J.M., and Rheinboldt, W.
[1] *Iterative Solution of Nonlinear Equations in Several Variables.* New York: Academic Press, 1970.
Panin, V.M.
[1] "Reshenie sistem nelinejnykh uravnenij dempfirovannym metodom Newtona pri uslovii prodolzhimosti (Solution of Systems of Nonlinear Equations by a Reduced Newton's Method Under a Continuity Condition)." In *Teoriya optimal'nykh reshenij*, pp. 39-51. Kiev: Izd-vo IK AN Ukrainskoj SSR, 1976.
Pearson, J.
[1] "The Discrete Maximum Principle." *Int. J. Control*, vol.11, no.2 (1965).
[2] "On Variable Metric Methods of Minimization." *Comp. J.*, vol.12, no.2 (1971): 171-178.
Pietrzykowski, T.
[1] "An Exact Potential Method for Constrained Maxima." *SIAM J. Numerical Analysis*, 16 (1969): 299-304.
Polak, E.
[1] *Computational Methods in Optimization. A Unified Approach.* New York: Academic Press, 1971.
[2] "A Modified Secant Method for Unconstrained Minimization." *Math. Programming*, vol.6, no.4 (1974): 264-280.
Polak, E., and Teodoru, I.
[1] "Newton Derived Methods for Nonlinear Equations and Inequalities." In *Nonlinear Programming*, pp. 255-277. 2nd ed. Edited by O. Mangasarian, R. Robinson, and R. Meyer. New York: Academic Press, 1975.
Polyak, B.T.
[1] "Iterative Methods Using Lagrange Multipliers for Solving Extremal Problems with Constraints of the Equation Type." *U.S.S.R. Comput. Maths. Math. Phys.*, vol.10, no.5 (1970): 42-52.
[2] "The Convergence Rate of the Penalty Function Method." *U.S.S.R. Comput. Maths. Math. Phys.*, vol.11, no.1 (1971): 1-12.
Polyak, B.T., and Tret'yakov, N.V.
[1] "The Method of Penalty Estimates for Conditional Extremum Problems." *U.S.S.R. Comput. Maths. Math. Phys.*, vol.3, no.1 (1973): 42-58.
Pontryagin, L.S., Boltyanskij, V.G., Gamkrelidze, R.V., and Mishchenko, E.F.
[1] *Matematicheskaya teoriya optimal'nykh protsessov* (Mathematical Theory of Optimal Processes). Moskva: Fizmatgiz, 1961.
Powell, M.J.D.
[1] "A Method for Nonlinear Constraints in Minimization Problems." In *Optimization*, pp. 283-298. London: Academic Press, 1969.
[2] "A New Algorithm for Unconstrained Optimization." In *Nonlinear Programming.* Edited by J. Rosen, O. Mangasarian, and K. Ritter. New York: Academic Press, 1969.
Powers, W., and Shich, C.
[1] "Convergence of Gradient-Type Methods for Free Final Time Problems." *AIAA J. Numerical Analysis*, vol.14, no.11 (1976): 1598-1603.

Propoj, A.I.

[1] "O printsipe maksimuma dly diskretnykh sistem upravleniya (On the Maximum Principle for Discrete Control Systems." *Avtomatika i telemekhanika*, no.7 (1965): 1177-1187.

[2] *Elementy teorii optimal'nykh diskretnykh protsessov* (Fundamentals of the Optimal Discrete Process Theory). Moskva: Nauka, 1973.

Pshenichyj, B.N.

[1] "An Algorithm for the Solution of a Non-linear Problem of Optimum Control." *U.S.S.R. Comput. Maths. Math. Phys.*, vol.5, no.2 (1965): 94-102.

[2] "Newton's Method for the Solution of Systems of Equalities and Inequalities." *Math. Notes of the Academy of Sciences of the USSR*, vol.8, no.5 (1970): 827-830.

[3] *Vypuklyj analiz i ekstremal'nye zadachi* (Convex Analysis and Extremal Problems). Moskva: Nauka, 1980.

Pshenichnyj, B.N., and Danilin, Yu.M.

[1] *Chislennye metody v ekstremal'nykh zadachakh* (Numerical Methods for Extremal Problems). Moskva: Nauka, 1975.

Razdol'skij, A.R.

[1] "Skhema resheniya zadachi nelinejnogo programmirovaniya pri unimodal'noj tselevoj funktsii (A Scheme for Solving Nonlinear Programming Problems for a Unimodal Objective Function)." *Izvestiya Akademii Nauk SSSR, ser. Tekhnicheskaya kibernetika*, 4 (1973).

Razumikhin, B.S.

[1] *Fizicheskie modeli i metody teorii ravnovesiya v programmirovanii i ekonomike* (Physical Models and Methods of Equilibrium Theory in Programming and Economics). Moskva: Nauka, 1975.

Rheinboldt, W.

[1] *Methods for Solving Systems of Nonlinear Equations.* Philadelphia: Soc. Ind. Appl. Math., 1974.

Rockafellar, R.T.

[1] *Convex Analysis.* Princeton, NJ: Princeton University Press, 1970.

[2] "A Dual Approach to Solving Nonlinear Programming Problems by Unconstrained Optimization." *Math. Programming*, no.5 (1973): 354-373.

[3] "Augmented Lagrange Multiplier Functions and Duality in Nonconvex Programming." *SIAM J. Control*, vol.12, no.2 (1974): 268-285.

Rosen, J.

[1] "The Gradient Projection Method for Nonlinear Programming." *J. Soc. Ind. Appl. Math.*, vol.9, no.4 (1961): 514-532.

Rybashov, M.V.

[1] "Metod differentsial'nykh uravnenij v zadache otyskaniya ekstremuma funktsij s pomoshchyu analogovykh vychislitel'nykh mashin (The Method of Differential Equations for Finding Extrema of Functions Using Analogue Computers)." *Avtomatika i telemekhanika*, 5 (1969): 181-194.

[2] "Nepreryvnye algoritmy prodolzheniya resheniya konechnykh uravnenij, zavisyashchikh ot parametrov (Algorithms for Continuation of Solution of Finite Parametric Equations)." *Avtomatika i telemekhanika*, 4 (1975): 11-18.

Schuller, Günther

[1] "On the Order of Convergence of Certain Quasi-Newton Methods." *Numer. Math.*, vol.23, no.1 (1974): 181-192.

Shamanskij, V.E.

[1] *Metody chislennogo resheniya kraevykh zadach na ETsVM* (Methods for Numerical Solution of Boundary Value Problems on a Digital Computer). Kiev: Naukova Dumka, 1966.

[2] "Ob odnoj modifikatsii metody Newtona (On a Modification of Newton's Method)." *Ukrainskij mathem. zhurnal*, 19 (1967): 133-138.

Shanno, D.

[1] "Conditioning of Quasi-Newton Methods for Function Minimization." *Math. Comput.*, 24 (1970): 647-656.

Shanno, D., and Phua, K.

[1] "Numerical Comparison of Several Variable Metric Algorithms." *J. Optim. Theory Appl.*, vol.25, no.4 (1978): 507-518.

Shatrovskii, L.I.

[1] "On a Numerical Method for Solving Problems of Optimum Control." *U.S.S.R. Comput. Maths. Math. Phys.*, vol.2, no.3 (1963): 511-514.

Shepilov, M.A.

[1] "Nepreryvnye analogi metoda shtrafov dlya zadach vypuklogo programmirovaniya (Continuous Analogs of the Penalty Function Method for Convex Programming Problems)." *Ekonomika i matematicheskie metody*, vol.11, no.1 (1975): 130-141.

Shor, N.Z.

[1] *Metody minimizatsii nedifferentsiruemykh funktsij i ikh prilozheniya* (Methods of Minimization of Nondifferentiable Functions and Applications). Kiev: Naukova Dumka, 1979.

Skalkina, M.A.

[1] "O svyazi mezhdu ustojchivost'yu reshenij differentsial'nykh i konechnoraznostnykh uravnenij (On the Relationship Between the Stability of Solutions of Differential Equations and the Stability of Solutions of Finite-Difference Equations)." *PMM*, vol.19, no.3 (1955): 287-294.

Skarin, V.D.

[1] "The Method of Penalty Functions for Nonlinear Programming Problems." *U.S.S.R. Comput. Maths. Math. Phys.*, vol.13, no.5 (1973): 108-123.

Smol'yakov, Eh.R.

[1] "Printsip maksimuma dlya zadach s fazovymi ogranicheniyami (The Maximum Principle for Problems with State Constraints)." In *Issledovanie operatsij*, pp. 136-155. Moskva: VTs AN SSSR, vyp. 2, 1971.

Spedicato, E.

[1] "Quasi-Newton Methods for Nonlinear Unconstrained Minimization: A Review." In *Toward Global Optimization*, pp. 191-207. Edited by L. Dixon. Amsterdam: North-Holland, 1978.

[2] "Computational Experience with Quasi-Newton Algorithms for Minimization Problems of Moderately Large Size." In *Toward Global Optimization*, pp. 209-213. Edited by L. Dixon. Amsterdam: North-Holland, 1978.

Spedicato, E., and Greenstadt, J.

[1] "On Some Classes of Variationally Derived Quasi-Newton Methods for Systems of Nonlinear Algebraic Equations." *Numer. Math.*, vol.29, no.4 (1978): 363-380.

Stoer, J.

[1] *Einführung in die Numerische Mathematik*. Berlin Heidelberg New York Tokyo: Springer-Verlag, 1976.

[2] "On the Relation Between Quadratic Termination and Convergence Properties of Minimization Algorithms." *Theory Numeric. Math.*, vol.28, no.3 (1977): 343-366.

Strongin, R.G.

[1] *Chislennye metody v mnogoekstremal'nykh zadachakh* (Numerical Methods in Multiextrema Problems). Moskva: Nauka, 1978.

Tabak, D., and Kuo, B.
[1] *Optimal'noe upravlenie i matematicheskoe programmirovanie* (Optimal Control and Mathematical Programming). Moskva: Nauka, 1975.
Thomas, S.
[1] *Sequential Estimation Techniques for Quasi-Newton Algorithms*. Cornell Univ., 1975, TR 75-227.
Tikhonov, A.N.
[1] "Sistemy differentsial'nykh uravnenij, soderzhashchie malye parametry pri proizvodnykh (Systems of Differential Equations Containing Small Parameters in the Derivatives)." *Matematicheskij sbornik*, vol.31, no.3 (1952): 576-586.
Tornheim, L.
[1] "Convergence of Multipoint Iterative Methods." *J. Assoc. Comput. Mach.*, vol.11, no.2 (1964): 210-220.
Tret'yakov, N.V.
[1] "Metod shtrafnykh otsenok dlya zadach vypuklogo programmirovaniya (The Penalty Function Method for Convex Programming Problems)." *Ekonomika i matemat. metody*, vol.8, no.5 (1972): 740-751.
Ul'm, S.Yu.
[1] "Extension of Stefensen's Method for Solving Nonlinear Operator Equations." *U.S.S.R. Comput. Maths. Math. Phys.*, vol.4, no.6 (1964): 159-165.
[2] "Ob obobshchennykh razdelennykh raznostyakh (On Generalized Divided Differences)." *Izvestiya Akademii Nauk Estonskoj SSR*, ser. phiziko-matemat. nauk, vol.16, no.1 (1967): 13-26.
Vasil'ev, F.P.
[1] *Chislennye metody resheniya ekstremal'nykh zadach* (Numerical Methods for Solving Extremal Problems). Moskva: Nauka, 1980.
Vasil'eva, A.B., and Butuzov, V.F.
[1] *Asimptoticheskie razlozheniya reshenij singulyarno vozmushchennykh uravnenij* (Asymptotic Expansions of Solutions of Singularly Perturbed Equations). Moskva: Nauka, 1973.
Velichenko, V.V.
[1] "A Numerical Method for Solving Optimal Control Problems." *U.S.S.R. Comput. Maths. Math. Phys.*, vol.6, no.4 (1966): 34-50.
[2] "O zadachakh optimal'nogo upravleniya dlya uravnenij s razryvnymi pravymi chastyami (On Optimal Control Problems for Equations with Discontinuous Right-hand Sides)." *Avtomatika i telemekhanika*, 7 (1966): 20-30.
[3] "Sposob opredeleniya uslovnogo minimuma funktsij mnogikh peremennykh (A Technique for Determining the Minimum of Multivariable Functions)." *Avtomatika i telemekhanika*, 2 (1967): 171-172.
[4] "K zadache o minimume maksimal'noj peregruzke (On the Problem of the Minimum of the Maximal Overload)." *Kosmicheskie issledovaniya*, vol.10, no.5 (1972): 700-710.
Venets, V.I., and Rybashov, M.V.
[1] "The Method of Lyapunov Functions in the Study of Continuous Algorithms of Mathematical Programming." *U.S.S.R. Comput. Maths. Math. Phys.*, vol.17, no.3 (1977): 64-73.
Voevodin, V.V.
[1] *Linejnaya algebra* (Linear Algebra). Moskva: Nauka, 1980.
Volkonskij, V.A.
[1] "Optimal'noe planirovanie v usloviyakh bol'shoj razmernosti (Optimal Planning of Large-Scale Systems)." *Ekonomika i matematicheskie metody*, vol.1, no.2 (1965): 195-219.

Wilson, R.
[1] *A Simplicial Algorithm for Concave Programming*. Boston: Harvard University Press, 1963.
Wolfe, P.
[1] "The Secant Method for Simultaneous Nonlinear Equations." *Comm. Assoc. Comput. Mach.*, vol.2 (1959): 12-13.
Yakovlev, V.M.
[1] " O diskretnom printsipe maksimuma (On a Discrete Maximum Principle)." *Problemy kibernetiki*, pp. 247-257. Vyp. 34, 1978.
Zangwill, W.
[1] "Non-linear Programming Via Penalty Function." *Management Science*, vol.13, no.5 (1967): 344-358.
[2] *Nelinejnoe programmirovanie: Edinyj podkhod* (Nonlinear Programming: A Unified Approach). Moskva: Sovetskoe radio, 1973.

INDEX

V.F. Dem'yanov, and L.V. Vasil'ev
Nondifferentiable Optimization
1985, approx. 350 pp.
ISBN 0-911575-09-X Optimization Software, Inc.
ISBN 0-387-90951-6 Springer-Verlag New York Berlin Heidelberg Tokyo
ISBN 3-540-90951-6 Springer-Verlag Berlin Heidelberg New York Tokyo

V.P. Chistyakov, B.A. Sevast'yanov, and V.K. Zakharov
Probability Theory For Engineers
1985, approx. 200 pp.
ISBN 0-911575-13-8 Optimization Software, Inc.
ISBN 0-387-96167-4 Springer-Verlag New York Berlin Heidelberg Tokyo
ISBN 3-540-96167-4 Springer-Verlag Berlin Heidelberg New York Tokyo

B.T. Polyak
Introduction To Optimization
1985, approx. 450 pp.
ISBN 0-911575-14-6 Optimization Software, Inc.
ISBN 0-387-96169-0 Springer-Verlag New York Berlin Heidelberg Tokyo
ISBN 3-540-96169-0 Springer-Verlag Berlin Heidelberg New York Tokyo

V.A. Vasilenko
Spline Functions: Theory, Algorithms, Programs
1985, approx. 280 pp.
ISBN 0-911575-12-X Optimization Software, Inc.
ISBN 0-387-96168-2 Springer-Verlag New York Berlin Heidelberg Tokyo
ISBN 3-540-96168-2 Springer-Verlag Berlin Heidelberg New York Tokyo

V.F. Kolchin
Random Mappings
1985, approx. 250 pp.
ISBN 0-911575-16-2 Optimization Software, Inc.
ISBN 0-387-96154-2 Springer-Verlag New York Berlin Heidelberg Tokyo
ISBN 3-540-96154-2 Springer-Verlag Berlin Heidelberg New York Tokyo

A.A. Borovkov, Ed.
Advances In Probability Theory:
Limit Theorems For Sums of Random Variables
1985, approx. 400 pp.
ISBN 0-911575-17-0 Optimization Software, Inc.
ISBN 0-387-96100-3 Springer-Verlag New York Berlin Heidelberg Tokyo
ISBN 3-540-96100-3 Springer-Verlag Berlin Heidelberg New York Tokyo

Continued on page 560

TRANSLATIONS SERIES IN MATHEMATICS AND ENGINEERING

V.V. Ivanishchev, and A.D. Krasnoshchekov
Control of Variable Structure Networks
1985, approx. 200 pp.
ISBN 0-911575-05-7 Optimization Software, Inc.
ISBN 0-387-90947-8 Springer-Verlag New York Berlin Heidelberg Tokyo
ISBN 3-540-90947-8 Springer-Verlag Berlin Heidelberg New York Tokyo

A.N. Tikhonov, Ed.
Problems In Modern Mathematical Physics and Computational Mathematics
1985, approx. 500 pp.
ISBN 0-911575-10-3 Optimization Software, Inc.
ISBN 0-387-90952-4 Springer-Verlag New York Berlin Heidelberg Tokyo
ISBN 3-540-90952-4 Springer-Verlag Berlin Heidelberg New York Tokyo

N.I. Nisevich, G.I. Marchuk, I.I. Zubikova, and I.B. Pogozhev
Mathematical Modeling of Viral Diseases
1985, approx. 400 pp.
ISBN 0-911575-06-5 Optimization Software, Inc.
ISBN 0-387-90948-6 Springer-Verlag New York Berlin Heidelberg Tokyo
ISBN 0-387-90948-6 Springer-Verlag Berlin Heidelberg New York Tokyo

V.G. Lazarev, Ed.
Processes and Systems In Communication Networks
1985, approx. 250 pp.
ISBN 0-911575-08-1 Optimization Software, Inc.
ISBN 0-387-90950-8 Springer-Verlag New York Berlin Heidelberg Tokyo
ISBN 3-540-90950-8 Springer-Verlag Berlin Heidelberg New York Tokyo

TRANSLITERATION TABLE

R	E	R	E
а А	a	р Р	r
б Б	b	с С	s
в В	v	т Т	t
г Г	g	у У	u
д Д	d	ф Ф	f
е Е	e	х Х	kh
ё Ё	e	ц Ц	ts
ж Ж	zh	ч Ч	ch
з З	z	ш Ш	sh
и И	i	щ Щ	shch
й Й	j	ъ Ъ	"
к К	k	ы Ы	y
л Л	l	ь Ь	'
м М	m	э Э	eh
н Н	n	ю Ю	yu
о О	o	я Я	ya
п П	p		